Flames

Their structure, radiation and temperature

Flames

Their structure, radiation and temperature

A. G. GAYDON, D.Sc., F.R.S.
*Emeritus Professor of Molecular Spectroscopy
Imperial College, London
and formerly Warren Research Fellow
of the Royal Society*

AND

H. G. WOLFHARD, Dr. rer. nat.
Institute for Defense Analyses, Arlington, Va.

FOURTH EDITION

LONDON
CHAPMAN AND HALL

A Halsted Press Book
John Wiley & Sons, New York

First published 1953
Second edition (revised) 1960
by Chapman and Hall Ltd
11 New Fetter Lane, London EC4P 4EE
Third edition (revised) 1970
Fourth edition (revised) 1979
© 1979 A. G. Gaydon and H. G. Wolfhard
Typeset by Fletcher & Son Ltd, Norwich
and printed in Great Britain by
Richard Clay (The Chaucer Press) Ltd
Bungay, Suffolk
ISBN 0 412 15390 4

All rights reserved. No part of this book may be reprinted, or reproduced or utilized in any form or by any electronic, mechanical or other means, now known or hereafter invented, including photocopying and recording, or in any information storage or retrieval system, without permission in writing from the publisher.

Library of Congress Cataloging in Publication Data

Gaydon, Alfred Gordon.
    Flames, their structure, radiation, and temperature.

    "A Halsted Press Book."
    Bibliography: p.
    Includes indexes.
    1. Flame. I. Wolfhard, Hans G., joint author. II. Title.
QD516.G28   1978   541'.361   78-16087
ISBN 0-470-26481-0

    Distributed in the U.S.A. by Halsted Press, a Division of John Wiley & Sons, Inc., New York

Preface to First Edition

The science of combustion has now become so large, and embraces so many other aspects of science that it is no longer practical adequately to cover the whole field in a single book, or indeed for two authors to deal with all sides of the subject. The aim here is to give a fairly advanced discussion of a part of the field, that concerned with stationary flames, with the emphasis on the physical rather than the chemical viewpoint.

Bone & Townend's *Flame and Combustion in Gases* (1927) gives a clear account of many aspects of flame structure, and is valuable historically, but is now out of date. More recent books, *Explosion and Combustion Processes* by Jost, and the revised edition of *Combustion Flames and Explosions* by Lewis & von Elbe give a very good general introduction to the subject of combustion, especially the chemical aspects. There are, however, many parts of the subject which are not completely covered for various reasons. Among the subjects on which we have especially concentrated are the measurement of flame velocity, the theories of flame propagation, the method of carbon formation in flames, flame radiation, the measurement of high flame temperatures and ionisation in flames. Many of these topics were considered by Lewis & von Elbe to be in too fluid a state or too controversial, and this is indeed to some extent true, but nevertheless they are among the most interesting. Some fields, such as the chemical kinetics of combustion processes, cool flames, minimum ignition energies, quenching distances and flame stability which have been very ably discussed in the books mentioned are not fully dealt with here, only sufficient mention being made of them to preserve the continuity of this book. Rather surprisingly there is little overlap with *Spectroscopy and Combustion Theory* (Gaydon, 1948) although our recent spectroscopic work on low-pressure flames and on the structure of flat diffusion flames is included here. The new book by Thring *The Science of Flames and Furnaces* gives a good account of the heating of furnaces by large industrial flames, but has little in common with the problems of laboratory flames discussed here.

Our aim is to present as clearly as possible the underlying physical processes occurring in flames. We fully realise, of course, the need for quantitative measurements, but have avoided purely mathematical discussion;

indeed we have little enthusiasm for abstract mathematical treatments of combustion, these usually involving many unknown and often unknowable parameters.

We should like to express our sincere thanks to Sir Alfred Egerton, F.R.S. for his encouragement and stimulation. We are also grateful to a number of persons for helpful discussion on various parts of the subject and for co-operation in obtaining photographs and diagrams. We should like especially to thank Drs. J. Barr (Glasgow), H. P. Broida (U.S. Bureau of Standards), J. H. Burgoyne (Imperial College), H. Behrens (Weil a/R. Germany), J. E. Garside (Leeds), W. Gohrbandt and B. P. Mullins (National Gas Turbine Est.), H. Hahnemann, W. G. Parker (Royal Aircraft Est.), B. Lewis and G. von Elbe (U.S. Bureau of Mines), W. A. Simmonds (Gas Research Board) and K. Wohl (Delaware). One of us (A.G.G.) is indebted to the Royal Society for the award of a Warren Research Fellowship, and the other (H.G.W.) to the Chief Scientist, Ministry of Supply, for permission to publish. We also wish to thank the Editor of *Endeavour* for help in supplying the blocks for the colour plates.

Chemical Engineering Department,   A.G.G.
Imperial College,
London, S.W.7.

Royal Aircraft Establishment,   H.G.W.
Farnborough,
Hants.

May 1952

Preface to Fourth Edition

The first edition of this book appeared in 1953 and it was revised in 1960 and 1970. This further revision has been prompted by the steady demand. Although the lay-out and chapter headings have remained practically unchanged for all editions, revision in each case has been thorough. In the preface to the third edition we said that knowledge of flames was then more certain and quantitative so that the book was in more definitive form; this has been partly true, but the development of new techniques, such as laser-Doppler anemometry, laser-Raman spectroscopy and computer handling of data, have led to appreciable advances in our detailed knowledge of flames. This fourth edition appears deceptively similar to the third. The problem has been to bring the book up to date without losing the basic clarity and emphasis on a simple physical understanding of flame processes. To this end we have added over 200 new references, deleting over 60 old ones, and have also inserted 17 new figures and two plates, with some extra sections on new techniques and a more thorough revision of some parts, such as flame propagation, structure of turbulent flames and soot formation. We have still endeavoured to keep the emphasis on the basic physical principles, avoiding detailed chemical processes or too much formal mathematics, but we have been able to include some extra numerical data, e.g. for reaction rates.

There are some fairly recent books on various aspects of combustion such as *Flame Structure* (Fristrom & Westenberg, 1965), *Fundamentals of Combustion* (Strehlow, 1968), *Electrical Aspects of Combustion* (Lawton & Weinberg, 1969), *Industrial Flames. I. Measurements in Flames* (Chedaille & Braud, 1972), *Combustion Technology* (Palmer & Beer, 1974), *Spectroscopy of Flames* (Gaydon, 1974) and *Introduction to Combustion Phenomena* (Kanury, 1975), but the emphasis in all these is different and they do not cover in any detail the subjects discussed here.

A minor problem has been the recent change to International System (S.I.) units. Most of the work on flames so far published has been in c.g.s. or more practical units. While a change here to S.I. would appear desirable, there are difficulties in making a simple conversion. Thus if a measurement is made at 10 torr pressure, using an actual mercury manometer, is it realistic to say it was made at 1·333 kN m^{-2}? Should work at 1 atm. be described as at 101·3 kN m^{-2}? We have also left most thermochemical data in kcal/mole (1 kcal = 4·184 J) and flame speeds in cm/sec rather than m/s, but have changed temperatures from °K to K and inches to mm.

The revision for this edition has been done by one of us (A.G.G.) as the other (H.G.W.) has been unable to spare the time.

We are grateful to Dr D. R. Ballal and Dr P. A. Marsh for providing prints for the new Plates 21 and 22.

Chemical Engineering Department,   A.G.G.
Imperial College,
London, S.W.7.

Institute for Defense Analyses,   H.G.W.
Arlington,
Virginia.

Contents

Preface to First Edition   *page* v

Preface to Fourth Edition   vii

I INTRODUCTION   1
Flame – Stationary flames – Propagating flames – Physics or chemistry?

II PREMIXED FLAMES   7
The Bunsen burner – The Méker burner – Laminar premixed flames – The Smithells separator – Turbulent flames – Flame stability – Quenching – Low-pressure flames and their structure – Spontaneous ignition and cool flames – Ignition by external means – Limits of flame propagation – The position of luminous zones – Polyhedral and cellular flames – The transition between premixed flames and diffusion flames.

III FLOW VISUALISATION AND FLAME PHOTOGRAPHY   40
The gas flow from a burner – The measurement of the gas flow – Laser-Doppler anemometry – The photography of the flame cone – Particle-track photography – Schlieren photography – Interferometry – Lasers and holography – The flame thrust, and its effect on cone shape – The tip of the flame.

IV MEASUREMENTS OF BURNING VELOCITY   58
The burning velocity – Gouy's flame-area method – Modified flame-area and cone-angle methods – The nozzle method – The particle-track method – Methods of locating the flame front – Rectified flat-flame methods – The flame-thrust method – The soap-bubble method – The spherical-bomb method –

Page

The revised tube method – Comparison of burning velocity measurements – The temperature dependence of burning velocity – The pressure dependence of burning velocity – The influence of hydrocarbon structure on burning velocity – The influence of mixture strength, diluents and additives – The burning velocity of turbulent flames.

V THE STRUCTURE OF THE REACTION ZONE AND ITS RELATION TO FLAME PROPAGATION 92

Introduction – The preheating zone – The ignition point – The reaction zone; radical concentrations – The reaction zone; its thickness in relation to burning velocity – The relation between burning velocity, temperature and radical concentration – The effect of additives on burning velocity, flammability limits and spontaneous thermal ignition – Early thermal and diffusion theories of flame propagation – Combined thermal and diffusion theories – The concept of excess enthalpy – The concept of flame stretch – Reaction rates – The formation of oxides of nitrogen – Flame propagation in turbulent gases.

VI DIFFUSION FLAMES 146

The height and shape of small flames – The candle flame – The Wolfhard–Parker flat diffusion flame – Spectroscopic study of structure – Counter-flow diffusion flames – Comparison of diffusion with premixed flames – Turbulent flames – Thin flames – The Coanda burner – Furnace flames – Combustion of droplets and dusts – Flames on wicks; mixed fuels – Combustion of liquids at a free surface.

VII FLAME NOISE AND FLAME OSCILLATIONS 177

Flame noise – Jet noise – Singing flames – Combustion chamber oscillations – Flame flicker and other instabilities – Sensitive flames – The influence of sound on premixed flames.

VIII SOLID CARBON IN FLAMES 195

Introduction – Diffusion flames – Premixed flames – The effect of pressure – The effect of temperature – Explosion flames – The effect of non-metallic additives – The effect of metallic additives – Electrical effects – The nature of carbon particles and deposits – The equilibrium between carbon and the burnt gases – Detection of intermediate products – Isotope

Page

tracer studies – The pyrolysis of hydrocarbons – The problem of the formation and growth of soot particles – The carbon-black process.

IX RADIATION PROCESSES IN FLAMES 238

The nature of radiation – The equipartition of energy – Processes of electronic excitation – Conditions for radiative equilibrium – Chemiluminescence – Departures from equilibrium in flame gases – Radiation and scattering from solid particles – Calculations of radiative heat transfer.

X FLAME TEMPERATURE. I. MEASUREMENT BY THE SPECTRUM-LINE REVERSAL METHOD 268

Flame temperature? – The sodium-line reversal method – Corrections for reflection loss and for change of brightness temperature with wavelength – Effect of complex flame structure – Use of other lines for reversal work – Background sources – Adaptations for non-visual recording and for time-resolved studies – Accuracy and reliability – The Kurlbaum method – Results for interconal gases – Results for reaction zones.

XI FLAME TEMPERATURE. II. OTHER METHODS OF MEASUREMENT 291

Brightness and emissivity methods – The colour-temperature method – The line-ratio method – Effective translational temperature by Doppler broadening – Rotational temperatures – Vibrational temperatures – Laser-Raman scattering – Refractive-index methods – α-particle and X-ray methods – Sonic methods – Hot-wire methods – Ionisation and electron temperatures.

XII FLAME TEMPERATURE. III. CALCULATED VALUES 323

Systematic calculation of composition for a given temperature – Calculation of temperature – Values for equilibrium constants, enthalpies, etc. – Discussion of the effect of varying mixture strength – The effect of pressure – Temperatures and compositions of some typical flames.

Page

XIII IONISATION IN FLAMES   340

Measurement of ionisation – Ionisation in local thermodynamic equilibrium (LTE) – The role of chemical reactions in the ionisation of metal additives – Ionisation in hydrocarbon flames – Further investigations of ion reactions in flames – The effect of electric fields on flames – Ionisation in rocket exhausts.

XIV COMBUSTION PROCESSES OF HIGH-ENERGY AND ROCKET-TYPE FUELS   372

Flames supported by oxides of nitrogen – Nitric acid flames – Nitrate and nitrite flames – Propellant flames – Halogens as oxidisers – Ozone – Metal alkyl flames – High-energy fuels – Combustion of solid fuels – Ignition and flammability of solids – The burning characteristics of solids – Supersonic combustion.

XV FLAME PROBLEMS   401

    REFERENCES   406

    AUTHOR INDEX   427

    SUBJECT INDEX, with definitions of symbols and values of physical constants, units and conversion factors.   439

List of Plates

COLOUR
(*between pages* 10–11 and 20–21)

1. Premixed and diffusion flames of ethylene.
2. Smithells separator, low-pressure flame.

MONOCHROME
(*between pages* 210–211)

3. Rectified flat flame, with cool flame below; air entrainment into diffusion flame.
4. Carbon formation in premixed flames.
5. Cellular and polyhedral flame structures.
6. Examples of flame photography.
7. Structure of turbulent flames.
8. Effect of pressure and 'vitiation' on diffusion flames.
9. Direct and schlieren pictures of diffusion flames.
10. Instantaneous schlieren photographs of turbulent diffusion flames.
11. Spectra of flat diffusion flames.
12. Schlieren pictures of candle and burning kerosene spray.
13. Height to turbulence in flames; flat diffusion flames; starting vortices in jets.
14. Photographs of vibrating flames.
15. Carbon particles in flames and carbon deposits.
16. Spectra showing abnormal excitation and OH rotational and vibrational anomalies.
17. Flames of dusts and rocket-type fuels.
18. Spectra of flames containing oxides of nitrogen.
19. Spectra of CH_4/NO_2, CH_3NO_2 and CH_3NO_3 flames.
20. Spectra of methane/chlorine trifluoride flames.
21. Turbulent flame spread, for turbulent burning velocity.
22. Phase-contrast electron micrographs of soot and carbon black.

Chapter I

Introduction

Flames

We all have a pretty clear idea of what is meant by a flame, but it is very difficult to give the word a precise meaning. Although a few flames, such as that of hydrogen burning with clean dust-free air, may be practically non-luminous, we generally associate the emission of light with flames. We shall see that both the quantity and quality of this light may differ remarkably from that of mere hot gases, and the causes of this will form a part of our examination of the nature of flame. The light of the flame is also useful in locating the position and form of the flame front, which is important for the determination of burning velocity and quenching distance; the limitations of this method of examining the flame and the use of other optical methods such as those of shadow and schlieren photography will be considered. The emission and absorption of radiation by the hot flame gases also offers several possible methods of measuring the gas temperature and studying the state of equilibrium in the reacting gases. Indeed the emission of light is one of the most characteristic properties of a flame, and it is natural that we should seek to use this light to learn as much as possible about flame processes.

Another general property of ordinary flames is that of a rapid temperature rise, usually to a temperature of over 1400 K. We shall deal fairly fully with the problems of the measurement and calculation of the final temperature of the burnt gases, and also the temperature gradient through the reaction zone of the flame. Although most flames are very hot, there are some exceptions such as the well-known glow of oxidising phosphorus and the 'cool flames' of hydrocarbons and certain other organic vapours which may give luminous reactions at temperatures between 200°C and 400°C; the steady glow obtained during such controlled reactions can hardly be called a flame, but the pulses of luminosity which sweep through the reacting mixture, which are accompanied by fairly rapid changes in chemical composition, although not in temperature, probably do deserve the name of flame. These cool flames will not, however, be considered in detail in this book.

Again, flames are generally associated with oxidation processes. There are, however, some other reactions, such as those of fluorine and other halogens with hydrogen and hydrocarbons, which may not involve any oxygen at all and yet quite clearly merit the term flame. The flames of fluorine with hydrocarbons are indeed very similar to those of organic fuels with oxygen, being very brilliant and giving the same characteristic banded spectra of diatomic molecules such as C_2. We shall consider some of these flames which are not supported by air or oxygen in Chapter XIV.

True flames are usually associated with highly exothermic reactions between gases. The combustion of finely divided solid particles, e.g. of aluminium, or sprays of liquids, may, however, also give rise to flames. While this book is mainly concerned with small laboratory-scale flames of burning gases, we shall also refer to the burning of oil sprays and the combustion of solid particles; in the latter case the radiation may be relatively important because it may play a bigger part than usual in heat transfer and maintenance of the flame.

Stationary flames

This book is devoted almost entirely to the study of stationary flames, as opposed to propagating or explosion flames. Generally these are the flames which are of greatest industrial importance for heating and other purposes, and also are more suitable for making fundamental scientific measurements. This class of flame may be considered for convenience as of two general types. In the first the fuel burns as it is brought into contact with the air. On a small scale the combustion processes are then mainly determined by the rate of inter-diffusion of air and fuel, and we shall refer to these as *diffusion flames*. In larger flames of this type the mixing may be due to turbulence and other movements of the gases rather than to diffusion, and in these cases the problems of flame stability and size are mostly of an aerodynamic nature. Industrial flames of this large type have been discussed in the books by Thring (1962) and by Chedaille & Braud (1972); here we shall be more interested in the structure of the simple diffusion flames. The second general type of stationary flame is that where the fuel and air or oxygen are premixed; we shall refer to these briefly, but a little loosely, as *premixed flames*. The best example of this type of flame is the common Bunsen flame, in which the premixed gases flow up a burner tube at a rate which exceeds the normal burning velocity of the mixture, a steady flame being maintained above the burner top.

The premixed flames are most used for domestic heating, in gas fires and for cooking. They have also been the subject of much more scientific investigation than diffusion flames because they can be used to give information about a number of fundamental properties of the gas mixture such as its burning velocity and temperature. We shall discuss these first, in Chapters II and III, and the measurement of burning velocities, in Chapter IV; the discussion of theories of flame propagation, in Chapter V, will be mainly concerned with premixed stationary flames.

Diffusion flames include simple gas jets burning in air or oxygen and flames on wicks, in which the heat transfer from the flame causes a steady production of flammable vapour. With diffusion flames there is a steady change in chemical composition as one passes through the flame, and there are few physical constants which can be measured. The rate of combustion is determined by rates of diffusion and mixing; there is no burning velocity to measure. The temperature distribution through the flame and its radiation are of interest, but although regions of a diffusion flame often give temperatures near the theoretical maximum for a stoichiometric premixed flame there is no true final flame temperature unless the air supply as well as the fuel supply is restricted, which is not usually the case. In certain enclosed systems such as furnaces it may be possible to define the temperature of a diffusion flame, or rather of the burnt gases after the reactions are complete. There are, however, a number of interesting features about the structure of diffusion flames which have recently come to light, and these are discussed in Chapter VI. Thus it is found from spectroscopic studies that in hydrocarbon/oxygen diffusion flames the hydrocarbon is decomposed thermally before it comes in contact with any oxygen.

Although the division into premixed flames and diffusion flames is convenient, it cannot be insisted on too rigidly. Thus in a diffusion flame with both air and fuel moving upwards, the flame would lift off the burner if it were not for the existence of some mixing near the burner rim, causing a flame which can propagate back against the gas stream and maintain the flame. As the pressure is reduced, this region of premixing near the base of a diffusion flame becomes larger until the whole flame may, at low pressure, become indistinguishable from a premixed flame. Similarly, rich premixed flames of the Bunsen type, although giving clear inner cones, are dependent on the outer diffusion flame, or outer cone, for their stability. This is often not fully realised; many apparently stable premixed flames will lift off the burner if enclosed. It seems that oil sprays may approximate to the conditions of either premixed or diffusion flames according to the size of the oil droplets. Some phenomena which occur in both premixed and diffusion flames are described in Chapter VIII (the formation and combustion of solid carbon particles) and Chapter VII (flame vibrations, flame noise and the effects of sound on flames).

Propagating flames

Expanding spherical explosion waves, either in closed vessels or by the soap-bubble technique, are of some interest for measurements of burning velocity (Chapter IV) and also sometimes show interesting cellular structure, like some flat stationary flames. Generally, however, the propagation of explosion flames in tubes and in closed vessels will not be dealt with here.

A flame propagating down a tube usually proceeds at first with fairly uniform speed, this being determined by the fundamental burning velocity, the flame area and the expansion of the hot gases behind the reaction zone. As the flame proceeds the gas expansion thrusts the flame front forward with increasing velocity so that the flame assumes a long sausage shape of larger area. As the flame proceeds down the tube its velocity thus increases, and this may be further accelerated by turbulence, which enlarges the area of the reaction zone. Depending on the size of the tube, and thus the Reynolds number, the flame may reach high speeds. The moving gas thus acts like a piston on the unburnt portion of the gas, and the resulting pressure pulse may sharpen up to a shock wave. If this shock wave is strong enough, the temperature behind the shock front may be high enough to cause rapid chemical reaction and a *detonation* may be formed. The flame front is then propagated by the shock wave which travels at a speed, usually between 1500 and 3000 m/s, which depends on the velocity of sound under the conditions of high temperature and pressure which occur behind the front. For gases initially at 1 atm, pressures up to 20 atm may be produced behind the detonation, and up to 100 atm if the detonation is reflected at the closed end of the tube; hence the dangerously destructive effects of detonations. Early treatments of detonation processes are given in the books by Bone & Townend (1927) and Lewis & von Elbe (1961); a simple treatment of the initiation and propagation mechanism is given by Gaydon & Hurle (1963), and a more thorough treatment of the detailed structure of detonations by Strehlow (1968). It should be mentioned here, however, that explosion pressures can in some cases be higher than detonation pressures. If no shock wave is set up in the closed tube, then the unburnt gas is compressed to high pressures before the combustion occurs. Thus the combustion process occurs at pressures higher than the initial pressure and this raises in turn the final pressure. In contrast, in a detonation no compression of the unburnt gas takes place once the detonation has started (Wolfhard & Seamans, 1962).

The most important application of closed-vessel explosions is of course in the internal-combustion engine. This has, indeed, been the subject of a tremendous amount of research and development. Generally the problems are mainly of a technical nature, and the actual flame in the cylinder of an internal-combustion engine is rather rarely studied. There are some interesting problems of a chemical nature, such as knock, which is due to pre-inflammation of the charge ahead of the main flame front and may, within limits, be inhibited by chemical additives like lead tetra-ethyl. These effects are linked with studies of thermal ignition, the production of cool flames and two-stage ignition processes and hardly come within the scope of this study of flames. They are described in a number of existing books (Lewis & von Elbe, 1961; Minkoff & Tipper, 1962).

Physics or chemistry?

The subject of combustion falls, of course, between physics and chemistry. It has, however, usually been considered more as a branch of chemistry, and from the earliest days of the phlogiston theory until the more recent theories of chain reactions and chemical kinetics, those studying combustion have made major contributions to the science of chemistry. In recent years, though, the emphasis in combustion work has been shifting more and more towards the physical processes.

In the low-temperature region, for work on cool flames, for determination of ignition temperatures and for limits of flammability for mixtures which are on the border-line of self-propagating combustion, the chemical processes are probably the most important.

For practical purposes, limiting flame conditions are of overriding importance in fire prevention and the assessment of explosion hazards. The book, however, concerns itself with the heat release and radiation of hot flames. The release of energy by the chemical reaction is the beginning of it all, but the governing factors of flames and their radiation are mainly physical. For diffusion flames the rate of chemical reaction is usually immaterial; it is the rate of inter-diffusion and, for larger flames, the rate of mixing which are the rate-determining processes. The aerodynamics of the system (onset of turbulence, entrainment of air, etc.) are most important for determining the flame size and stability. We also encounter problems in heat transfer by conduction and radiation. In premixed flames the general problems are similar, but in this case the rate of chemical reaction does determine the burning velocity, although the propagation also involves the rate of heat transfer ahead of the flame front and the diffusion of active species. The stability of flames may also be affected by vibrations due to resonance effects (e.g. singing flames) or external noise (sensitive flames), while flame noise is becoming an increasing problem.

Many of the methods used to study flames involve physical measurements. Thus studies of flame speed involve those of flame photography and lead to some tricky optical problems. The measurement of temperature is done largely by optical methods and is described in Chapters X and XI;

these follow Chapter IX which deals with the emission of radiation by flames.

In considering the detailed structure of the reaction zone and the adjustment of the equilibrium after the rapid initial heat release, we are brought in contact with detailed molecular processes such as lag in equipartition of vibrational and electronic excitation energy with that in other forms. These are discussed in Chapter IX. There is also some interest in the causes and effects of the high ionisation in flame gases; these are examined in Chapter XIII.

Astrophysicists have made staggering progress in their study of stellar atmospheres from quantitative measurements of the emission spectra. In fact it might almost be true to say that we know more about processes in some stellar atmospheres than we do about processes in a Bunsen burner! Surely with a controllable laboratory source it should be possible to learn from emission and absorption spectra a tremendous amount more than we do at present. The purely spectroscopic considerations are discussed in *The Spectroscopy of Flames* (Gaydon, 1974), and there is comparatively little overlap between that book and this. Results on the high level of electronic excitation in the reaction zones of organic flames and on the spectra and structure of diffusion flames are included here. We also discuss here the effects of departures from radiative equilibrium.

In earlier editions of this book we said that little progress had been made in applying these astrophysical-type techniques to the study of flames, but in recent years there has indeed been a lot of good quantitative work on flame radiation, using measurements of emission, absorption and scattering; new techniques, such as the use of lasers, have greatly contributed to these studies.

Chapter II

Premixed Flames

The Bunsen burner

The familiar laboratory burner was invented by Bunsen around 1855. It resulted in a major change in the gas industry. Previously, simple flames of the diffusion type had been used; these were luminous and sometimes even smoky and tended to form carbon deposits on surfaces in contact with the flame; their effective temperatures were rather low. The clear premixed flames give much more intense combustion, have a higher effective temperature, give better heat transfer, and are relatively free from sooting troubles. The simple principle of the Bunsen burner is now incorporated in many gas appliances such as cooking stoves and domestic gas fires.

The gas issues from a small nozzle or orifice and entrains some air (primary air) injector-wise. The mixture of gas and primary air passes up the burner tube at a speed which is sufficient to prevent the flame striking back down the tube. The mixture thus burns at the top of the burner, the combustion being assisted by the surrounding (secondary) air. This secondary air plays an important part in stabilising the flame; without it the limits between striking back and blow-off are surprisingly narrow.

In the simple type of Bunsen burner (Fig. 2.1) the gas issues from a nozzle shaped orifice about 1·5 mm in diameter. The air which is entrained enters through either one or two adjustable air holes. Even with the air holes fully open, the amount of air entrained is usually well below that for complete combustion of the gas (i.e. below the stoichiometric amount), and in some cases may be less than half the amount required. For good air entrainment the air hole must be of sufficient size; Eiseman (1949) says it should be at least 1·25 times the area of the burner mouth. The proportion of air entrained can also be increased by reducing the size of the gas orifice, but this reduces the size and power of the flame unless the gas pressure is increased. The amount of air which can be entrained is limited by the pressure of the gas supply; the maximum velocity of the gas jet cannot exceed $(2 \times \text{pressure of gas supply}/\text{gas density})^{\frac{1}{2}}$, where the velocity is measured in cm/sec and the pressure in dyne cm^{-2}. The momentum of the gas in the jet must then be shared with the entrained air. Since the gas and air mixture must flow up the tube fast enough to prevent the flame flashing back, there is a maximum possible air entrainment for a given burner and gas supply pressure. The theory of air entrainment and flow conditions in the burner have been given by Lewis & Grumer (1948) and Lewis & von Elbe (1961).

There are limits to the size of Bunsen burners. On the small side the flame tends to blow-off, and the minimum size of a burner on which a flame can be stabilised is related to the quenching distance. The maximum size is set by the increasing tendency to flash-back with large burners. As we have seen there is a maximum gas velocity for a given supply pressure, and in order to obtain sufficient velocity at all points in the tube to prevent flash-back the average velocity must be increased for bigger burners. The best dimensions will depend on the nature of the gas supply. For manufactured gas, which usually contains 40% to 50% of hydrogen by volume and has a high burning velocity, a normal size for a Bunsen is 10 mm diameter, with a maximum size of up to 20 mm. With natural gas, consisting mainly of methane which has a low flame speed, larger burners are possible, and the burner is often designed so that it opens out to a wider diameter at the mouth. The natural gas has to be supplied at higher pressure, too, to give it sufficient momentum for adequate air entrainment.

The stability of Bunsen-type flames for various fuels, including pure gases and manufactured town's gases, has been examined by Fuidge, Murch & Pleasance (1939). The stability limits for hydrogen are rather low, because of the high flame speed which causes a tendency to flash-back. Methane has good flexibility but with a tendency to blow-off; flash-back rarely occurs. Carbon monoxide is poor. Ethylene has wide stability limits but tends to give luminous smoky flames. The addition of a little hydrogen decreases the tendency of methane to blow-off. Also it is found that small amounts of unsaturated hydrocarbons are valuable in improving the stability for manufactured gas; they also make the flames more easily visible, which may be a safety advantage for domestic use. Inerts, such as nitrogen, are also of some value for gas with a high hydrogen content because they assist air entrainment since the momentum of the jet is proportional to the square root of the gas density.

To obtain good mixing and allow time for the turbulence created by the gas jet to die down, we obviously require a long burner (*see* page 14), but if it is too long it will cause unnecessary resistance to flow and reduce air entrainment. For natural gas flames Eiseman says the length should be about six times the diameter at the mouth. For manufactured gas, with higher flame speed and higher flow rates, a longer burner is presumably required, probably at least ten times the diameter. The orifice must be accurately aligned to give the best stability and the best burners contain an adjustable orifice as well as adjustable air ports.

The flame produced on a Bunsen burner is too familiar to need much description. With maximum aeration there is a roughly cone-shaped inner luminous region which is blue or blue-green in colour. This is surrounded

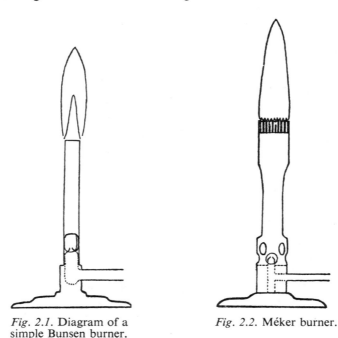

Fig. 2.1. Diagram of a simple Bunsen burner.

Fig. 2.2. Méker burner.

by a paler blue-violet sheath of flame, usually referred to as the outer cone, although it is not actually conical in shape. The ordinary Bunsen flame is too unsteady to be suitable for detailed flame studies. The structure of Bunsen-type flames will be discussed in later sections.

The Méker burner

The chief limitations of the simple Bunsen are its inadequate entrainment of air and the tendency to flash-back with large diameter burners. In the Méker burner* rather greater air entrainment is usually achieved, giving a higher temperature, and larger burners are possible. The general principle is the same as for the Bunsen, but a deep grid across the mouth prevents

* This type of burner was developed by G. Méker in France. In 1965 we were pleased to get a letter from M. Méker himself and to learn that the burners are still marketed by the firm of G. Méker & Co. at Courbevoie.

flash-back. This grid is usually of metal (most frequently nickel) but may be of a refractory ceramic. In addition to this grid, Méker burners usually have a venturi throat (*see* Fig. 2.2) above the gas orifice to increase air entrainment. The best size for the throat corresponds to about 40% of that of the burner mouth (Eiseman, 1949), and the air holes should have an area at least 2·25 times that of the mouth. The size of the holes in the grid is probably not very critical. If they are too large there will be a risk of flash-back through the holes for fast-burning mixtures, and if they are too small and close there is a tendency for the normally separate flame cones to fuse together and lift off the grid surface to give a more Bunsen-shaped cone. The central apertures in the grid may give slightly different combustion conditions to those in the outer cones or in a Bunsen burner as there is no secondary air reaching this region. This secondary air is important in preventing lifting of the flame, and with deficient air supply there is always a tendency for the flame cones in the centre of the grid to become unstable. This is particularly so for methane or natural gas as fuel.

Laminar premixed flames

The Bunsen flame, in which the air is entrained by the gas jet, is not very suitable for systematic study. The flame is usually unsteady because of turbulence; the mixture strength can only be varied over a limited range on the fuel-rich side, and stoichiometric and weak mixtures cannot be studied; the flow-metering of the air supply is difficult. It is therefore easier to study flames in which the gases are mixed at a T-piece after flow-metering. The burner tube must be sufficiently long to enable good mixing and steady flow conditions to be established. It is usually advisable with air flames to use a tube at least 30 cm long, and for fast-burning mixtures much longer tubes may be necessary (*see* page 14). The optimum burner diameter will again depend on the flame speed, but is best kept as large as possible without the flame becoming turbulent. For H_2 and C_2H_2, burners of only a few millimetres can be used. For CH_4, burners up to about 2 cm may be utilised without difficulty. For flames with oxygen, much smaller burner diameters must be used. In order to prevent turbulence a burner with a nozzle mouth is best, although this increases the risk of the flame striking back violently; commercial welding burners, which have a small robustly-built mixing chamber, are often convenient for studying premixed flames with oxygen.

For a hydrocarbon, such as ethylene, burning with air, the simple diffusion flame of pure ethylene burning in the surrounding air at the top of a circular burner is luminous (i.e. yellow) due to emission by incandescent carbon particles; this is shown in Plate 1c. With a little air added to the ethylene the flame becomes less luminous, the yellow region contracting

PLATE 1

(c) Diffusion flame of ethylene in air.

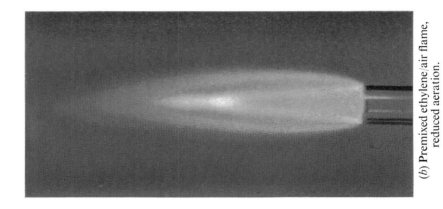

(b) Premixed ethylene/air flame, reduced aeration.

(a) Premixed ethylene/air flame, fairly high aeration.

and an outer blue-violet sheath developing. With slightly more air the inner cone becomes obvious (Plate 1b) with some residual yellow luminosity at the tip. For acetylene, on the other hand, the carbon luminosity appears to start from the base of the inner cone and extend upwards as a luminous cylinder. In the case of ethylene and some other fuels this yellow carbon region appears to start within the inner cone, just below the tip. With ethylene the luminous tip just disappears with about half the air supply required for complete combustion; on equilibrium considerations it should go earlier, at about one-third of the required air supply, but the flame is actually quite yellow with this mixture strength. After the disappearance of the carbon the inner cone becomes brighter, shorter and blue-green in colour (Plate 1a) and the outer cone also contracts. As we approach the stoichiometric mixture the inner cone becomes less bright and more blue-violet and the outer cone becomes shorter and less distinct. Beyond the stoichiometric point the outer cone disappears, while the inner cone becomes paler and more violet. The burnt gases above the inner cone emit a pale bluish-grey radiation. Beyond a certain mixture strength the flame blows-off and cannot be relighted.

All hydrocarbons and organic fuels, including methane, methyl alcohol and formaldehyde give this general type of flame with a well-marked inner cone, although methyl alcohol and formaldehyde do not show the luminous tip due to carbon formation; the diffusion flames of these two gases are clear bluish. Flames of hydrogen, on the other hand, if pure, are practically non-luminous, and there is no obvious inner cone. There is an inner, cone-shaped, dark region but no increase in visible luminosity in the flame front. The OH radiation in the ultra-violet is, however, stronger in a rather diffuse region at the base of hydrogen-air flames. With carbon monoxide the flame is bright blue and the inner cone is fairly well-defined, consisting of a relatively thick layer of increased luminosity. The formation of clearly defined, bright, inner cones is not restricted to organic fuels. The ammonia/oxygen flame and the hydrogen/nitrous oxide flames among others also give well-defined inner cones.

Flames of hydrocarbons and organic fuels with oxygen generally have a similar structure to the flames with air, but are much brighter, especially at the inner cone. With rich mixtures, just slightly weaker than those giving carbon luminosity, these flames have a luminous mantle or sheath surrounding the inner cone. This mantle shows a banded spectrum (due to bands of radicals like C_2, CH and CN) superimposed on a continuum due to carbon or incipient carbon particles. It may be a few millimetres thick, becoming larger and whiter in colour as the mixture is made more fuel-rich, until it finally merges with the carbon luminosity. We shall see in the following sections that the inner cone corresponds to the flame front or reaction zone, and the outer cone to the diffusion flame where excess oxidisable constituents (mainly CO and H_2) burn, and this mantle above the inner cone for rich mixtures may be considered as a very hot region where there is an equilibrium concentration of radicals like CH and C_2 which can give thermal radiation in the visible region. Unfortunately this mantle is not very easy to study separately as it is usually only obtained with flames which are so rich that they are difficult to stabilise without the support of secondary air. In some flames of acetylene with oxygen we have found that if secondary air is excluded the mantle spreads much higher up the flame, suggesting that its normal limitation to a few millimetres around the inner cone is due to inward diffusion of either air or combustion products from the outer secondary combustion. The mantle is only observed well in very hot flames, not in the cooler flames with air. For ethylene/air the mantle is usually just visible, but is difficult to distinguish from the onset of carbon formation; for acetylene/air there may be some glow just above the tip of the inner cone commencing at about the mixture strength for which carbon formation starts from the base of the cone. These mantles are especially noticeable in flames with N_2O or NO as oxidiser.

The Smithells separator

The separation of the inner from the outer cone of a Bunsen-type flame was achieved independently by Smithells and by Teclu around 1891 (Smithells & Ingle, 1892). They found that if a glass tube was fitted over the top of a Bunsen burner, then by reducing the gas flow the flame burning on top of the glass tube could be made to strike back and burn on the Bunsen tube. The flame then appeared as two separated flames, the inner cone resting on the top of the Bunsen, and the outer cone continuing to burn at the top of the glass tube.

These separated flames are best studied on concentric glass or quartz tubes of adequate length, the gases being premixed at the bottom of the inner tube after flow-metering. The best dimensions for the tubes again depend on the fuel, but for many hydrocarbons a convenient diameter for the inner tube is around 1 cm and around 2 cm for the outer tube. The gap at the bottom of the outer tube between the inner and outer tubes is usually closed with a ring of cork or rubber so that the outer tube may be slid up and down to vary the distance between the cones.

Plate 2a shows ethylene and air burning at a separator; the inner tube had in this case an oval section, so that the inner cone is not the typical shape.

With the aid of the separator the general structure of rich Bunsen-type flames at once becomes apparent. The inner cone is the initial flame front due to combustion of the fuel with the primary air, and the outer sheath of flame is due to the further burning of the mixture when it comes in contact with the secondary air. The inner cone is thus the true flame of the premixed fuel and air. The outer cone is a diffusion flame.

A striking feature for these rich mixtures revealed by the Smithells separator is that the hot gases between the inner cone and the outer cone, which we shall refer to as the interconal gases, are practically non-luminous. They do, however, emit strongly in the infra-red. All the visible light comes from the two reaction zones.

Chemical analysis of the interconal gases shows that they consist mainly of CO, H_2, CO_2, H_2O and N_2. The hydrocarbons are usually almost completely broken up in their passage through the inner cone. The composition of the interconal gases varies, of course, with the initial mixture strength, the amount of CO and H_2 decreasing and the amount of H_2O and CO_2 increasing as more air is added. Table 2.1 gives some of the original analyses from Smithells & Ingle (1892).

TABLE 2.1

Analyses of interconal gases for flames with air in a Smithells separator: per cent by volume

Fuel	C_2H_4	CH_4	C_6H_6	Coal-gas	C_5H_{12}
Diameter outer tube	29 mm	29	19	19½	20
Diameter inner tube	20 mm	20	8	12	13
CO_2	3·6	6·8	13·1	4·2	7·0
CO	15·6	4·5	5·0	8·8	7·9
Hydrocarbons	1·3	—	0·6	—	—
H_2	9·4	3·9	0·64	9·3	5·4
H_2O	9·5	17·6	7·7	16·0	13·1
N_2	60·6	67·2	73·1	62·0	66·2

Thus it seems that the interconal gases are fairly near an equilibrium mixture of CO, H_2, CO_2 and H_2O, with N_2 as diluent. It is possible that there may be some minor departures from equilibrium. Nitric oxide appears to pass through the inner cone of some flames without being decomposed very much. It is possible that a little methane may also get through with flames of relatively low temperature, and the above analyses do show that in some cases there are small amounts of hydrocarbons. The possibility of other forms of disequilibrium, such as the presence of excess free atoms or abnormally high ionisation will be discussed in later chapters.

If the inner and outer cones are separated by some distance (perhaps 15 cm or more) then it is possible to blow out the outer cone. If, however, the tube length is short so that the inner and outer cones are relatively close then it is impossible to extinguish the outer cone alone. This effect is probably caused mainly by the cooling of the interconal gases in a long tube.

Turbulent flames

A turbulent flame can usually be recognised by its hissing sound, and by the thickening of the flame front and rounding and shortening of the tip of the flame.

The state of flow in a long tube is either laminar or turbulent. In the laminar case the flow of all volume elements is parallel, but for turbulence the velocities have components normal to the average flow direction. For laminar flow the gas tends to develop a parabolic distribution of velocity across the burner, but for turbulent flow the velocity near the walls is much higher than in the laminar case. The state of flow is characterised by a dimensionless quantity known as the Reynolds number, R_e, where

$$R_e = v \cdot d/v,$$

where v is the average velocity, d is the diameter of the tube (or in the general case a characteristic length) and v is the kinematic viscosity (= viscosity/density).

For $R_e < 2300$ the flow in a tube supporting a flame is always laminar and for $R_e > 3200$ it is usually turbulent (Mache, 1943). For the range between R_e 2300 and 3200 small random flow fluctuations tend to be damped out, but the flow is unstable to larger fluctuations. The flame thus tends to alternate abruptly between the laminar and turbulent forms with longer or shorter periods of laminar flow according to whether R_e is nearer the lower or higher limit; the frequency of alternation may be of the order of a second.

For laminar flow we have a thin flame front which is between conical and slightly bell-shaped. For values of R_e above the turbulent limit but not too large, the centre of the flame becomes blurred, the rim remaining steady. With increased turbulence the whole flame front becomes very thick and blurred; this whole volume in which the primary reaction occurs may be referred to as the flame brush; Plate 7a shows the appearance of such a brush (from Bollinger & Williams, 1949). With very fast flows the flame becomes still more rounded and shaped rather more like the end of a bolt, the diameter at the base often exceeding that of the tubes; for Bunsen-type flames with air, however, the flame blows off long before this stage is reached; it is usually necessary to hold such highly turbulent flames with a pilot light near the rim of the burner.

In order to get fully developed turbulence or pure laminar flow in a tube, we require a certain minimum length, L_m. The value of L_m depends on Reynolds number, R_e, and tube diameter d. Various values for the relationship have been given, but, roughly,

$$L_m = 0.05 \, R_e \cdot d.$$

Even for a simple Bunsen flame, R_e is of the order 1000, so for a 1 cm diameter tube, the minimum length for a fully established laminar flow pattern with a parabolic velocity distribution is around 50 cm, much longer than the normal Bunsen burner. For established turbulent-flow patterns, at still higher R_e, extremely long tubes would be necessary. Turbulence can be created by the introduction of a grid in the flow, although a long tube is still required to give a steady mean velocity profile. Laminar flow can be maintained to R_e rather above 2300 by using smooth tubes and taking care not to introduce too much swirl at the injector. If laminar flow is established through a wide tube with a short nozzle-like outlet it is possible to maintain laminar flow through the nozzle to much higher Reynolds numbers, even up to $R_e = 40\,000$.

The structure of the reaction zone of a flame under turbulent-flow conditions is discussed more fully at the end of Chapter V.

Flame stability

The Bunsen-type flame is only stable within certain flow limits, the stability limits. The flame is held from striking back by the gas flow. The flame front will always adjust itself so that at any point in the flame front the component of the gas flow normal to the flame front is equal to the normal burning velocity at that point. A full discussion of burning velocity measurements and variations will be given in a later chapter, but roughly we may say that the burning velocity is nearly constant over most of the flame front but falls sharply to a much lower value near the cool rim of the burner, due to quenching effects at this rim. For a long tube we have a parabolic velocity profile in the gas stream as it emerges from the burner, the velocity being zero at the walls and rising towards a maximum in the centre. Owing to this zero velocity at the walls a flame would always strike back along the walls if it were not for their quenching effect. With a slow flow of gas, the flow velocity at some little distance from the walls will be below the burning velocity and the flame will strike back. With increased gas flow the flow will exceed the burning velocity for all points and the flame will rise until it takes up a position above the burner rim where the burning velocity and flow velocity are just equal. This is possible because the gas flow lines will diverge slightly, giving a slower flow just outside the burner, while the burning velocity itself increases as the distance from the rim increases. A stable flame is thus possible. With further increase of gas flow, however, the flame may be lifted so high that appreciable inter-diffusion with secondary air occurs near the burner rim, thus reducing the burning velocity near the rim. If this reduction of burning velocity exceeds the increase due to the reduced quenching effects, then the flame will continue to rise and will blow off. This is the blow-off limit. A flame on a nozzle lifts off more easily than one on an ordinary burner because of the higher flow velocity near the rim, as the boundary-layer thickness is decreased, and it may strike back down the centre rather than at the walls.

The stability of a flame thus depends mainly on conditions near the burner rim, especially on quenching effects by the burner walls. The basic theory has been given by Lewis & von Elbe (1961) and developed and applied also to turbulent flames by Khitrin, Moin, Smirnov & Shevchuk (1965). Janisch & Günther (1973) have reproduced a good particle-track photograph showing the gas flow lines and velocity near the burner rim, and also the position of the luminous flame front. The locus of the first temperature rise ahead of the flame front, usually used in defining the burning velocity, is never normal to the flow direction but the luminous reaction zone, where the heat release rate is near its maximum, is normal to the flow at one point just above the burner rim. In addition to wall quenching by heat loss and perhaps radical recombination on the surface, the burning velocity may also be reduced by interdiffusion with the surrounding air which changes the mixture strength. The possibility of a fall in burning velocity due to flame stretch (using the Karlovitz concept, *see* page 126) has also been considered by Reed (1967; 1971) and Edmondson & Heap (1969). Lawton & Weinberg (1969) have discussed the mathematical conditions for blow-off and flash-back in terms of the gradients of the burning velocity and of the flow velocity with distance from the burner rim.

Apart from stabilisation at the mouth of an open tube, flames may also be stabilised on gauzes, as with the Méker burner. It is also possible sometimes to stabilise an inverted flame on an object such as a wire or rod held in a flame which would otherwise blow off. These inverted flames are sometimes useful for studying special effects, and examples of this may be found in Plate 4*c* and *d* and even more pronounced in Plate 6*b*. Flame holders are also essential in many large combustion units, such as rocket motors.

Quenching

We have seen that the quenching effect at the walls is very important in determining flame stability. Some very elaborate theories of quenching and ignition processes have been put forward. Here we shall confine ourselves to a brief discussion of the methods of measurement of quenching distances and the physical processes involved.

The simplest case of quenching is the two-dimensional case where a flame propagates between parallel plates. The distance between the plates may then be referred to as the quenching distance. There are two fairly straightforward methods of measurement. Blanc, Guest, von Elbe & Lewis (1947) studied spark ignition at the centre of a pair of parallel glass plates whose separation was adjustable. With sufficiently powerful sparks a flame develops in the immediate neighbourhood of the spark, but will only propagate if the plates are separated by more than the quenching distance. In the other method (Friedman, 1949) a rectangular burner, in which the width is variable, is used. A steady flame is established above the burner and then the gas flow is suddenly stopped. The flame will then either strike back or quench, according to whether the distance between the burner sides is greater or less than the quenching distance. In both these methods there are some possible slight errors due to gas movements, but on the whole they probably give a fairly reliable value for the quenching distance. Berlad (1954) found that this distance was roughly inversely proportional to pressure.

In another method of studying quenching one burns a very small flame on a burner and measures the diameter below which the flame will not strike back. Alternatively and more conveniently one adjusts the ambient pressure at which a given burner diameter no longer allows flash-back. This gives us the *quenching diameter* for a certain ambient pressure. It is obviously related to the *quenching distance*. However we now have a two-dimensional quenching instead of the one-dimensional quenching between the parallel plates. We may point out that it is possible to run a flame *above* the burner top for a burner of slightly smaller size than the quenching diameter. The method has the great advantage that we can also measure the minimum gas flow and burning velocity at the quenching point. The quenching diameter is probably about 1·4 times the quenching distance.

These measurements of quenching diameter are best discussed in relation to our work on low-pressure flames. The burner, etc., are described in the next section. Here we shall discuss the stability regions for a flame on a given burner. For a certain gas flow we may have a stable flame and this will have two pressure limits of stability, the blow-off point and the point at which the flame strikes back. If we go to a lower mass flow, then we can find the limits again, and these will generally be closer. Below a certain mass flow the limits converge and beyond this we cannot maintain a flame. In the tip of the stability region we have the smallest possible flame.

Fig. 2.3 shows a group of stability regions for stoichiometric acetylene/ air on different burners. The flame is quite steady near the limiting point and its features can be studied. The thickness of the visible flame front is much greater than for full-strength flames at the same gas pressure, and the burning velocity, which is practically equal to the average gas flow up the burner, is much less than for the full-strength flame. For C_2H_2/air there seems to be a limiting burning velocity of about 70 cm/sec compared with about 150 cm/sec for the full-strength flame.

For various hydrocarbons burning in stoichiometric proportions with

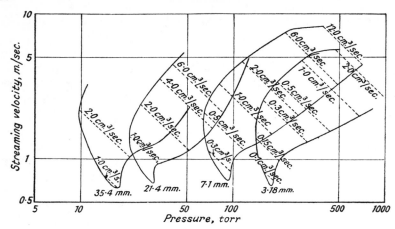

Fig. 2.3. Stability regions for stoichiometric acetylene/air. The average streaming velocity of the cool mixture in the burner is plotted, logarithmically, against the pressure for blow-off or strike-back. The burner diameter is indicated below each curve, and the mass flow, expressed as a volume of fuel at S.T.P., is indicated against the diagonal broken lines.

air, quenching distances, δ_q, are about 2 mm. The faster acetylene/air and hydrogen/air flames have δ_q about 0·5 mm. Hydrocarbon/oxygen flames have δ_q down to about 0·3 mm, but acetylene/oxygen is only about 0·15 mm. We have noted the approximate inverse relationship with pressure; Friedman & Johnston (1950; 1952) made more careful measurements and found that the pressure dependence varied from $P^{-0.8}$ to $P^{-1.15}$, being around $P^{-0.9}$ for many stoichiometric or rich hydrocarbon/air flames, but with oxygen it was nearer $P^{-1.1}$. We note that generally fast burning mixtures have smaller quenching distances than slow burning mixtures, so that δ_q increases rapidly as we approach the limits of flammability.

The quenching effects may be due either to heat loss to the walls or to removal of active centres by diffusion to the walls. The quenching distance does not seem to vary appreciably with the nature of the surface. We might expect that if the effect was thermal it would depend on the thermal conductivity of the walls, but actually this does not appear to be so because the heat capacity of solids is so much greater than that of the gases. On a radical-diffusion theory we might expect differences between various surfaces, say between quartz and metal (cf. the experiments of David and colleagues on bare and coated wires in flames, page 318); such effects of the nature of the surface are, however, unimportant. Recombination of free atoms and radicals increases the heat flux to the surface, but Cookson & Kilham (1963) suggested that the recombination occurred in the cool boundary-layer rather than on the surface. Friedman & Johnston (1950) found that heating the walls did reduce the quenching distance, which is roughly inversely proportional to the square root of the absolute temperature. Detailed temperature and composition profiles round a cooled heat sink in an ethylene flame, and a study of the gas-flow pattern by particle-track photography (Tewari & Weinberg, 1967) have shown that H-atom diffusion to the sink is very important in quenching; surface recombination $H + O_2 = HO_2$ is postulated as a likely mechanism for H-atom removal.

Potter (1960) reviewed work on quenching, and Rozlovskii & Zakaznov (1971) have extended the treatment. For a dominantly thermal method of flame propagation the burning velocity, S_u, and the thermal diffusivity of the initial mixture, κ_0, are important, and the Peclet number $P_e = S_u \kappa_0 / \delta_q$ apparently has a value of 65 within a factor of about 2.

These quenching effects are made use of in flame traps. It is often necessary for safety reasons to prevent a flame striking back down a tube. For slow-propagating flames with air or near the limits of flammability, it is fairly easy to stop a flame. A fine gauze, as in the well-known miner's safety lamp, or a piece of rolled-up gauze plugged into the tube, will often suffice. There is a critical flame speed above which a particular gauze will not arrest the flame; this critical speed is approximately inversely proportional to the width of the mesh of the gauze (Palmer, 1959). For very fast flames, such as oxy-acetylene, it is, therefore, very difficult to prevent the passage of a flame, especially if detonation is once established. Egerton, Everett & Moore (1953) have reported some experiments with sintered metal plugs, but even with these it is found difficult to check a detonating explosion with certainty. For details of the mechanisms by which flames propagate through narrow channels see Wolfhard & Bruszak (1960) and Rozlovskii & Zakaznov (1971).

Low-pressure flames and their structure

In the preceding section we have pointed out that the quenching diameter increases as the pressure is reduced, so that larger tubes are required to maintain flames at low pressure. The stability regions for low-pressure flames of C_2H_2 and air (stoichiometric) have been given in Fig. 2.3. From the minima of these stability regions it is possible to determine a number of points at which a flame can be run in minimum conditions for the various burner diameters. Fig. 2.4 shows such graphs, plotted logarithmically over a very large pressure range, for various stoichiometric gas mixtures. It will be seen that the limiting pressure is almost exactly inversely proportional to the burner diameter for all the flames. Thus the secret of obtaining flames at very low pressure is to use very large burners.

Since the flame speed does not vary much with pressure, this implies that

FLAMES

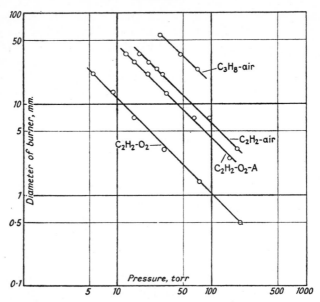

Fig. 2.4. Dependence of burner diameter on pressure for limiting flames of various stoichiometric mixtures.

the mass rate of burning tends to increase roughly inversely with pressure, and hence very high pumping rates, of the order of cubic metres per minute, are necessary. For most fuels we have been able to run flames down to a few torr pressure (Wolfhard, 1943; Gaydon & Wolfhard 1950c). For C_2H_2/O_2 we have maintained a stable flame at pressures rather below 1 torr.

It is clear that the term 'limiting pressure' which is often used has no meaning except in relation to the burner size. Curves showing limits of flammability against pressure, are of little value unless the dimensions of the vessel and ignition energy are stated.

A diagram of the burner which we have used for work at very low pressure is shown in Fig. 2.5. The premixed gases enter at the bottom of a wide burner of Pyrex glass or, for intermediate pressures, of Pyrex with graded seal and a quartz tip. The burner tube is enclosed in a large (5 litre) Pyrex vessel which is continuously evacuated from the top through a valve by a powerful rotary oil-pump. The flame is ignited by a discharge between two electrodes from a 2 kVA transformer; the electrodes were waxed in through side tubes in the vessel. The flame was viewed through quartz windows, which are omitted, for simplicity, from the diagram. The limit to the pressure which can be reached depends mainly on the pumping speed. We have been able to use burners up to 10 cm diameter, giving flames of most fuels

PLATE 2

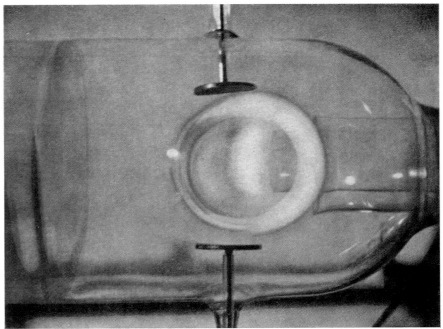

(b) Premixed oxy-acetylene flame at a pressure of about 3 torr showing the structure of the thick flame front.

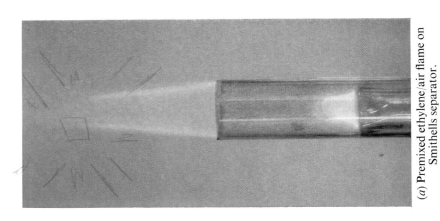

(a) Premixed ethylene/air flame on Smithells separator.

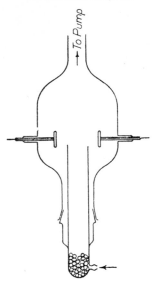

Fig. 2.5. Low-pressure burner.

with oxygen down to a pressure of around 1/100 atm. The burner tube is usually partly filled with glass beads; these serve to even out the gas flow and also, by adjusting the quantity of beads, it is possible to exercise some control over flame vibrations; these vibrations were sometimes troublesome and occurred most strongly when there was resonance between the volume of the vessel and the burner tube acting as an organ pipe. In later work a corrugated metal matrix near the top of the burner tube has been successfully used to stabilise the flow, and the outer Pyrex bulb has been replaced by a water-cooled metal vessel.

These enclosed flames do not, of course, possess an 'outer cone', and the appearance of the inner cone or reaction zone is modified by the much greater thickness of the flame front. A colour photograph of a low-pressure flame in this type of burner is shown in Plate 2b. By control of the gas flow the flame can be varied from a conical or hemispherical form to a flat disc.

The main importance of these low-pressure flames is in the greatly increased thickness of the flame front, which makes detailed examination of it possible. This is particularly the case when the flat disc-shaped flame is used, as it is then possible to examine the light from various parts of the reaction zone without it having to pass through the surrounding interconal gas and the outer cone as in ordinary conical flames at atmospheric pressure. The thickness of the luminous reaction zone increases as the pressure is reduced so that it is roughly inversely proportional to the pressure.

With hydrocarbon/oxygen mixtures the most noticeable feature of these low-pressure flames is that they are relatively green at the base and around the edge, and rather more blue or blue-violet towards the top of the reaction zone. This is seen in the colour photograph. The separation is not complete, but the spectrum shows clearly that the C_2 bands are strongest rather low in the flame and the CH bands, which give the blue-violet colour, occur rather higher. This is generally true for all hydrocarbons and similar fuels for weak, stoichiometric or moderately rich mixtures. For very rich mixtures the structure is rather different and we have observed flames in which the blue-violet CH radiation commences below the green C_2 radiation. Methane does not show the separation into green base and blue top as well as other fuels, but mixtures of methane and hydrogen do. Some pretty effects are obtained with flames supported by nitrous oxide; in some of these we have quite a rainbow effect, with a greenish base emitting C_2 and NH_2 bands, a violet upper part with CH and CN emission, and above this a yellowish-green glow due to the NO + O reaction. The spectra of many of these low-pressure flames have been discussed in detail by Gaydon & Wolfhard (1947; 1949d).

The general character and also the spectra of these low-pressure flames are very similar to those of corresponding flames at atmospheric pressure, and there is little reason for thinking that there is any major difference in the chemical processes or method of propagation. Hsieh & Townend (1939) and E. C. W. Smith (1940a) were very struck with the intense green colour of some flames, especially for rich mixtures of ethylene, at reduced pressure. Rich ethylene/air flames are very green at atmospheric pressure and we have not noticed any very striking change at very low pressure; at low pressure, with reduced gas density, the onset of luminosity due to carbon formation is rather less troublesome, especially with relatively small flames near the quenching limit, and thus it may be possible to observe the flames to rather richer, and therefore greener, conditions.

With these flames with the very thick reaction zone, which may be up to as much as a few centimetres, compared with a small fraction of a millimetre at atmospheric pressure, the flame tends to have a more rounded shape. Conical flames with a sharp tip are no longer observed. At best, the tip is very rounded, and usually the whole flame is more spherical or dome-shaped. This is presumably because we are always working rather near the limiting conditions; flames near limiting conditions at 1 atm probably have a similar structure but because of their small size are less easy to see.

So far there does not seem to be any sign of the existence of a true low-pressure limit to combustion. At some extremely low pressures, loss of energy by radiation might become increasingly important and cause a limit, but at present the limitations are practical, due to pumping speeds.

Spontaneous ignition and cool flames

When a combustible mixture is heated, chemical reactions commence, and since these are normally strongly exothermic, the mixture tends to heat up spontaneously, so further accelerating the rate of the chemical reaction. When the rate of self-heating exceeds the heat loss to the vessels walls the mixture will ignite, i.e. explode. In practice it is found that there is, for any particular fuel/air mixture, a fairly well-defined temperature at which this spontaneous thermal ignition occurs.

We can understand that with a particular mixture in a reaction vessel, the rate of heating by chemical reaction may rise exponentially with temperature, as $\exp(-E/kT)$ where E is the Arrhenius activation energy, k is the Boltzmann constant and T the absolute temperature, while the heat loss from the gas to the walls will vary linearly with the temperature difference $(T_{gas} - T_{wall})$; thus a sharp ignition temperature is to be expected. What is surprising is that the ignition temperature is usually so well-defined and does not vary much with vessel dimensions or fuel/air ratio. Obviously it is not a true physical constant, dependant only on the gas mixture, but because of the large activation energy E and thus the dominance of the dependence on temperature, the spontaneous ignition temperature is of practical importance. The following are some values taken from a book by Mullins (1955):

Spontaneous ignition temperatures, in °C.

	In air	In oxygen
Hydrogen	576	560
Carbon monoxide	609	588
Methane	632	556
Ethane	472	—
Propane	493	468
Ethylene	490	485
Acetylene	305	296
Benzene	690	662
Ethyl ether	343	178

Although these ignition temperatures are practically useful, there is, of course, some dependence on vessel dimensions and the nature of the wall surface, and especially on gas pressure. The subject is mainly within the realm of chemical kinetics, as radical chain mechanisms are often involved, and the balance between branching and terminating steps depends in a complicated way on surface reaction and partial pressures of the gases (see, for example, Cullis, Fish & Gibson, 1965; Cullis & Foster, 1977).

When premixed gases are quickly admitted to a heated vessel at controlled temperature, there is usually a fairly definite temperature, the ignition temperature, above which ignition occurs. Below this temperature some slow combustion may occur, and around this ignition temperature there is usually an appreciable delay, the induction period, before ignition occurs, this induction period usually decreasing as the temperature is raised. The ignition temperature depends on the experimental conditions, especially on surface effects, the presence of catalysts or inhibitors and on pressure. For many fuels, such as the higher hydrocarbons, aldehydes and ethers, the curve of ignition temperature against pressure takes a peculiar form. A typical curve, from Newitt & Thornes (1937), for propane/O_2 is shown in Fig. 2.6. Over a certain range of pressure and temperature the well-known cool flames may be seen to traverse the mixture. These cool flames and thermal ignition studies have been the subject of extensive experiments by Townend, Newitt and others and are fully described in books dealing with the more chemical aspects of combustion.

The pale, cool flames may readily be distinguished from normal flames by their spectrum, which shows bands due to formaldehyde instead of the usual C_2 and CH bands (*see* Gaydon, 1974). They are associated with the presence of peroxides and the formation of formaldehyde in the mixture.

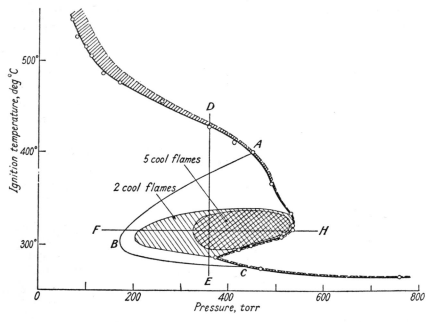

Fig. 2.6. The spontaneous ignition temperature for an equi-molecular propane/oxygen mixture, as a function of pressure. The regions in which cool flames occur are indicated.

It has been found possible in some cases (Topps & Townend, 1946; Spence & Townend, 1949) to stabilise these cool flames in heated conical tubes, and sometimes a second-stage cool flame can be stabilised as well. (Sheinson & Williams, 1973; Ballinger & Ryanson, 1971). In slow-burning mixtures on a flat-flame burner, Thabet (1951) and Egerton & Thabet (1952) have found a thin, cool flame lying a considerable distance below the normal flame; this is shown in Plate 3a. Agnew & Agnew (1965) have used a flame of this type, on a modified burner, for chemical study of cool flame processes. In considering the chemical processes of ignition and in relation to knock, or 'pinking', in internal-combustion engines these cool flames and two-stage ignition phenomena are very important (Halstead et al., 1975), but they will not be discussed further here. The cause and mathematical interpretation of the periodicity of these cool flame oscillations have also received a lot of attention (for example, Gray & Yang, 1969; Depoy & Mason, 1971; Halstead, Prothero & Quinn, 1971). Kapila & Lundford (1977) have discussed the relative positions of the reaction zones in two-stage combustion.

Ignition by external means

The lighting of a stationary flame may be achieved in a number of ways, e.g. by a hot wire, a spark or another flame. It may sometimes happen that although it is possible to maintain a stable flame on a burner, it is not possible to light it directly, because the gas movements associated with the sudden ignition cause the flame to strike back. These effects are often troublesome with the low-pressure flames, and can best be overcome by lighting a flame of a more stably burning mixture, and then gradually changing the mixture composition or flow rate until the desired flame conditions are reached.

A certain minimum energy is always required to ignite a flammable mixture. The case of spark ignition has been fully investigated by Lewis & von Elbe (1961). The energy dissipated in the spark may be assumed to be close to that of the electrical energy available, $\frac{1}{2}CV^2$, where C is the capacity and V the voltage. Ignition is most readily obtained when the separation between the spark electrodes is around that of the quenching distance for the mixture; long sparks are less efficient because the available energy is dissipated through a long thin channel through the gas, but if the electrodes are too close there will be appreciable heat loss from the gas to the electrodes. At atmospheric pressure, hydrocarbon/air mixtures have minimum ignition energies of about 0·2 to 0·3 millijoules; ammonia/air requires 8 mJ (Verkamp, Hardin & Williams, 1967), but hydrogen/air requires only 0·065 mJ. Ignition energies in oxygen are much lower, of the order 0·002 mJ for hydrocarbons.

This subject of minimum ignition energies has become of increased importance in Britain because of the change from manufactured town's gas to natural gas (methane) which has a higher minimum ignition energy and a much narrower flammability range, as shown in Fig. 2.7. The whole subject has been excellently reviewed by Sayers *et al.* (1971). Some recent values for hydrocarbons have been given by Moorhouse, Williams & Maddison (1974); they have discussed corrections to the usual assumption that the energy of the spark discharge is $\frac{1}{2}CV^2$ and have made measurements over a range of pressures and initial pressures and shown that generally the ignition energy varies approximately as the inverse square of the temperature and the inverse square of the pressure.*

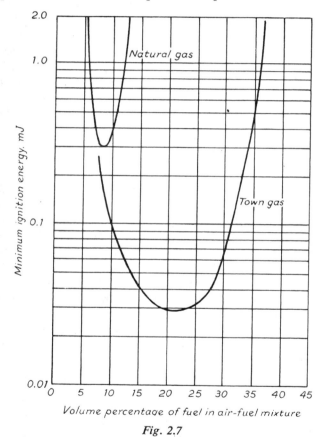

Fig. 2.7

* There appears to be a misprint in their paper which gives $E_i = AP^\alpha T^\beta$ with $\alpha = 2$ instead of -2.

It might at first sight seem that for combustion by a chain branching mechanism, it should be possible to obtain ignition from an infinitely small ignition source. In practice, except for spontaneously flammable mixtures, the chain branching reactions require an activation energy, and the loss of heat from a very small volume exceeds the heat generated by reaction in that volume. It seems, moreover, that it is not sufficient to bring a minimum volume up to the self-ignition temperature; it must be brought up to some temperature nearer to that of the burnt gases in the propagating flame.

We have seen that for any gas mixture there is a certain quenching distance. For ignition we may assume that a certain minimum volume of gas, related to the quenching distance, i.e. a sort of 'quenching volume', must be brought up to flame conditions, so that the amount of energy required will depend on the amount of gas in this minimum volume. We know that the quenching distance is nearly inversely proportional to pressure, P. The minimum volume will thus be proportional to $(1/P)^3$, and since the density is proportional to P, we may expect the mass of gas to be heated to be proportional to $1/P^2$. Experiments confirm that for instantaneous ignition the minimum energy required varies as $1/P^2$.

Minimum ignition energies for oil sprays have been measured by Rao & Lefebvre (1976); values depend on air velocity and drop size, being lowest (down to 0·3 mJ) for small drops. In turbulent air flow, both for oil sprays and for gas mixtures there is an increase in ignition energy (de Soete, 1971; Ballal & Lefebvre, 1977). The minimum energy increases with turbulence intensity due to an effective increase in thermal conductivity, and may also be related to the concept of 'flame stretch'.

For ignition by a continuous source, we must also take into account the time factor in supplying the necessary energy to the minimum volume. Both heat and particle diffusion will propagate with an inverse square law. Thus for diffusion $x^2 = 2\,Dt$ where D is the diffusion coefficient and t the time to travel a distance x. Consider ignition at 1 atm and at 1/10 atm. We find that x^2 changes by a factor of 100 but D by only 10, so that we have ten times the time available to supply the required minimum energy. Thus we should expect that the pressure dependence for the minimum ignition energy would be less for a continuous electric discharge or a hot wire than for instantaneous ignition.

For ignition by a hot wire, obviously the wire temperature should exceed the spontaneous ignition temperature, but the ignition condition will also depend to some extent on other factors, such as the flow rate of gas mixture past the wire. For fast flows the gas will not remain near the wire long enough to reach the ignition temperature. In stagnant or nearly stagnant gas, slow reactions may take place which slowly build up an inert sheath of partial combustion products round the wire; thus there is often an optimum, rather slow, flow rate for ignition. The wire temperature has to be higher for a thin wire than for a thick one. A mathematical relationship between wire temperature, activation energy, heat flux and other parameters has been derived by Adomeit (1963; 1965), who has also studied ignition by a suddenly-heated rod, and shown that delay times up to 60 msec occur before ignition. With bare wires, catalytic effects may assist ignition.

Gas mixtures can also be ignited by shock waves. The temperature to which the mixture is raised by the shock can be calculated precisely if the shock speed is measured (Gaydon & Hurle, 1963). In ordinary shock-tube experiments the duration of shock heating is usually short, from 100 μsec to at most a few msec, and ignition temperatures determined in this way are usually rather above spontaneous thermal values and correspond to values at short induction times; for this reason they are of some interest for comparison with the conditions in a flame front. Shock ignition temperatures are usually a little higher than spontaneous ignition values, although for methane/oxygen the shock value, around 1360 K, is much higher than the spontaneous ignition point of 829 K (Gaydon, Hurle & Kimbell, 1963). Some early reports of very low ignition temperatures in shock waves were apparently due to diaphragm opening processes or to the production of hotter reflected shocks at surface irregularities. It may be noted that shock waves give homogeneous gas-phase ignition, whereas more conventional measurements of spontaneous thermal ignition are connected with surface ignition.

A recent addition to ignition sources is the laser. The very parallel beam of light from a laser can be brought almost to a point focus and in pulsed lasers the energy density momentarily attained at the focus is fantastically high. Usually gas mixtures are very transparent to visible light, and hence radiation does not cause heating. It is, however, found experimentally that focused laser pulses do cause ignition; the detailed mechanism of laser breakdown is still somewhat obscure, but the initial process may involve multiple photon absorption followed by ionisation and cascade build up of electron concentration. Weinberg & Wilson (1971) used a Q-switched ruby laser and made measurements of transmitted light intensity with and without breakdown, so deducing the energy absorbed in the laser 'spark' kernel; they also took schlieren streak photographs of the developing kernel, which was often very elongated. At reduced pressure, 0·1 to 0·5 atm, minimum ignition energies were in fair agreement with those obtained with ordinary sparks, although the curves of ignition energy against pressure crossed those from the data of Lewis and von Elbe. Strong laser 'sparks' produce shock waves which readily cause spherical detonations instead of deflagration.

Limits of flame propagation

For mixtures of fuels with air or oxygen there are certain limits of composition over which flame propagation may occur, and outside which it is not possible to have a self-sustaining flame. If an attempt is made to ignite a mixture just outside the limit by a powerful ignition source, such as a condensed spark, combustion may be initiated and a flame may start, but quickly dies out. In some cases when a small premixed flame is run in a nearly flammable atmosphere a luminous cap may be maintained above the premixed flame. These 'flame caps' were described very well by Bone & Townend (1927). In determining the limits of flame propagation, or limits of flammability, the dimensions of the vessels used are important as the quenching distances near the limits are very big.

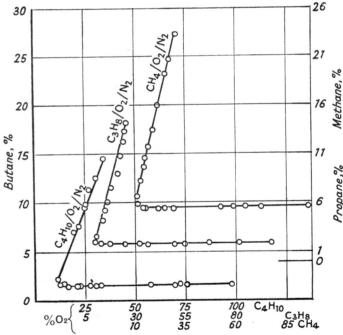

Fig. 2.8. Flammability limit curves for butane, propane and methane with oxygen, with nitrogen as diluent, the N_2 amount being expressed as $100 - (\text{fuel} + O_2)$. [*From Egerton & Thabet* (1952)]

It seems that for many flames the limits correspond to a certain minimum final flame temperature. For methane this is around 1400 K for both rich and weak limits in air and also for mixtures limited by the addition of inert diluents. Egerton & Thabet (1952) have studied the limits for various fuels and diluents and given clear diagrams, which are reproduced in Fig. 2.8. The horizontal branches of the curves may be regarded as lean limits, and the vertical ones as rich limits.

The appearance of all hydrocarbon flames near the lean limit is similar, with a blue reaction zone, the spectrum of which shows CH and OH bands and some CO-flame spectrum. At the rich limit, the flame is still blue, but reactions do not seem to be complete as the blue zone is usually followed by a dark zone of appreciable thickness and then by a luminous zone due to radiation from carbon particles. This dark zone and the carbon zone are shown above the cellular flame front in Plate 5*b*.

In mixed fuels of similar type the limits can roughly be predicted from Le Chatelier's rule.* This is just a mixing rule. Thus if components 1, 2, etc., have their lean limits (expressed as a percentage) at N_1, N_2, etc., and their percentage proportions in the mixture are P_1, P_2, etc., then the mixture will have its limit at

$$L = 100/\left(\frac{P_1}{N_1} + \frac{P_2}{N_2} + \ldots\right).$$

Burgess & Wheeler (1911) found that the lean limit (in %) multiplied by the calorific value (in kcal/mole) is constant at about 1100 for most fuels. This would imply that all lean limits correspond to about the same final flame temperature. A more exact evaluation is given by Egerton & Powling (1948) whose curve is shown in Fig. 2.9.

There are some effects of pressure on the limits, and in some cases these effects are difficult to understand. At atmospheric pressure methane has an upper (i.e. rich) limit at about 15% but at 400 atm the limit moves to 46% (Berl & Werner, 1927); a calculation of the temperature (at constant pressure) for this mixture, even assuming no liberation of free carbon, leads to a temperature of not more than 400°C, a great contrast to the normal value of above 1100°C. Some endothermic compounds like acetylene and ethylene oxide will burn without any oxygen at all at high pressure.

The reason for the existence of rather sharp limits of composition outside which a flame will no longer propagate is still something of a mystery. On most simple theories one would expect the burning velocity of a mixture to fall smoothly to zero at the limit. Measurements indicate that the burning velocity is actually finite, usually between 2 and 5 cm/sec at the flammability limits. Heat losses by radiation from the burning mixture seem unable to account for the phenomenon. Linnett & Simpson (1957) sugges-

---

* In some mixtures (e.g. propene with methane), however, selective thermal diffusion may modify the local stoichiometry and cause failure of this rule (Barnes & Fletcher, 1974).

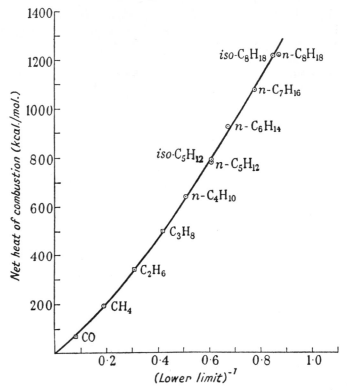

Fig. 2.9. Heat of combustion plotted against the reciprocal of the lean limit, expressed as a percentage.

ted that convective effects are important near the limits and that the rising of hot exhaust gases has a significant effect on flame stability for slow burning mixtures. Levy (1965) has made an optical study of flames propagating in tubes at near the fuel-lean limit of flammability and shown that these buoyancy effects are indeed important and that a bubble of hot exhaust gases may continue to rise up the tube even after the flame itself has failed.

One view was that if the hot gas bubble rose faster than the flame velocity then the flame would be unable to keep up with the bubble and would extinguish, but this view is probably rather naïve.

Lovachev (1971) has now developed the theory of the effect of convection on flammability limits. Simple theory, ignoring convection, would predict limits set by radiation heat losses involving extremely low burning velocities at the limits, with correspondingly large minimum ignition energies, large minimum critical volumes and very long induction times. However, the rise of the hot, roughly spherical, flame kernel increases convective heat losses to the surrounding unburnt mixture and this quenches the slowly developing flame. Lovachev has put this into mathematical form and, using reasonable assumptions, has made some trial calculations for lean hydrocarbon/air mixtures which would give a burning velocity of 7·1 or 5·5 cm/sec at the limit. He comments that inhibitors, although having little effect on the burning velocity, do narrow the flammability limits because they reduce the rate of heat release. Cool flames, with a much lower final flame temperature, cause less convection and should have wider flammability limits.

The limits vary, of course, with initial gas temperature and become very wide at high temperature and as the spontaneous thermal ignition temperature is approached they widen so that practically any mixture will react. Lloyd & Weinberg (1975) have shown that flames well outside the normal flammability range can be maintained by preheating the incoming fuel mixture. They have designed a spiral burner (Fig. 2.10) in which the outflowing hot products preheat the incoming mixture and found that methane, which has a normal weak limit around 5% can be burnt down to 1%. The flame temperature at the centre of the spiral is about 1400 K, the usual value near the flammability limit. An extension of this work (Jones, Lloyd & Weinberg, 1978) has involved considering various types of recirculation device and has put the subject of combustion in heat exchangers on a more quantitative basis.

Thus it appears that limits of flammability may not be fundamental properties of a gas mixture. However, in practice they are fairly well de-

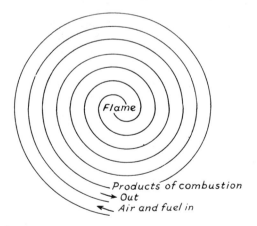

Fig. 2.10. Spiral burner

fined and knowledge of them is important for safety considerations, to enable one to predict whether or not a mixture is likely to be explosive. Determination of such limits is often made by experiments to see whether a flame will propagate through long tubes. Because of the buoyancy effects, propagation upwards in a vertical tube is somewhat easier than downward propagation. Table 2.2 shows some flammability limits for upward and downward propagation for various fuels with air; these are taken from collected data by Lewis & von Elbe (1961). Tables 2.3 and 2.4 show addi-

TABLE 2.2

Flammability limits (in fuel percentage) for upward and downward propagation, for various hydrocarbons with air

Fuel	Propagation direction	Lower mixture limit	Upper mixture limit
Methane	Upward	5·35	14·85
	Downward	5·95	13·35
Ethane	Upward	3·12	14·95
	Downward	3·26	10·15
Pentane	Upward	1·42	8·0
	Downward	1·48	4·64
Benzene	Upward	1·45	7·45
	Downward	1·48	5·55

TABLE 2.3

Flammability limits (fuel percentage) for mixtures with air at 1 atm.

Fuel	Lower limit	Upper limit	Fuel	Lower limit	Upper limit
Hydrogen	4·0	75·0	Acetylene	2·5	80·0
CO (moist)	12·5	74·0	Ethylene	3·1	32
Ammonia	15	28	Methyl alcohol	7·3	36
Cyanogen	6·0	32	Ethyl alcohol	4·3	19
Methane	5·3	15·0	Acetaldehyde	4·1	57
Propane	2·2	9·5	Ether	1·9	48
Butane	1·9	8·5	Ethylene oxide	3·0	80

TABLE 2.4

Flammability limits for mixtures with oxygen

Fuel	Lower limit	Upper limit
Hydrogen	4·0	94
Deuterium	5·0	95
CO (moist)	15·5	94
Methane	5·1	61
Ethane	3·0	66
Ethylene	3·0	80
Ether	2·0	82

tional data for upward propagation for other fuels with air or oxygen. Slightly wider limits are obtained when measurements are made with a steady tent-shaped premixed flame burning in the gas mixture under investigation (Sorenson, Savage & Strehlow, 1975). The book by Spiers (1955), *Technical data on Fuels*, tabulates, among other things, values of spontaneous ignition temperature, flammability limits and flash points; see also *Handbook of Laboratory Safety* (Steere, 1967).

The position of luminous zones

The formation of luminous carbon particles in flames forms the subject of Chapter VIII. In this early consideration of flame structure we shall briefly examine the part of the flame front in which the luminosity, due to carbon, occurs. Generally, with a few exceptions, we shall find that after passage through the flame front the gases are in fairly good chemical equilibrium. We should therefore expect carbon formation to show up, if at all, fairly uniformly throughout the interconal gases, commencing immediately behind the flame front. This does not, however, always happen (Behrens, 1950b). In some cases this appears to be due to real departures from chemical equilibrium, and in other cases to thermal diffusion effects.

Premixed flames of heavy hydrocarbons, like benzene, octane, heptane, etc., show luminosity especially at the tip of the inner cone (*see* Plate 4a); visually it can be seen that carbon formation actually occurs below the blue-green inner cone (Plate 1b). In some cases, as the conditions for polyhedral or cellular flames are approached, the carbon formation may be seen to occur in streaks. The occurrence of this luminosity at the tip seems to be due to thermal diffusion effects. Before the heavy hydrocarbon is cracked thermally into soot and H_2 it will tend to diffuse in the temperature gradient towards the cooler region, and thus accumulate towards the tip of the flame. This occurs most strongly with heavy hydrocarbons as fuel, but also to some extent with ethylene (*see* Plate 1b), acetone, ether, etc.

Acetylene gives luminous, and even smoky, flames very readily. Here the luminosity comes more from the main body of 'interconal' gases, there being if anything less carbon at the tip of the inner cone and a definitely increased luminosity from around the base of the inner cone, as shown in Plate 4b. Carbon formation can also be studied in inverted flames, in which the inverted cone is stabilised by a wire or other object. Behrens (1950b) used flames in which the central region was inverted in this way. For benzene the luminosity then appears weak at the downward pointing tip (Plate 4c) and strongest from the annular rim where the flame inverts; this supports the view that the effect is due to selective thermal diffusion in the temperature gradient rather than merely by change of mixture strength due to entrainment of air from outside the flame gases. For acetylene with the partially inverted flame (Plate 4d) luminosity now occurs most strongly in the centre.

The lower saturated hydrocarbons, methane, ethane and propane should not, in equilibrium, form solid carbon particles in premixed flames, because the mixtures which should set carbon free lie outside the limits of propagation. The formation of carbon in flames of this type probably occurs because we are really not dealing with pure premixed flames but with an intermediate stage between premixed and diffusion types. The change, as air is progressively added to a diffusion flame burning on a tube, is interesting. The pure diffusion flame shows a blue-violet outer sheath, separated by a thin, dark region from a blue-green zone, which is strongest at the base of the flame and shows C_2 and CH bands in its spectrum; this blue-green zone is followed by the luminous carbon zone. As air is added the blue-violet region remains fairly stationary, but the blue-green zone separates from it and moves towards the centre, taking the carbon zone with it. As the air supply is increased further, the blue-green zone strengthens and elongates and finally closes at the top, forming the inner cone of the premixed flame. At this stage the luminosity commences below the tip of the blue-green zone but extends through and above it. Although flames of this type possess an apparent true inner cone and look like premixed flames, they are dependent on the outer secondary air supply for their stability and maintenance. The cones cannot be separated with a Smithells separator. The occurrence of carbon in such flames is not therefore a true indication that carbon can occur in a pure premixed flame.

In some cases, however, carbon formation in premixed flames definitely does occur in conditions for which it would not be expected from equilibrium considerations. Egerton & Thabet (1952) using large flat flames with the rectified flow pattern (*see* page 72) have found luminosity in premixed propane/air and methane/air; this only started some distance above the visible reaction zone. For oxy-acetylene mixtures carbon formation is only observed for mixtures* richer than $\lambda = 0.4$, which is just about where it should commence. For ethylene/air flames we have, however, observed that the luminous tip forms at about $\lambda = 0.52$, although on equilibrium considerations it should not have appeared until a much richer mixture, about $\lambda = 0.34$, was attained.

Polyhedral and cellular flames

The majority of ordinary flames have a uniform continuous flame front, but under some conditions the flame front tends to break up. This type of

* λ is the mixture strength, defined as the actual amount of O_2 present divided by that amount required for the stoichiometric mixture. It is the reciprocal of the equivalence ratio ϕ.

effect seems to occur for mixtures which are near the limits of flame propagation, near the lean limit for hydrogen and methane and near the rich limit for the higher hydrocarbons and similar fuels. The breaking-up of the flame front shows itself in a number of phenomena; before the end of the last century Smithells & Ingle (1892) commented on the petal-like formation of some benzene flames, and this structure is now well known and is usually referred to as a polyhedral flame; spherical explosion flames may show non-isotropic propagation and may even, as fully discussed by Lewis & von Elbe (1961), have a pimply or cellular surface; large flat flames of near-limit mixtures may break up into a series of cells or cusps, as observed by Markstein (1949; 1951) and Egerton & Thabet (1952); upward propagating flames may break up into threads or tubes, with only partial combustion of the mixture.

It is probable that all these phenomena are related and have the same basic cause – differential rates of diffusion of the fuel and the molecules supporting combustion. Jost (1944) has discussed the possibility of spatial periodicities in reacting systems. Let us examine conditions in a lean hydrogen/oxygen mixture just outside the limit of flammability. Suppose we have a locally high concentration of hydrogen with ignition started at this point: Then owing to the combustion the temperature will rise and the concentrations of hydrogen and oxygen will fall, causing diffusion of both H_2 and O_2 towards this region. Owing to the higher rate of diffusion of H_2 it will diffuse in faster and the mixture strength will shift locally towards the stoichiometric point, thus facilitating further combustion. Since the mixture as a whole is incapable of sustaining combustion, local pockets of flame of the type considered will be restricted to a limited number. The pockets of combustion will tend to repel each other because each pocket requires a certain zone of influence from which to draw more hydrogen.

Upward propagating flames of H_2/O_2 can be obtained down to a limit of only 4·1% of H_2 (Goldmann, 1929) for which mixture the calculated final flame temperature is only about 350°C, which is lower than the self-ignition temperature for H_2. This type of combustion is very far from complete, because at the limit the flame propagates as a single tube or thread through the mixture with only about one-tenth of the available hydrogen being burnt. With slightly less lean mixtures several threads of flame, each with a cusp-shaped flame front, may travel up through the mixture.

The importance of diffusion in this upward propagation in lean-limit H_2 flames was proved by Clusius, Kölsch & Waldman (1941) who showed that in mixtures of hydrogen and deuterium a greater fraction of the hydrogen was burnt and more of the D_2 remained unburnt. For H_2 they found a lower limit of only 3·85% H_2, but for D_2 the lower limit was at 5·65%; the ratio 5·65 to 3·85 is 1·46 which agrees quite well with the ratio of the diffusion coefficients which should be $\sqrt{2}$ or 1·41. This agreement is so close that for this particular case there can be no doubt that the instability is due to selective diffusion effects.

Behrens (1950b) made experiments on $H_2/O_2/CO_2$ mixtures on a burner and studied the dependence of the number of flame threads on mixture strength (*see* Plate 5a). He also obtained stationary flames of similar mixtures burning above a sintered glass matrix; the flame threads were relatively short and the flame front was broken up into a number of cusps (*see* Plate 5d). Similar flames were obtained with lean CH_4/H_2/air mixtures (*see* Plate 5e).

Similar flames have been obtained with rich mixtures of other hydrocarbons. Markstein (1949; 1951) burnt rich propane/air at the top of a wide tube and then reduced the flow so that the flame entered the tube and remained stationary. A multitude of flame cusps developed and some relations were deduced for their dimensions. Markstein found that the diameter was inversely proportional to pressure and in his experiments $P \cdot d = 800$, where the pressure P was measured in millimetres Hg and the diameter d in centimetres. The greater the molecular weight of the hydrocarbon, the smaller the cusp diameter was found to be. These cellular flames are best displayed in flat flames using the rectified flow-pattern burner of Egerton & Powling (*see* page 72). Egerton & Thabet (1952) found cellular flames for lower-limit mixtures of methane and coal-gas, and for rich-limit mixtures of methane, propane and butane. The occurrence of the structure at both limits for methane is surprising. Plate 5b and c shows the cell structure obtained with the flat-flame burner. The fascinating changes of structure and the movements of the flame and also the effects of various additives to the mixture, are described in detail by Thabet (1951).

There can be no doubt that differential diffusion effects are responsible for some of the phenomena, and Markstein (1959) has indeed shown by chemical sampling from the ridges and valleys that there are variations in gas composition. It must, however, be remembered that a break-up into cellular structure may be obtained due to instability from a variety of causes; such structures have been observed as the result of the convective forces in clouds, of the forces due to shrinking in drying mud, and in the cellular pitting of the pole of a carbon arc due to instability in thermal conductivity (Steinle, 1939). Prandtl (1942) has obtained cellular structure due to heat convection in a liquid and his plate, which we reproduce (Plate 5f), closely resembles that observed in flames. We think that most of these instabilities in the flame front are due to diffusion, but they may also be influenced by heat convection and changes of density due to both heating and diffusion. The phenomena with hydrogen occur readily with upward propagating flames, but we are not aware of any similar observations with

H$_2$ for downward propagation. The observation of structure for methane flames at both limits is not easy to explain on the simple diffusion picture; the diffusion of one gas into another is not very simple as it involves both mass and collision cross-section; also we have both concentration gradients and temperature gradients influencing the diffusion. Thus, while feeling that selective diffusion is the main cause we must bear in mind the possibility of other factors contributing to the instability of the flame front, at least in some cases.

The polyhedral flames appear to be a special case of the cusp-shaped or cellular flames. Apart from their early observation by Smithells, they have been examined by Smith & Pickering (1929) and many later workers. Rich Bunsen-type flames of hydrocarbons, especially benzene, show a faceted structure, with luminous carbon appearing from the tip of the flame and from the edges of the facets. From three to seven sides are usually observed, the number depending on the mixture strength, fuel, burner size, etc. The flames frequently appear to rotate rapidly and only settle into steady polyhedral flames for correct mixture strengths and flow rates. Plate 4a shows a photograph of a polyhedral benzene flame. In some cases these flames have open tops, i.e. the inner cone does not close at the top, but just dies out. Polyhedral flames, with open tops, occur also for weak mixtures of hydrogen (Broida & Kane, 1952).

The transition between premixed flames and diffusion flames

The distinction between flames of premixed gases and flames due to diffusion at an interface between different gases is both convenient and basic. In the first case we have essentially an explosion wave travelling through a mixture in which fuel and oxygen are in intimate contact. In the diffusion flame we have no true propagation or burning velocity and, as we shall see in Chapter VI, in many cases the fuel is decomposed thermally before it reaches the oxygen, and fuel and oxygen are always separated by a wedge of intermediate combustion products.

However, in practice many stationary flames on open burners do come rather between these two ideal conditions. The ordinary Bunsen flame is, as we have seen, really a double flame, the inner cone being the flame front propagating through the mixture against the gas stream, and the outer cone being a diffusion flame. The two zones of the flame obviously have a strong influence on each other. Many quite stable Bunsen-type flames will lift off if enclosed so as to prevent the secondary combustion of the outer cone. Also in separated flames, in a Smithells separator, we have seen that the outer cone, or diffusion flame, may be extinguished if the cones are well separated, but not if they are close together.

In rich premixed flames burning in open air, the entrainment of secondary air and its diffusion into the flame gases has a much greater effect than is commonly supposed. Plate 3b gives a vivid impression of entrainment of air, as shown by particle tracks. Flames of rich mixtures may form an inner cone, and thus enable the measurement of an apparent burning velocity, but if the supply of secondary air is reduced and then cut off, the form of the inner cone changes and then the flame blows-off. In the rich flames of premixed gases with oxygen, such as an oxy-acetylene welding flame, we have already commented on the existence of a brightly luminous mantle of fairly considerable thickness just above the inner cone; this is sometimes known as the 'acetylene feather'. In open flames this mantle has a fairly sharp upper edge, but if the outer air is excluded and the flame can be prevented from blowing-off, then this mantle becomes much larger and its top becomes indefinite in position. It seems that its limit is determined by the influence of the surrounding air on the gas composition well into the flame.

A possible criterion for a true premixed flame is that the proportion of hydrogen (or other fuel) to oxygen n_H/n_O, shall not change between the unburnt and burnt gases. Even this would not rule out possible influences of the outer diffusion flame by heat transfer.

In the simple diffusion flame of a jet of gas burning in air, the main combustion is, of course, of the diffusion type. However, at the very base of the flame, fuel and air must be in direct contact and by interdiffusion form a small volume of premixed gases. This is normally shown by the emission of C_2 and CH bands in the spectrum from this region and by the absence of carbon formation. We still do not have a very clear physical picture of the various factors governing the stability of diffusion flames but it seems to us that there must be a sort of flame velocity in this premixed zone at the base which causes the combustion processes to propagate downwards along the flame front against the gas stream, thus preventing the flame from lifting. In diffusion flames at low pressure it is very noticeable that the blue-green region at the base, in which there is some premixing of gases by diffusion before ignition, becomes much more important, and at very low pressures (say below 20 torr) there is practically no visible difference between a premixed oxy-acetylene flame and a diffusion flame (Wolfhard, 1956).

The distinction between diffusion flames and premixed flames may also tend to break down when the flow conditions are turbulent. In a highly turbulent diffusion flame the vortices and irregular movements cause entrainment and mixing of air and fuel before ignition. This is especially the case with lifted turbulent flames; these show a blue-green base whose spectrum is similar to that of the inner cone of a premixed flame. These lifted turbulent diffusion flames also show less carbon formation. Even acetylene under these conditions does not give a smoky flame.

Chapter III

Flow Visualisation and Flame Photography

Before we can seriously discuss the measurement of burning velocities it is necessary to consider the gas-flow pattern and various other influences on the shape of the flame. We shall be concerned especially with optical techniques for observing the position of the flame and studying its structure. Since the earlier editions of this book there have been great advances, and especially the introduction of lasers has revolutionised many of these techniques. We have not been able fully to discuss these new sophisticated methods but we have added a new section on Laser-Doppler anemometry and added a number of references to recent researches. The older work described here still provides basic information on which to build an understanding of the shape and structure of simple flames.

The gas flow from a burner
The velocity distribution of issuing gas across the radius of a circular burner depends on burner length and diameter and on mass flow. Above the burner the gas may be observed by schlieren photography to remain as a coherent jet for an astonishing distance; this may not be true for heavy gases like butane which fall under gravity in an irregular manner. It is generally believed that for some distance, provided the Reynolds number is not too big, the velocity distribution in the jet will remain the same as at the burner mouth. Actually schlieren pictures only show where the refractive-index gradient is greatest and give no information about the velocities. In the jet the momentum will be the only value which will remain constant. Viscous drag within the jet will tend to cause a decrease in the velocity in the centre and an increase near the rim. Still air from the surroundings will also be sucked into the jet, broadening and slowing it. Plate 3*b* gives a vivid impression of the entrainment of air into a flame. So far an exact solution of the flow pattern in a jet has only been developed for flow from an infini-

tesimally small hole; Fig. 3.1 shows the broadening of a jet from a very small hole. For faster flows the shearing forces produce a complex vortex pattern at the interface between the issuing gas and the surrounding air, and at still faster flows this develops into random turbulent motions (*see* pages 134, 188). For ordinary flames on small burners the flow pattern at the position of the flame front probably deviates very little from that at the burner mouth. For more recent work see Fristrom (1956) and Fristrom & Westenberg (1965).

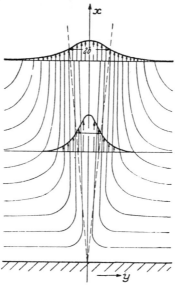

Fig. 3.1. The broadening of a laminar jet issuing from a small hole (*after* Schlichting)

There are two types of flow pattern which can easily be realised experimentally: (*a*) the parabolic velocity profile, and (*b*) the uniform velocity profile. The parabolic form is always obtained in sufficiently long circular tubes for Reynolds numbers below about 2300; the required length is $l > 0.05\, d\,.\,R_e$, where d is the diameter and R_e the Reynolds number. For a fast flow in a 2-cm tube a length of over a metre is necessary. For the parabolic pattern, the velocity u at a distance r from the centre is

$$u = \frac{2V}{\pi R^2}\left(1 - \frac{r^2}{R^2}\right),$$

where R is the radius and V is the volume flow of gas. The velocity at the centre is thus twice that for uniform flow. The velocity is zero at the burner wall.

The uniform type of flow pattern can be realised by discharging the gas through a nozzle. The gas in the container before the nozzle is practically at rest, and receives everywhere the same acceleration due to the pressure drop through the nozzle; the velocity is thus uniform except for a boundary-layer along the walls. The design of the nozzle is not very critical, but a convenient shape is described by Hahnemann & Ehret (1943). The walls should be highly polished to keep down the thickness of the boundary-layer; the thickness of this layer is proportional to $1/\sqrt{R_e}$, and it is therefore best to use fast flows. Fortunately, for nozzles, it is possible to reach much faster flows (sometimes up to $R_e = 40\,000$) before turbulence sets in.

The velocity in the central uniform part, away from the boundary-layer, cannot easily be determined from the overall gas-flow because of the effect of the boundary-layer. As long as the flame is not burning the velocity may be measured from the Pitot head; if $\triangle P$ is the pressure difference between that before the nozzle and outside, ρ is the gas density and u_i is the initial velocity of gas before it reaches the nozzle, then the velocity u is given by

$$\triangle P = \tfrac{1}{2}\rho \, (u^2 - u_i^2).$$

Frequently u_i^2 is so small that it may be neglected. For flames, the back pressure due to the flame interferes with the measurement of the Pitot head and it is best to make a calibration of gas flow against velocity (from Pitot head) without the flame being lit. For a fuller account of Pitot-tube measurements and probe design see Chedaille & Braud (1972).

For slow flames the velocity is too small to be measured by its Pitot head, and the velocity must be determined from the gas flow and diameter, after allowing for the boundary-layer. Tollmien (1931) gives the formula

$$\frac{\delta}{d} = 1\cdot 73 \sqrt{\frac{v}{u \times d}}$$

where δ is the effective thickness of the boundary layer, d the diameter and v the kinematic viscosity. We then have

$$u = \frac{4V}{\pi(d - \delta)^2}$$

V being again the volume flow.

For very slow flows, a uniform flow pattern can also be realised by using a long, straight burner fitted with a honeycomb matrix; this matrix is conveniently made by winding together into a spiral two layers of nickel foil, one of these layers being corrugated and the other smooth (Powling, 1949; Levy & Weinberg, 1959). The uneven flow due to the boundary layer at the walls in the honeycomb will be smoothed out if there is a little distance between its top and the flame.

The measurement of the gas flow

The regulation of the gas flow is usually done with a reducing valve, to give constant pressure, and then a needle valve. Flows are usually compared using capillary flow-metres; the flow through the capillary should be kept laminar, as unsteady readings are obtained if the flow is turbulent; near the onset of turbulence there is sometimes a sort of hysteresis effect due to delay in change from one flow form to the other. For fast flows it is therefore necessary to use wide, short capillaries, and the calibration curve is then often so far from linear that the sensitivity is reduced. Thus for very large flows annular restrictions made by sealing one glass tube into another are to be preferred to capillaries, as a larger flow can be attained. The U-tube for measuring the differential pressure should be of fairly wide bore to reduce interference from surface tension forces at the meniscus. Xylol or dibutylphthalate are better for filling the U-tube than water as the capillary forces are lower. The viscosity of a gas varies rapidly with temperature and for accurate work the capillary of the flow-meter should be enclosed in a thermostat. For flames at other than atmospheric pressure, or if there is a big back pressure, a needle-valve must be near the capillary, the valves being adjusted to give a constant pressure in the flow-meter. This method also eliminates errors due to change in atmospheric pressure.

Rotameter flow-meters, in which a small rotating float is lifted part of the way up a slightly coned vertical tube, are also convenient.

The flow-meters are usually calibrated most conveniently by the soap-bubble method. The principle is to time the rate of travel of a bubble along a tube of known bore. Alternatively, of course, the gas may be collected for a known time over water in a measuring vessel of known volume. In both of these methods the gas should be saturated with water after passing through the flow-meter before measurement; the volume flow is then corrected for the partial pressure of water vapour added. This type of method cannot be used for gases like NH_3 which are very soluble. For these the best method is probably to expand the gases, after flow-metering, through a needle-valve into a large evacuated vessel. If the volume of the vessel is known and the rate of pressure rise is measured the flow can be determined. The method is especially suitable for vapours which can only be handled at low pressure.

Laser-Doppler anemometry

This recently developed technique is proving most valuable for the study of flame gases because it enables measurement of very localised gas velocities, at the intersection of two narrow pencils of light from a laser. Usually it is necessary to seed the flame with small particles which scatter the light; particles of magnesium oxide, titanium oxide or silica (obtained by introducing atomised silicone oil) have been used. In luminous flames the soot particles may be made use of. However, in some cases 'phase objects' such as turbulence balls or boundaries between hot and cold gas which produce sharp variations in refractive index can be utilised (Schwar & Weinberg, 1969; Weinberg & Wong, 1975; Hong, Jones & Weinberg, 1977).

If a monochromatic light beam, of frequency v_1 is scattered through an angle θ by a particle moving with velocity v, the frequency v_2 of the scattered light is

$$v_2 = v_1(1 + v \sin \theta/c).$$

If the scattered light is allowed to interfere with an unscattered reference signal, then the beat frequency is

$$v_b = v_1 v \sin \theta/c = v \sin \theta/\lambda$$

where c is the velocity of light and λ is the wavelength.

A simple system, shown in Fig. 3.2, is to split the laser beam so that one part is scattered from a particle approaching the beam, and the other

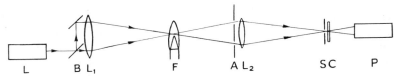

Fig. 3.2. L, laser. B, beam splitter. L_1 and L_2, lenses. F, flame, seeded with particles. A, aperture stop. P, photomultiplier. S, small stop or slit at focus. C, colour filter.

part is scattered from the receding particle. The two scattered beams are then brought to focus on the cathode of a photomultiplier. The beat frequency is then

$$v_b = (2v_1 v/c) \sin \theta = (2v/\lambda) \sin \theta.$$

The two beams produce interference fringes which move across the cathode surface. The effect may be regarded either as a moving interference fringe system or as a Doppler beat frequency. The photomultiplier output may be displayed on an oscilloscope or processed electronically to yield the beat frequency.

The method is entirely dependent on the highly monochromatic nature of light from a continuously running gas laser; argon-ion and helium-neon lasers have been used successfully. Various refinements to the basic method shown in Fig. 3.2 have been introduced (e.g. the use of a diffraction grating as beam splitter, and schlieren-type systems for studying phase objects) and these are discussed in recent papers. In addition to measuring local gas flows down to very low velocities in laminar flames, the method has been adapted to the study of fluctuation velocities in turbulent flow, and the results will be discussed when we come to discuss the structure of turbulent flames. Recent papers on flame anemometry are by Durst & Whitelaw (1971), Durst, Melling & Whitelaw (1972), Durão & Whitelaw (1974), and Jones, Schwar & Weinberg (1971).

The method may well become very important for detailed study of velocity distribution and velocity fluctuations in larger practical flames such as those in furnaces and gas turbines, and, for example, some measurements on flames on burners with swirl have been made (Baker, Bourke & Whitelaw, 1973; Chigier & Dvorak, 1975).

The photography of the flame cone

Measurements of flame cones are usually made from photographs. For hydrocarbons the inner cone is usually sufficiently bright to enable photographs to be taken with exposures of less than a second. It is essential to use plates giving good contrast. The measurements are complicated by the fact that the flame has a finite thickness and has not got sharp edges. Thus a very awkward problem arises if we wish to determine from the photograph the position of a reference line in the flame or the flame thickness (Hübner & Kläukens, 1941). The emission from the inner cone is mainly from C_2 and CH radicals, and, fortunately, for this radiation the self-absorption is negligible and so we are spared further complications from this cause. Fig. 3.3 shows the intensity distribution in the image for two

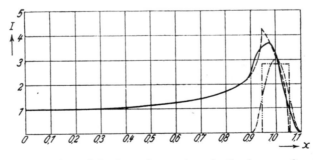

Fig. 3.3. Representation of the intensity contour in the image of a luminous cylinder as a function of the distance x from the axis of the cylinder.

– · – · – density D of radiating molecules in the object
――― resulting intensity I in image
$\Bigg\}$ for $D = D_0(r_a-r)^2(r - r_i)$

– – – – density D of radiating molecules in the object
– – – – resulting intensity I in image
$\Bigg\}$ for $D = $ constant

assumed distributions of light intensity through the reaction zone, both having the same half-breadth. It can be seen that the points of maximum contrast in the image shift appreciably. The eye cannot easily assess the true intensity contour, but quickly recognises the position of maximum contrast, where the rate of change of blackening with distance is greatest. The intensity maximum in the image is very near the inner edge of the visible flame cone; the intensity falls to zero at the outer edge.

The measurement of the thickness of the inner cone is especially difficult if an accuracy better than about 20% is required. It then becomes necessary to calculate the true intensity distribution in the cone from that in the image with the Abel integral equation, in a similar way as was done for an arc by Hörmann (1935). The intensity distributions for Fig. 3.3 were derived assuming an optical system of very small aperture. In practice, cameras of fairly large aperture are often used, and this may introduce further serious complications.

A source of error in these direct studies of flame light is the optical distortion caused by the strong refractive-index gradients in the flame. Weinberg (1963) has devoted a short chapter to distortion of flame luminosity by this effect. It is particularly important for observations of the thickness of the reaction zone, and Weinberg's calculations suggest that an error of the order 1 mm is to be expected in observations of reaction zone thickness, using slow flames on an Egerton-Powling burner or faster flames on a slot burner. With a conical Bunsen-type flame the apparent increase in reaction-zone thickness due to optical distortion is, however, less, being around 0·03 mm for a flame of 3 mm diameter and burning velocity of 240 cm/sec. These effects tend to become more important at high pressure, but are negligible for low-pressure flames.

Particle-track photography

A useful way of studying the gas flow pattern is by particle-track photography using intermittent illumination of the particles. In early work on fairly fast flames by Neubert (1943), Lewis & von Elbe (1943) and Andersen & Fein (1949) magnesium oxide particles were used and the intermittent illumination was provided by either a long-duration photo-flash bulb and sectored wheel or by a controlled spark system. In later work with slower gas flows (Levy & Weinberg, 1959; Pandya & Weinberg, 1964) particles of Bentonite, a finely divided colloidal clay with particle sizes down to 4μm, have been used, and these have been illuminated at 10 msec intervals by a high-pressure mercury lamp running on the 50-cycle mains.

The region of flame being studied may be restricted by control of the illuminating light beam and by using lenses of high aperture and thus small depth of focus. The Tyndall scattering is strongest in the forward direction, but the best location of particles is obtained by viewing at right angles to the illuminating beam; in practice a compromise position for the camera is usually used.

In calculating the gas flow velocity it may be necessary to make a small correction because the particles are falling freely through the gas; the downwards velocity due to free fall can readily be calculated by Stoke's law. For 4μm Bentonite particles at 100°C the velocity due to falling is 0·11 cm/sec, decreasing to 0·06 cm/sec at 400°C and still less as the viscosity rises at higher temperature. The larger MgO particles (Andersen & Fein quote 20 and 40μm) have an appreciable rate of fall and are unsuitable for measuring slow flows. Weinberg and colleagues have selected the less bright, and therefore smaller, particles of Bentonite for their measurements.

The particles may be introduced in various ways. Andersen & Fein placed the charge of MgO particles between two wads of steel wool and by-passed part of the gas flow through. For some purposes it is sufficient to lay a little powder somewhere in a slow part of the supply stream and just to tap the tube before the photograph is taken.

Some examples of particle-track photography are shown in Plate 6b, c and d.

Schlieren photography

'Schliere' (plural schlieren) is a German word meaning the inhomogeneous regions in otherwise homogeneous matter. These schlieren can easily be observed when salts are dissolved in water or when hot air rises above a hot body. Generally schlieren are any things which cause irregular deflection of light from a relatively small area. They may be caused either by change in refractive index or by change in the thickness of a transparent body such as a glass plate which then behaves like a lens. The nature of schlieren is best illustrated by a simple shadow experiment: let us have a point light source, a glass plate and a screen some distance behind it; if the plate is not of good quality we shall see dark shadows and regions of greater brightness on the screen. It is important to realise that only the shadows of the schlieren are in a straight line between source and the inhomogeneities causing them. The brighter regions are illuminated by excess light deviated from other regions, and it is not in general easy to associate these bright regions with the inhomogeneities causing them.

Schlieren arrangements have several advantages over normal flame photography. One of the greatest is that of exposure time. Usually it is not possible to take an 'instantaneous' photograph of a flame because the light is not sufficiently bright. For schlieren, photographs can be taken usually in 10^{-3} sec, and with spark sources down to 10^{-5} or even 10^{-6} sec, while with Q-switched lasers times down to 10^{-8} sec may be attained. It is thus possible to explore non-stationary conditions such as the formation of turbulence balls and vortices. We shall also see, when we come to discuss measurements of burning velocity, that the velocity is referred to the propagation into the unburnt mixture, and the flow lines are less disturbed in the relatively low-temperature region where the schlieren are formed than in the hot luminous region.

Schlieren photography has been extensively used in aerodynamics for the location of shock fronts and the study of the onset of turbulence. For flame studies the large temperature jump at the flame front produces a strong schliere and the flames are often smaller than aerodynamic flow regions, so that the large expensive mirrors used in aerodynamic work are not often needed for flame study. Many arrangements of mirrors, lenses and stops are possible for the study of flame schlieren and these have been arranged into various groups and reviewed by Weinberg (1963). Here a few of the simpler systems most suitable for the study of small flames are discussed.

(*a*) One of the simplest systems, used by Schardin (1942), is to place a coarse grating, such as a photograph of a series of equally spaced black and white lines, a short distance behind the flame and illuminated from behind, and to photograph flame and grating together with a small-aperture camera so that both are in adequate focus. The lines of the grating are deviated by the schlieren in the flame and can be measured to give semi-quantitative information about refractive-index deviations in the flame. Schardin also described a modification using several colours.

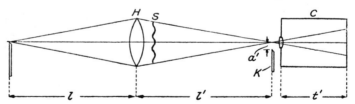

Fig. 3.4. Toepler's schlieren arrangement.
H = lens of focal length *f*. S = schliere (i.e. the flame).
K = knife-edge. C = camera.

(*b*) Toepler's arrangement, shown in Fig. 3.4, is probably the most important; this has been used so frequently that the term 'schlieren photography' has tended to mean use of Toepler's arrangement, to the exclusion of shadow and other forms of schlieren photography. An image of a convenient bright light source, such as an arc or Hg high-pressure lamp, is focused on to a very wide slit or just one jaw of a slit. The light from this is then focused with a lens H on to a knife-edge which is accurately parallel to the first edge so that the light is just, but only just, completely cut off by this second knife-edge. The test object (i.e. the flame) is placed directly behind the lens *H*, and a camera situated behind the knife-edge is focused on the test object. Without interference from the test object, all the light will be intercepted by the second knife-edge. Deviation of light due to changes of refractive index in the test object will cause some of the light to be deflected so that it passes the knife-edge, and the camera will then record a photograph of those regions of the test object which deflect the light in the right direction. To get all the inhomogeneities in the test object we must either turn it round or turn both knife-edges through 90°.

The lens *H* has to be of good quality. It must be achromatic, or else monochromatic light must be used; it is often convenient to use monochromatic light from a mercury lamp or a laser and suitable colour filter as source. The lens *H* may be replaced by a pair of telescope lenses, with the test object between them; this is especially suitable as in this way the lenses may be used in the arrangement for which they are designed, with a parallel beam of light between them through the test object.

Concave mirrors may be used instead of expensive lenses, but these mirrors give astigmatism in the image. This astigmatism may be overcome, however, by focusing so that the image is drawn out parallel to the direction of the knife-edge. It can be shown that the spherical aberration is least when the angle between the parallel beam and the light source is the same as that between the parallel beam and the camera (*see* Fig. 3.5).

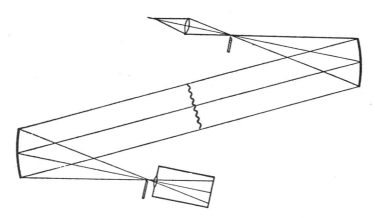

Fig. 3.5. Schlieren arrangement with concave mirrors.

The theory of arrangements for studying schlieren has been discussed by Schardin (1942) and by Weinberg (1963). Here it must suffice to stress some of the more important points. For visual observation the eye is placed directly behind the second knife-edge; the focal length of *H* should be as big as possible and the distance a' (Fig. 3.4) small. However, if a' is too small interference patterns, which disturb the schlieren, may be observed. For photography f should be as big as possible and t' as small as possible to obtain maximum sensitivity. However, these conditions may be incompatible with the need to keep a convenient magnification, say 1:1 of the test object. In this case the focal length of H should be as small as possible. For a large object it would not be possible to make t' and f small to get high sensitivity because the image deviation would get too big. Our purpose is to stress that many conflicting conditions have to be fulfilled and the choice of the best schlieren arrangement for maximum sensitivity needs careful consideration.

It is important to have the test object in accurate focus on the plate in the camera, otherwise strange deviations occur. Frequently it is possible to focus sharply on reflected or diffracted light (visible as a schliere) from the burner rim. Another method is to focus directly on the light from the flame with the schlieren light source off and the knife-edge temporarily removed.

As an alternative to using two knife-edges, a point source may be used and the second knife-edge replaced by a small hole, or by a small opaque stop; then any deflection of the light, irrespective of its direction, will be observed.

A quantitative evaluation of the schlieren image in the Toepler arrangement is possible, although not very convenient, by calibrating with a standard schliere, such as a lens of known small curvature.

(c) A modification of the Toepler set-up which is suitable for quantitative work is obtained by replacing the first knife-edge by a fairly narrow slit SL and the knife-edge by a stop like a coarse grating (Fig. 3.6). The image

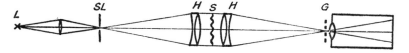

Fig. 3.6. Schlieren arrangement for quantitative measurements.
L = light source. SL = slit. H = lenses. S = schliere (i.e. the flame).
G = coarse grating.

of the slit is then focused into one of the gaps in the grating, so that without schlieren the test space appears uniformly illuminated. With a slight schliere, sufficient to displace the image to the first dark bar of the grating, the region of the schliere appears dark. Thus a series of dark contour lines will indicate the intensity of the schliere. By covering the clear lines in the grating-like stop with different colour filters and using colour photography it is possible to distinguish more easily between the different orders.

(d) Euler & Hüppner (1950) have used a small dispersing prism over an illuminated slit as source, and replaced the knife-edge by a second slit. The deviations due to the schlieren then appear as different colours passing the second slit, and may be observed visually without a camera.

(e) The simplest schlieren system is shadow photography and was first used by Dvorak in 1880. A point source of light, or a small pin hole with a light focused on to it, is used, and the shadow of the flame is thrown on to a screen or photographic plate. Maximum sensitivity is obtained with the flame midway between the source and the screen. The pattern is not easy to interpret. The dark regions tend to be in line with the source and schlieren causing them, but the superposition of bright areas of deflected light from other parts may complicate the image. Diffraction effects between the schlieren may also be troublesome, these being worst when the distance from the flame to screen is about a quarter of the source to screen spacing.

A slight modification is to use a lens so that the flame is in parallel light. Such a system, as used by Weinberg (1956b) for studying a flat (Egerton-Powling type) flame is shown in Fig. 3.7. If d is the distance of the centre of

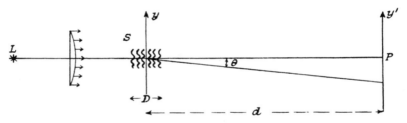

Fig. 3.7. Shadow arrangement for flat flame using parallel light. L is the light, S the schliere (flame) and P the plate or screen.

the flame to the screen, D is the thickness of the flame and y is the height, then, as can be seen from Fig. 3.7.

$$y' = y - \theta d$$

and, since $\theta = dn/dy \cdot D$, where n is the refractive index, it follows that

$$y' = y + \frac{dn}{dy} \cdot Dd,$$

and differentiating

$$dy' = dy + \frac{d^2n}{dy^2} \cdot Dd \cdot dy.$$

Thus, whereas the normal schlieren deflection depends on the first derivative of the refractive index dn/dy, the shadow tends to depend on the second derivative d^2n/dy^2. If the screen is placed very close to the flame, then this deduction is valid and the positions of maximum and minimum intensity can be related to maximum and minimum d^2n/dy^2 in the flame. At larger distances from the flame, crossing of rays from different parts of the flame complicates the shadow pattern and the above relationship does not hold. At sufficiently large d, a discontinuous increase in intensity is observed and this can be used as a reference point. For further discussion of the location of shadow and schlieren images see pages 69–72.

Interferometry

An interferometer is more expensive and more complicated to use than a schlieren set-up but it enables more quantitative results to be obtained. It enables the refractive index to be determined, whereas the schlieren method gives primarily the gradient of the refractive index.

To examine flames it is necessary to have a very big separation between the interfering light-beams so that the burner or combustion chamber may be inserted in one beam, without causing distortion of the mirrors by heat from the flame. A Mach-Zehnder interferometer (Mach, 1892; Zehnder, 1891) is commonly used; the optical system, consisting of two plane mirrors and two half-silvered plane mirrors, is shown in Fig. 3.8. The

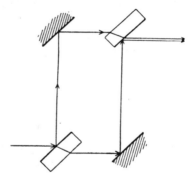

Fig. 3.8. Diagram of Mach-Zehnder interferometer.

system is illuminated by a source, such as a mercury lamp or laser, placed at the focus of a condensing lens. Behind the last mirror is another lens which focuses the light source into the camera, the camera being adjusted to focus on the flame, which is between the last two mirrors. The rather complicated adjustment of the instrument is described by Hansen (1940) and Weinberg (1963). The adjustment is much easier with the highly monochromatic light from a laser.

The change in refractive index due to the presence of the flame causes a shift in the system of interference fringes, and measurement of this shift quickly gives the product of the flame thickness and refractive-index change. If the thickness of the flame is measured (by direct, shadow or schlieren method, or with the interferometer itself) then the mean refractive index of the flame gases can be calculated. This refractive index depends mainly on temperature, but also on pressure and chemical composition. The possibility of determining both temperature and composition by using light of various wavelengths has been discussed by Olsen (1949). It seems that for normal flame studies, the interferometric method may be rather oversensitive; as a technique for aerodynamic studies it is very valuable and methods have been reviewed by Ladenburg & Bershader (1955).

An interferometric picture of a candle flame is shown in Plate 6e. The candle is, of course, a three-dimensional problem, the fringe displacement depending on the thickness of the flame as well as its refractive index.

An alternative to the Mach-Zehnder type of interferometer is a system of either two or four transmission diffraction gratings. Various arrangements are possible (Weinberg & Wood, 1959; Weinberg, 1963). In the one illustrated in Fig. 3.9, light from the source S is made parallel by a lens L and falls on the first grating D_1 so that the zero-order light continues straight on to the grating D_2, while the first-order diffraction is deviated to grating D_3. The first-order diffraction spectra from D_2 and D_3 then recombine at grating D_4, interference being produced between the first-order spectrum (produced in D_4) of the light received from D_2 and the zero-order

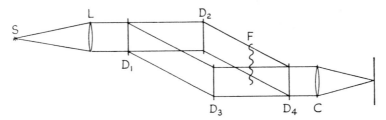

Fig. 3.9. Weinberg-Wood four-grating interferometer.

(in D_4) from D_3. The flame whose interference pattern is being studied is placed after the second grating, at F, and the fringe system is photographed by the camera lens C. For this type of arrangement, inexpensive replica gratings may be used and the alignment is easier than with a Mach-Zehnder instrument, but the system is more wasteful of light.

Lasers and holography

The laser is proving a valuable tool for combustion studies. The light from a laser possesses four special properties; it may be extremely intense, so that it can 'outshine' the brightest flame; it is extremely parallel, so that it may be focused down almost to a mathematical point, the practical size limit being set by the diffraction pattern; it is highly monochromatic and coherent (i.e. the waves from all emitting molecules are in phase) so that interference fringes may be obtained over long path differences; in some forms of laser it is possible, by what is known as Q-switching, or by the use of Kerr cells, to obtain controlled bright light pulses of very short duration, down to 10^{-8} sec. Gas lasers, such as the helium-neon laser, usually operate continuously and are more monochromatic. Solid-state lasers, such as the ruby laser, are less highly monochromatic, but give a higher intensity and are most suitable for pulsed light studies.

The most obvious application is in interferometry and in particular for taking instantaneous interferograms to show up fast gas movements and the behaviour of shock fronts and turbulent flow structure. Continuously running gas lasers are also helpful in aligning interferometers, although the absence of the zero-order white light fringe is sometimes a limitation. Lasers are also useful, because of the intensity and short duration, for instantaneous schlieren photography. Interference fringes are sometimes troublesome when using laser light for schlieren work, although this difficulty may be overcome. Oppenheim, Urtiew & Weinberg (1966) have developed laser-schlieren techniques for the study of the development of detonation waves and have taken photographs at 200000 frames a second with individual exposure times on each frame of only 10^{-8} sec. For a good review of the possibilities of laser techniques see Schwar & Weinberg (1969b).

Laser-Raman spectroscopy is mainly of use for temperature measurements and a section is now devoted to it in Chapter XI. A limitation of most lasers is that only a single spectrum line, or at best a few, are emitted. Tunable dye lasers can, however, cover a fair range of the spectrum; among their applications is the determination of pollutants (SO_2, NO_2, CO_2) in smoke plumes by absorption spectroscopy.

The laser is also making possible the use of holographic methods with visible light. If, for example, an interferometer is set up and a flame is placed in one beam, a complex system of interference fringes is produced. If a transparent positive photograph of this fringe system (hologram) is taken and then the flame is removed and the hologram is placed in the position of the original fringe system, then when the light is viewed through the hologram it appears as though the original flame were still present, since the wave pattern of the light after passage through the hologram is the same as that which would have been produced by the flame. One use of this type of hologram is to take an instantaneous record of a transitory phenomenon, such as a detonation, and then to use the hologram, with a continuously running gas laser as background, to study the structure of the phenomenon at leisure; schlieren and ray-deflection observations may be made on the hologram as though it were the original phenomenon. Another use of holography is by employing subtractive holograms to remove optical defects; thus if it is desired to study the interference or schlieren pattern of a flame travelling through an optically imperfect tube, then a hologram of the tube, without the flame, may be taken and a photographic negative of the hologram may be inserted, so that when the flame is studied the optical deviations produced by the tube are cancelled out. It has also been suggested (Schwar, Pandya & Weinberg, 1967) that inexpensive holograms of expensive optical components, like large schlieren lenses, might be used instead of the originals; this could be especially useful in the study of explosions where the optical components may be destroyed. An alternative is to replace lenses by 'point holograms', made by photographing the interference pattern from a divergent laser beam. A recent use is in the photography of droplets to study their instantaneous size, and thus their rate of evaporation, in burning oil sprays; three-dimensional records of droplets down to only a size of 20 μm have been obtained with an exposure time of 25 ns (Webster, Weight & Archenhold, 1976).

The flame thrust, and its effect on cone shape

Owing to the heating, expansion and acceleration of gases in the flame front there is a pressure difference across the flame front. It can be shown (Damköhler, 1940) that the pressure difference between that of the initial unburnt gas just inside the cone and that in the final condition, for the burnt gases above the cone is

$$P_i - P_f = \rho_i S_u^2 \left(\frac{\rho_i}{\rho_f} - 1 \right),$$

where ρ_i and ρ_f are the initial and final densities and S_u is the burning velocity as defined in the next chapter.*

As an example of the magnitude of this back pressure, we may consider a stoichiometric propane/oxygen flame for which $\rho_i = 1 \cdot 44 \times 10^{-3}$ and $S_u = 300$ cm/sec, and ρ_i/ρ_f is about 12; the pressure difference comes out at 1·45 cm of water. This is comparable with the Pitot head due to the gas velocity. A suitable velocity to obtain a good cone with this mixture would be around 2000 cm/sec. The Pitot head is

$$\tfrac{1}{2}\rho_i \text{ (gas velocity)}^2$$

which gives 288 N m^{-2} or a pressure of 2·9 cm of water.

* Edmondson and Heap (1969b) have pointed out that in conical flames this equation is not exact, the actual pressure being rather less because of the finite thickness of the flame and the consequent divergence of flow lines as the gases pass through the flame front. The ρ_i/ρ_f term should be multiplied by $(A_u/A_b)^2$ where A_u and A_b are the flame areas as measured at the positions before appreciable heat release (i.e. unburnt) and after complete heat release (burnt).

The flame thrust causes a deformation of the shape of the inner cone of a flame from its theoretical shape. Fig. 3.10 shows the simple theoretical shape of the cone for a parabolic velocity profile and the actual shape of a flame. The back pressure distorts the parabolic velocity profile in the region just below the cone and modifies the flame shape appreciably. As we approach the burner rim the loss of heat and active radicals to the wall causes the burning velocity and therefore the flame thrust in this region to fall. The gas is thus pushed out through the dead space between the burner rim and the flame front; this causes the well-known overhang of the flame. The effect of the back pressure is relatively greatest for slow initial gas flow which gives a wide-angle cone and big overhang. For small burners, when the burner diameter approaches the quenching distance, the effect may be so great that the flame may be higher near the rim than in the centre; in the most extreme case the flame may take the form of a hemisphere which is convex towards the burner.

The velocity v, with which the gas is pushed through the dead space may be calculated by treating the back pressure as equivalent to a Pitot head. Thus we obtain

$$\tfrac{1}{2}\rho_i \cdot v^2 = \rho_i \cdot S_u^2 \left(\frac{\rho_i}{\rho_f} - 1\right),$$

or

$$v = S_u \sqrt{2\left(\frac{\rho_i}{\rho_f} - 1\right)}.$$

For our example of propane/oxygen, this would correspond to a velocity of around 1400 cm/sec. Friction at the wall will obviously prevent the development of such high velocities, but it is clear that this back pressure will be sufficient to prevent diffusion of external gases through the dead space into the cone, as was assumed in some older theories. The outward flow of gas occurs not only in the dead space, but to some extent slightly higher up, where the burning velocity has not yet attained its full value.

The flow of gas through the dead space can be seen clearly in Plate 6*d*. Plate 6*b* shows the effect of back pressure on an inverted flame.

The tip of the flame

Fig. 3.10 shows, in addition to the flame overhang, the rounding at the tip of the flame. Owing to the thermal expansion as the gases pass through the flame front, the flow lines diverge outwards, away from the burner axis (*see* Fig. 4.6). At the tip, however, this divergence of flow lines must lead to a greater distance between lines, and hence to a lower flow rate, and, if the burning rate is constant, to an advance of the flame front and a rounding of the tip. Plate 6*d* and the reproduction by Fristrom (1957) show that this divergence of flow lines occurs below the luminous flame front. The round-

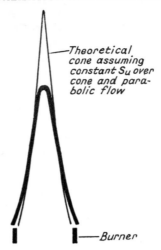

Fig. 3.10. Comparison of the actual and theoretical shape of the luminous inner cone of a propane/oxygen flame.

ing of the tip of the flame becomes particularly noticeable in low-pressure flames (Plate 2*b*), where the thickness of the reaction zone becomes comparable with the burner dimensions.

Although the divergence of flow lines is the main cause of the rounding at the tip, the rate of heat transfer and diffusion of active species will increase as the radius of curvature of the flame becomes comparable with the thickness of the reaction zone; this will cause an increase in the burning velocity (*see* Fig. 4.5, page 64) and thus contribute to the rounding of the tip. In extreme cases these effects may even cause the tip of the flame to dimple downwards; this occurs with low gas flows, giving a wide cone angle (Uberoi, 1954). Selective diffusion effects may also modify the gas composition, and hence the combustion rate, in the region of the flame tip; these effects have been discussed in the section on polyhedral flames, page 35.

Chapter IV

Measurements of Burning Velocity

The burning velocity

The burning velocity is the velocity with which a plane flame front moves normal to its surface through the adjacent unburnt gas. It is a fundamental constant of the gas mixture and is important both practically in the stabilisation of flames, and theoretically for theories of flame propagation. The difficulty of measuring the burning velocity is that a plane flame front can only be observed under very special conditions. In nearly all practical cases, the flame front is either curved, or it is not normal to the velocity of the gas stream.

In the ideal case, a thin plane flame front propagates into flowing gas as illustrated in Fig. 4.1. The most important relationship is

$$S_u = U \sin \alpha.$$

The flow direction changes due to the heating and chemical reactions in the flame front, and this flow line refraction (Lawton & Weinberg, 1969) is given by

$$\frac{\tan \beta}{\tan \alpha} = \rho_u/\rho_b$$

where ρ_u and ρ_b are the densities of the cold unburnt mixture and of the hot burnt products.

Methods of measurement may be divided into three groups: (a) study of burner flames, in which the flame is held stationary by a laminar counter flow of gas, (b) measurement of the speed of a flame travelling in a long tube, and (c) study of spherical explosion waves. In the burner methods, errors are caused by lack of uniformity of burning velocity due to curvature of the flame front, especially in the tip region, by quenching and air entrainment near the burner rim, and also by distortion of the flow lines in the preheating region, which has finite thickness. For travelling flames in tubes there are difficult corrections for the complex, and often varying, shape of the flame front, and sometimes problems occur due to gas movements ahead of the flame and to wall quenching. In spherical explosions the thermal expansion produces major movements of gas ahead of the flame and these must be allowed for in the calculations; the flame front is initially very curved and this may reduce the initial value of the burning velocity, and also combustion may not be complete near the centre of the developing flame kernel; in explosions in closed vessels the pressure rises as the flame expands and this causes some adiabatic heating and change in S_u.

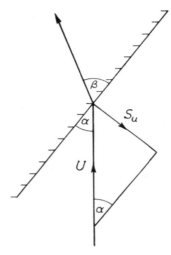

Fig. 4.1. Element of flame front propagating at the burning velocity S_u into cold gas mixture moving vertically upwards with velocity U. α is the angle between the flame surface and the initial gas flow direction. β is the angle between the flame surface and the final flow direction of the burnt products.

The burning velocity, as defined above, refers to laminar flow conditions. If the flow becomes turbulent this affects both the propagation process and the shape of the flame surface which is then no longer plane or smooth.

Gouy's flame-area method

Flame velocities are most frequently measured using a stationary flame on a burner with circular cross-section and we shall investigate conditions in a burner flame of this type. As a first approximation we assume that the burning velocity, which we shall call S_u, is constant over the whole surface of the stationary flame front. This flame front is not normal to the gas flow, but the component of the flow normal to the flame front is equal to the burning velocity. The conditions for an element of the flame front are illustrated in Fig. 4.1. Let the area of the burner mouth be A_0 and the average flow velocity in the burner mouth be v_0. The total volume flow of gas is therefore $A_0 v_0$. Let the total area of the flame front be A_f, moving with velocity S_u. Then

$$A_o \cdot v_o = A_f \cdot S_u,$$

or

$$S_u = \frac{A_o}{A_f} \cdot v_o.$$

It is therefore possible to determine the burning velocity by calculating the area of the flame front and measuring directly the gas flow and burner area. This method was developed by Gouy (1879) and is usually sufficient to give the burning velocity within $\pm 20\%$, and may be especially recommended for preliminary research on unknown mixtures.

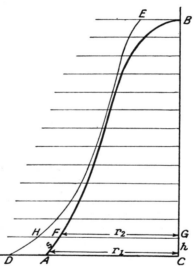

Fig. 4.2. Calculation of flame-front area. To measure the surface of the flame front AB, we divide it into a number of sections of equal height. The surface area of the section AF is $\pi s(r_1 + r_2)$ where s is the distance AF. The area $AFGC = \frac{1}{2} h (r_1 + r_2)$. If α is the cone half-angle at this point, $s = h/\cos \alpha$. If at each section we measure α and multiply the corresponding radius by $1/\cos \alpha$ we can plot a 'corrected' line DE. The area of a section $DHGC$ is then $\frac{1}{2} s(r_1 + r_2)$. If we determine the area $DEBC$ with a planimeter we obtain the sum $\Sigma \frac{1}{2} s_n (r_n + r_{n+1})$. If this is multiplied by 2π we have the surface area of the flame front.

The area of such a flame front can easily be calculated from a tracing or photograph using a planimeter. Fig. 4.2 shows the procedure. The cone is divided into a number of small sections. The surface area, a, of any such segment is $a = \pi s (r_1 + r_2)$. If α is the angle between s and the ordinate, the area $\frac{1}{2} h (r_1 + r_2)$, which can be measured with the planimeter, will give the area a when multiplied by $2\pi/\cos \alpha$.

If we determine α along the surface of the cone and multiply the corresponding radii by $1/\cos \alpha$, and then draw a second contour line as indicated in the figure, the enclosed area will be $\frac{1}{2} s (r_1 + r_2)$. The area obtained with the planimeter following the outer contour line has, therefore, only to be multiplied by 2π to give the surface area of the flame front.

Gouy's method does not depend on the establishment of any particular flow pattern in and above the burner. It does however suffer from two drawbacks. The tip of the luminous cone is rounded, not pointed; this curvature has been discussed at the end of the preceding chapter. The burning velocity is reduced at the base of the flame, due to heat loss and removal of active species at the burner wall, and the base itself is difficult to locate because the luminosity dies out gradually as the dead space is approached. The error due to the rounding of the tip is usually not very great because of the relatively small contribution of this region to the total surface area, but that at the base may be much more serious. These errors may be reduced by making the burner and flame as large as possible without the flame becoming turbulent. Also if a nozzle is used, to give a uniform velocity profile instead of the normal parabolic profile (*see* page 42), then the definition at the base of the flame is improved so that it may be located more easily and in addition we have the advantage that with this flow distribution we can go to higher Reynolds numbers and so use larger flames without trouble from turbulence. It seems that even with big burners, the influence of the base cannot be entirely eliminated, and values of S_u obtained by this method tend to be too low.

Another error, common, to some extent, to all burner methods, occurs with flames of rich mixtures. It seems that air entrainment tends to increase the burning velocity, especially towards the rim, and that this can have quite a major effect. Kimura & Ukawa (1962) studied the effect of changing the composition of the surrounding atmosphere; their results for a 6·5% propane/air mixture are shown in Fig. 4.3 for various burner diameters; for the 3-mm burner the measured burning velocity for an open flame in air was about 18 cm sec^{-1}; the true velocity in an inert atmosphere was probably about 9 cm sec^{-1}. This effect of entrainment is, of course, less marked for the larger diameter flames.

The flame front is not infinitesimally thin, as we have so far assumed. In slow flames it may be quite thick (*see* page 96). Thus a stoichiometric propane/air flame appears to have a luminous zone about 0·02 cm thick, with the preheating zone a further 0·07 cm. For rich and lean mixtures the burning velocity falls and the thickness of the reaction zone increases, that

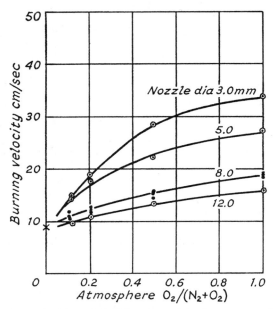

Fig. 4.3. Burning velocities of 6·5% propane/air flames as a function of the composition of the external atmosphere.   [*After Kimura & Ukawa*

of the preheating zone being of the order of 0·1 cm, which is by no means negligible compared with burner diameters which for this type of flame are commonly around 1 cm or perhaps less.

The question thus arises as to which part of the flame front should be selected to measure the flame area. For flames in which the front is normal to the flow direction there is no difficulty, but in ordinary burner flames the area of the flame front is not the same when measured from the outer boundary of the reaction zone as from the inner boundary. The outer boundary in the image of the flame is very badly related to the true outer boundary (*see* page 46), so that from this point of view the inner boundary is likely to be better. The burning velocity is defined with reference to a plane flame front; when the front is non-planar the best approximation is undoubtedly in terms of the velocity with which the position of the first temperature increase moves into the unburnt mixture; at all later positions in the reaction zone the velocity, referred to the gases at that point, is greater because of the thermal expansion. The position of the first temperature increase is difficult to define and in practice unmeasurable. The best approximation is provided by shadow or schlieren images, which are situated in a region of relatively low temperature (*see* page 70). They have the added advantage that the rounding of the tip practically disappears, and the base is better defined. We may stress again that the differences in area obtained by using different locations in the flame front only arise because the flame cannot be made so large that its thickness is negligible compared with the burner diameter.

Modified flame-area and cone-angle methods

A simplifying approximation is to assume that the reaction zone of a Bunsen-type flame is conical and to measure just the flame height, h, and the burner radius, r. The area of the flame surface is then $\pi r \sqrt{(r^2 + h^2)}$. Many measurements of burning velocity have been made by this method, and results by Jahn are reproduced by Lewis & von Elbe (1961). While it may be true that errors due to rounding of the tip and to overhang at the base tend partly to cancel out, the approximation is a crude one and we do not recommend it.

Measurement of height and radius of the flame cone is, of course, equivalent to measuring the cone angle, and we might equally use the simple relation (from Fig. 4.1)

$$S_u = U \sin \alpha$$

where α is the cone half-angle. This is indeed the basis of the nozzle method (*see* page 65). For flow up a long burner tube a parabolic velocity profile should be achieved and it can then be shown that the flame surface attains the correct angle α at a radius r given by

$$r/R = 1/\sqrt{2} = 0.707$$

where R is the radius of the burner. This method was used by Smith & Pickering (1936) and is the only method mentioned by Thring (1962).

However, the variation of cone angle with distance from the flame axis was examined by Garner, Long & Ashforth (1949; 1951) and Ashforth (1950) and this relation was found unsatisfactory. They used a long (140 cm) tube to give a parabolic velocity profile and compared the predicted cone angle with that obtained by photography of the luminous flame front. Fig. 4.4 shows that below $r/R = 0.3$ the cone angle (and therefore the burning velocity) was greater than predicted, and above 0.6 it was too small. These effects, although partly due to the increase in S_u near the tip and decrease near the rim, were attributed mainly to divergence of flow lines due to preheating and expansion of the gas before it reached the reaction zone.

To reduce this effect of flow-line divergence, Garner, Long & Ashforth also studied the flame shape by shadow photography and again calculated the burning velocity for various distances from the axis by the $S_u = U \sin \alpha$ method. In this case they found that the values of S_u tend to increase with r, in contrast with the variation found when the outer edge of the luminous cone was used. Their curves are shown in Fig. 4.5. The curves for the direct

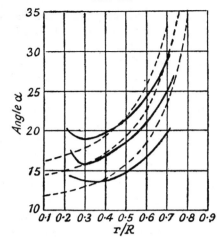

Fig. 4.4. Variation of cone half-angle, α, for three mass flows, as a function of fractional distance r/R from axis of burner.   [*After Ashforth* (1950)
— — — — calculated.
———— from observed luminous flame front.

and shadow photography cross at $r/R = 0.4$, and at this value the calculated and observed cone angles also agree (Fig. 4.4). Thus they suggested that measurements of cone angle at $r/R = 0.4$ combined with the calculated gas velocity at this radius* might be used to give the true burning velocity. It certainly appears that for the type of flame which they studied the direct and shadow cones are parallel at this radius and thus flow-line divergence

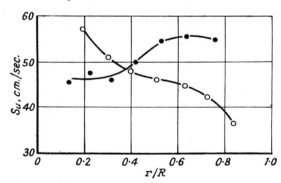

Fig. 4.5. Variation of burning velocity across flame, as function of fractional distance from axis of burner, r/R: from flame shadow — ●—●— and from outer edge of luminous cone — O — O — .

[*After Garner, Long & Ashforth* (1951)

* At $r/R = 0.4$ the gas velocity should be 1·68 times the mean flow velocity, calculated from the volume flow and burner area.

is small. The reliability of this $r/R = 0.4$ method has so far only been checked for a particular type of flame and might not hold for other fuels or burners.

Townend, Garside & Culshaw (1941–8) at Leeds have used Gouy's total area method, but have studied the effect of burner diameter. They found that the burning velocity determined from the inner edge of the luminous cone depends on the size of the burner because the base of the flame has a relatively greater influence with small burners. As the burner size is increased the influence of the base becomes less important and the burning velocity becomes constant and independent of burner size. The method is obviously better than using the Gouy method without regard to burner size or with the outer edge of the flame. It would, however, be better still to use schlieren photographs rather than the luminous cone.

It is particularly necessary to use large burners for flames at reduced pressure. In comparing burning velocities at different pressures the burner should be changed to keep its diameter inversely proportional to the pressure, or, more accurately, proportional to quenching diameter.

Other variations on burner methods have been used. Bunsen-type flames on a rectangular slot burner have the advantage that the tent-shaped flame is flat, thus avoiding some of the troubles due to curvature of the flame surface. Singer (1953) has described work with this type of burner, finding slightly lower values of S_u than with conical flames where the flame curvature increases heat transfer and the apparent value of S_u; these flames on slot burners have, however, a greater tendency to instabilities of the type which produce polyhedral flames. Singer also discusses the truncated-cone method, in which only the part of the flame away from the tip and the rim is used in the calculations. Another useful, more recent, review is by Andrews & Bradley (1972). Spherical flames supported by premixed gases flowing from a porous sphere have been studied by Fristrom & Westenberg (1965); the flame is then always normal to the flow lines, although the value for the flame area still depends on whether the flame's own light, its schlieren or its shadow is used to measure it; buoyancy effects on the flame shape are not, apparently, as troublesome as might be expected.

The nozzle method

In discussing the Gouy method, we have pointed out that there are advantages in using a nozzle instead of a simple burner tube. Moreover, the nozzle method may allow us to introduce some simplifications which make it unnecessary to calculate the flame area. Mache & Hebra (1941) were the first to realise that the uniform velocity profile gives the flame the form of a nearly perfect geometrical cone, at least in the central region. Thus if the gas velocity, U, is known, simple measurement of the cone half-angle, α, gives directly $S_u = U \sin \alpha$.

Nozzle flow patterns have been discussed in Chapter III. Mache & Hebra studied the velocity distribution and found that at 6 cm above a 1-cm burner the velocity distribution was already approaching the parabolic form again, but that flame cones of not too great height lay sufficiently well within the uniform velocity profile region. The design of the nozzle is not very critical, but the area contraction must be by at least a factor of 4.

Bartholomé (1949a; 1949b) refined this method by making allowance for the boundary-layer. His criterion was also to make the burner so big that the rounded tip occupied only a small part of the total cone length, so that the angle of the remaining straight part could be measured accurately. As an alternative to determining the velocity from corrected burner area and total flow, the velocity in the uniform region may be measured from the Pitot pressure as discussed in the previous chapter. A converging rectangular nozzle may be used on a slot burner (Kuehl, 1962).

An advantage of the nozzle method, using the slope, is that the choice of reference line in the flame front is less important as the luminous cone and schlieren are practically parallel. This method is likely to give the most trustworthy results for burning velocities, and is equalled only by the particle-track method, which is, however, very laborious. Generally the bigger the flame the better; for lean and stoichiometric mixtures it is only necessary to get a good straight part to the cone to measure α accurately, but for rich mixtures diffusion of external air may increase the burning velocity in small flames. The main disadvantage of the nozzle method, apart from the need to make the nozzle, is that the gas flows from a wide tube or container into a relatively narrow nozzle; if the flame strikes back into the wider part a serious explosion may occur. In practice it is difficult to prevent the flame striking back occasionally. A sintered porous plate in the wide part below the nozzle may serve as a flame trap, but will increase the danger if it fails to stop the flame; a safety precaution is to have a thin-walled section which will break first and reduce the destructive effects; glass should be avoided as far as possible.

The particle-track method

With conical flames on round burners we always have curved flame surfaces and a number of difficult three-dimensional problems arise, both in the detailed structure of the flame front itself and in its photography. To overcome these difficulties Lewis & von Elbe (1943) made an investigation using rectangular burners. For these it was not possible to calculate easily the flow pattern of the gas. A particle-track method was therefore devised in which small magnesium oxide particles in the gas stream were illuminated

MEASUREMENTS OF BURNING VELOCITY 67

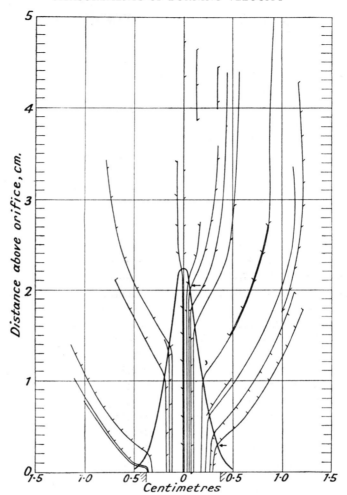

Fig. 4.6. Flow lines through the inner cone of a natural gas/air flame, from particle-track measurements using a rectangular burner. The particle tracks were interrupted at intervals corresponding to $1\cdot436 \times 10^{-3}$ sec.
[*From Lewis & von Elbe*

intermittently from the side (*see* page 46). The photograph of the track of the particle gives both its direction and velocity. Fig. 4.6 shows the tracks of particles in a narrow central section of the flame on the rectangular burner. Details of the complicated calculation of the burning velocity are given in the original paper. The results are shown in Fig. 4.7. It can be seen that over most of the flame front constant burning velocity does exist. The

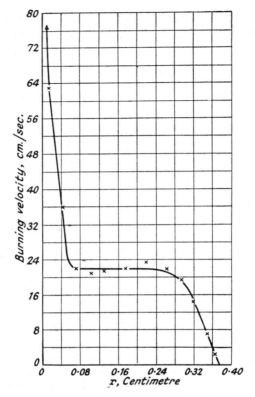

Fig. 4.7. Burning velocity across the flame as determined from the flow lines shown in Fig. 4.6; r is the distance from the axis of the burner.

[*From Lewis & von Elbe*

width of their burner, 0·755 cm, was rather small and the proportion of the flame front with uniform burning velocity would be greater if a larger burner were used.

The particle-track method is too laborious for regular measurement of burning velocities, but the results of Lewis & von Elbe are of great importance because they show that the burning velocity really is a genuine physical constant.

Andersen & Fein (1949) have also used a particle-track method, but with a circular nozzle burner, giving a straight-sided cone. They determined values of the burning velocity from the cone angle and the flow velocity as deduced by the particle tracks.

A possible weakness of the method is that the introduction of solid particles might, by catalytic effects at the surface, modify the combustion processes slightly and so alter the value of the burning velocity. Errors may also occur if the particles are too big to follow the gas flow accurately (*see* page 47).

Methods of locating the flame front
All the methods discussed so far require location of the flame front. The various methods of doing this, by direct photography of the luminous cone, by shadow photography and schlieren photography have been discussed in the previous chapter. We have seen that the reaction zone has a finite thickness and that the flow lines may diverge between the locus of the first temperature rise and the position of the luminous zone; as a result of this and of the photographic complications discussed in the previous chapter, the positions of the flame front as observed by direct-shadow and schlieren methods will not in general be parallel. Also owing to the conical shape of the flame surface there may be difficulty in interpreting the photographs of the flame. We shall therefore discuss the methods of flame photography again in relation to determination of the burning velocity.

Shadow photography. Light passing through the preheating zone of a flame is deflected downwards towards the cooler gas so that the shadow on a screen shows a dark zone above a brighter region. As the screen is moved away from the flame the separation between the bright and dark zones increases and the image becomes difficult to interpret. Obviously the bright zone is caused by the deflected light and its position on the screen is not simply related to the position in the flame at which the deflection occurs. The dark zone, however, is due to deflection of light from this zone and the dark shadow tends to be in a straight line with the source and the part of the flame causing the deflection; thus the dark shadows are, in first approximation easier to interpret. It is sometimes stated (e.g. Beams, 1955) that while schlieren measurements depend on the first derivative of the refractive index with respect to temperature, the shadow method depends on the second derivative. It might thus be expected that the shadow method would give information about a lower temperature region in the measurement of burning velocity. However, Weinberg's analysis (1956*b*) of the shadowgraph of a flat (Egerton-Powling type) flame has enabled us to clarify our ideas on this subject.

In the shadowgraph of the flat flame (*see also* page 51) there are two limiting conditions, one with the screen very close to the flame, and the other with the screen at a large distance. In the preheating zone the change in refractive index is due primarily to the temperature rise and not to change of chemical composition. For a gas such as air for which the thermal conductivity divided by the specific heat is approximately proportional to the absolute temperature (i.e. $k/c_p \propto T/T_0$) Weinberg has shown that with the screen close to the flame the bright region of the shadow is produced by a flame region at temperature T given by $T = 1 \cdot 14 \, T_0$, T_0 being the initial temperature. With $T_0 = 18°C = 291$ K this gives $T = 332$ K $= 59°C$. The dark part of the shadow corresponds to a flame region with $T = 2 \cdot 19 \, T_0$; for $T_0 = 291$ K this gives $T = 637$ K $= 364°C$. Thus the bright region of the shadow does indeed correspond to a flame region of very low temperature, but the dark region, which is easier to relate to its position in the flame, is produced by a region of higher temperature, higher indeed than the region giving the schlieren image. In practice the shadow obtained with the screen close to the flame is faint and difficult to see or photograph, while the interesting bright region moves and changes its intensity contour as the screen is moved away from the flame.

With the screen at a sufficient distance from the flame a sharp line in the shadow, due to a discontinuous increase in intensity, develops. Weinberg has shown that the flame region producing this is at temperature $T = 3/2 \, T_0$, which is (see next section) the same temperature as for production of the schlieren image. The discontinuity is displaced downwards from the flame region producing it. For the flat flame Weinberg has calculated the amount of this displacement in terms of the distance, flame thickness, burning velocity, thermal diffusivity and refractive index. For a conical flame this discontinuity in the shadow would be extremely difficulty to interpret, and since any expression would presumably involve a value of the burning velocity, it would hardly be suitable for measurement of this burning velocity.

Rather earlier, Grove, Hoare & Linnett (1950) made an analysis of the light pattern from an assumed infinitesimally thin cyclindrical flame front. They found that the light followed a caustic curve, and that the sharp edge did not coincide with the position of the flame front, but lay well inside it. This edge would therefore tend to give too a small flame area and thus too high a value for the burning velocity. Andersen & Fein (1950) studied the effect of varying the distance of the screen from the flame and extrapolated back to zero distance to locate the actual flame front. However, the assumption of an infinitely thin reaction zone and a cylindrical rather than a conical shape are obvious limitations. Using simple shadow photography, the bright regions of the shadow, although due to the important low-temperature region of the flame, are too difficult to interpret, while the dark shadows are easier to interpret but refer to a flame region which is hotter, and therefore less good, than that producing the schlieren image. Gilbert (1957) has used an adaptation of Weinberg's inclined slit technique; a shadow of the flame in parallel light was obtained with a grid of parallel diagonal wires interposed between the light source and the flame. Gilbert used the locus of the peak deflections of the wire images to determine the position of the flame. His photograph shows that diffraction patterns were rather troublesome. Although this method with a wire grid is better than direct-shadow photography, we think that ordinary schlieren photography is easier and better.

Schlieren methods. These methods give a focused image of the schlieren in the flame, and therefore give a better location of the change of refractive index. Plate 6a shows superposed schlieren and direct photographs of a flame. There is again some difficulty due to the different effective thickness of the cone at the top and base. This causes a greater deflection of the light passing through the base, and the schlieren in this region are much more clearly visible than near the tip of the flame. However, the position of the schlieren image is less affected than the edge between shadow and caustic curve.

Van der Poll & Westerdijk (1941) were the first to discover that the schlieren cone lies substantially inside the luminous cone. It was at first thought that this was because chemical reaction occurred below the luminous zone, but it is now clear that the schlieren deflections are caused by the temperature rise ahead of the reaction zone. Neglecting any change in composition, the refractive index, or rather $n - 1$, varies as $1/T$. Thus a change in temperature from say 100°C to 300°C has a vastly greater influence on the change of refractive index than a change from say 1500°C to 1700°C. Klaukens & Wolfhard (1948) made thermocouple measurements on the temperature distribution in the preheating zone of a flat flame and showed that the maximum schlieren deflection would be at around 200°C. Weinberg (1955; 1956a) has made a detailed theoretical study of the location of the region in the flame giving the maximum schlieren deflection. If the thermal conductivity k is assumed to be a simple function of temperature

$$k = KT^a,$$

where K and a are constants, it can be shown that the temperature for maximum schlieren deflection, T_s is given by

$$T_s = T_0 (a + 2)/(a + 1),$$

where T_0 is the initial gas temperature. Assuming that the conductivity was independent of temperature would thus give the schlieren image at twice room temperature. Actually for a gas such as air the conductivity is very nearly directly proportional to temperature over the interesting range, i.e. $a = 1$, so that

$$T_s = 1\tfrac{1}{2} T_0.$$

Thus for $T_0 = 291$ K ($= 18$°C) we obtain $T_s = 436$ K $= 163$°C. Allowing for some slight increase in T_0 due to heating of the burner by flame radiation we see that the value is very close to the experimental position of the schlieren at 200°C. These calculations were first made for a flat flame but have now been shown to apply approximately to conical flames as well. The schlieren thus serve very well to locate a flame zone at relatively low temperature such as is required for the measurement of burning velocity. Conan & Linnett (1951) find that the flow lines just begin to deviate from parallel flow at the position of the schlieren image, and expand further as they approach the luminous zone.

The burning velocity can be determined from the schlieren cone either by cone angle or by the total-area method. Conan & Linnett use the first method, and find that S_u is constant over an appreciable length of the cone and is independent of flow rate and the magnification of the optical system. The total-area method might be preferable if there is any doubt about the flow pattern because of back pressure. The maximum of the schlieren image should be taken as reference line as only this is independent of the geometry of the arrangement. We should like again to stress a point. Very careful focusing of the flame into the camera must be made before the schlieren are produced, since in the schlieren set-up the flame will always appear sharp because nearly parallel light is used. If this adjustment is not done accurately very queer focusing effects may occur.

Rectified flat-flame methods

Egerton & Powling (Powling, 1949; Egerton & Thabet, 1952; Badami & Egerton, 1955) have succeeded in running flat flames of slow-burning mixtures, with the burning velocity just balanced by the flow velocity of the gas (*see* Fig. 4.8). The burner is around 6 cm in diameter and is jacketed by a stream of nitrogen. Within the burner is a metal matrix which produces a uniform velocity profile. The flame takes the form of a flat disc. Its stabilisation seems to be effected by a very slight divergence of the flow lines just above the burner. Without this the flame would either blow-off or strike-back and it would not be practical to hold it steady. Particle-track photographs by Levy & Weinberg (1959) have shown a toroidal vortex which plays a part in anchoring the flame; there also appears to be an appreciable divergence of the flow lines, which, unless corrected for, will cause some error in measurement of burning velocity.

The measurement of the burning velocity thus presents no difficulty as it is nearly equal to the flow velocity. As a refinement the total area of the disc may be measured. This method should give really reliable results, but even here there are a number of limitations. It has only been found possible to obtain stable flames of this type with very slow-burning mixtures, roughly between 5 and 9 cm/sec. Also, to get the flame really flat the honeycomb-shaped matrix has to be fairly close to the flame, between 0·8 and 1·5 cm below it; for reasons not yet fully understood, this matrix becomes heated, sometimes to 200°C. The gases are therefore preheated by the matrix as well as by the flame front, and we no longer have a completely natural flame.

The method is useful for mixtures near the limits of flammability, which is just where other methods are very difficult to apply because of the low burning velocity and big thickness of the reaction zone producing disturbances in the flame shape. The greatest possibilities of the technique are,

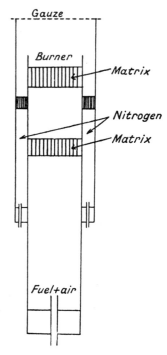

Fig. 4.8. Burner for flat flame with rectified flow pattern.

however, for dealing with cool flames and for producing flames with flat and very thick reaction zones which can be studied optically and chemically. Corbeels (1970) has noted some interesting effects on burning velocity in two-stage flames; if a cool flame develops well below the main hot flame the burning velocity in the main flame is reduced, but at the point where the cool and hot flames coalesce the value of S_u shows a marked increase.

Flat flames of faster burning mixtures can be stabilised over a cooled porous plate (Botha & Spalding, 1954; Yumlu, 1967). The heat loss from the flame reduces the burning velocity, so that the flame front takes up a position at some distance from the plate. The distance varies with the gas flow. The heat extracted by the cooled plate is measured by the flow and temperature rise of the cooling water. The flame speed for this flat flame is, of course, equal to the gas-flow rate. If, for various flows, the gas-flow rate is plotted against the heat extraction per unit volume of gas burnt, the extrapolation back to zero heat extraction will give the burning velocity of the mixture. By this method, it has been possible to measure the burning velocity of propane/air mixtures over the entire range of mixture strength. In some cases, as the gas-flow rate is raised to near S_u, the flame develops a cellular structure, leading to an apparent increase in burning velocity; in making the extrapolation back to S_u, the points for cellular flames must be ignored.

Later studies (Edmondson, Heap & Pritchard, 1970; Pritchard, Edmondson & Heap, 1972; Gunther & Janisch, 1972) have shown that these flat-flame methods are still subject to appreciable error due to edge effects and tend to give low values of S_u. Errors are reduced by using large diameter tubes and a better value of S_u can be obtained by studying the effect of burner size and extrapolating to infinite size.

The flame-thrust method

We have seen (page 53) that the back pressure of a flame, due to the acceleration of the flame gases as they pass through the reaction zone, depends on the burning velocity. For slow flames the back pressure is too small to measure easily, thus for natural gas/air it will only be at most about 0·01 cm of water. For really fast flames, however, it will be appreciable and may be used to determine the burning velocity. One needs to know the pressure difference between that in the burner and the surrounding atmosphere and the ratio of the densities of the burnt and unburnt gases. For slow flames there is no difficulty, in the large burners, in providing a pressure point to measure the pressure, but the pressure difference itself is very small. For very fast flames, very small burners must be used, and there may be some practical difficulty in measuring the pressure without disturbing the flow of gases in the burner. The measured back pressure will tend to give an average flame velocity over the flame front, including the dead-space area, and in this respect will suffer the same disadvantage as the total-area method. However, the form of averaging may be different; the dead space and base of the flame will be weighted high by virtue of their relatively large area, but as the pressure depends on the square of the velocity this dependence will favour, in the weighting, the parts of the flame front where the velocity is greatest.

The method requires a knowledge of the densities of the unburnt and burnt gas. The practical determination of this is difficult if not impossible. It may be obtained by thermodynamic calculation, as outlined in Chapter

XII. It is necessary to determine the change in mole number, allowing for dissociation in the burnt gases, and the final flame temperature.

Some measurements by this method have been made by Manson (1945; 1948) and by von Elbe & Mentser (1945); the latter found a back pressure of 5·5 cm water for oxy-acetylene, and their values for S_u agree with those obtained by Gouy's method.

The very sensitive pressure-measuring apparatus developed by Payne & Weinberg (1962) for studying electric-wind effects in flames, may have a use in studying small flame thrusts. It consists of a large diameter floating 'bell' of mica connected by a small tube to the pressure probe; the bell is connected to a balance arm and the movement is measured with an optical lever; pressures of around 10^{-6} atm or 10^{-3} cm water, may be measured.

The soap-bubble method

A method which is especially important for cross-checking results from burner experiments is that using soap bubbles. A combustible mixture is blown into a bubble and ignited centrally. The surface of the bubble expands freely as the explosion proceeds, thus ensuring constant pressure. The velocity measured directly is, of course, a spatial velocity, the expansion of the burnt gases causing a bodily movement of the gases through which the flame front is progressing. The method has been investigated by Linnett and co-workers, who have used both direct and schlieren photography to record the position of the expanding flame front. Khitrin (1936) has shown that the flame front is spherical and can be recorded with a revolving-drum camera. The burning velocity is the spatial velocity divided by the expansion ratio of the bubble. This ratio is $(D/d)^3$, where d and D are the initial and final values for the diameter of the bubble; these have to be measured separately.

The method gives directly the velocity of the propagating flame, corrected only for the expansion, and should therefore give accurate results. There are however some reservations. The gases must necessarily be moistened by evaporation from the soap solution and, especially for carbon monoxide, this may alter the burning velocity, although experiments have actually been made with CO. In some cases gases may be absorbed in the film; however Linnett* showed that this effect is small for acetylene/air flames.

The greatest experimental limitation is the determination of the initial and final size of the bubble, which must be known very accurately because the ratio appears in the calculation as the cube. The final size is difficult to measure. The expansion ratio can also be calculated assuming the gases reach the theoretical flame temperature, but comparison of measured and calculated expansion ratios often reveals serious discrepancies. Thus for

* Linnett, Pickering & Wheatley (1951); Pickering & Linnett (1951).

10% acetylene/air the calculated value is 9·17 and the measured is only 7·8. This is probably due to loss of heat to the electrodes used to ignite the flame. This heat loss means that it will not be a full-strength flame. Also the burning velocity will tend to be reduced because the flame develops against a curvature. This is the reverse of the effect at the tip of a burner flame where the curvature assists heat and radical transfer.

When the bubble of explosive mixture is surrounded by air, afterburning of any excess fuel with this air may cause trouble. Thus for propylene/air flames Gray, Linnett & Mellish (1952) found that for lean mixtures there was good agreement between soap-bubble determinations of burning velocity and those determined by a burner method, but that for rich mixtures the soap-bubble values were too low. The afterburning apparently affects the determination of final size. Strehlow & Stuart (1953) made similar observations on ethylene in which argon instead of air was used to surround the bubble; for slightly rich ethylene/air in air the burning velocity was measured as 55·3 cm/sec, and in argon as 65·8 cm/sec.

The soap bubbles used by Linnett were not very big. We have seen that for burner flames it is necessary to use large flames to reduce errors due to heat loss to the rim, effects of curvature at the tip and those due to the reaction zone being of finite thickness. Similar considerations must also apply to the soap bubble method. The method is thus more suitable for fast flames; for slow flames the time for heat losses will be greater and the thickness of the reaction zone will be larger. There is even the possibility with slow flames that the flame front may not remain spherical; the interesting cellular structure of some flames is described in Chapter II. The method is certainly not to be recommended for work at low pressure because the increase in thickness of the reaction zone will cause the effect of curvature to become more important, and also heat losses will be more serious. However, at high pressure we have at present very little knowledge of the variation of burning velocity. Such measurements are usually difficult to make because of the need for elaborate safety precautions. The soap-bubble method would appear relatively easy, safe and reliable at high pressure. Non-aqueous soap bubbles of glycerine with a detergent have been used (Strehlow & Stuart, 1953); these permit measurements on dry gases.

High speed cine photographs (Simon, 1959) of fast (2500 cm/sec) methane/oxygen/nitrogen flames show that the flame front is no longer smooth, the irregularities in the flame surface probably increasing its area and the apparent flame speed for these fast flames. For very slow flames, convective rise of the hot burnt gas prevents accurate measurement with a drum-camera; this effect is likely to be more serious with a horizontal slit to the camera, than with a vertical slit.

The spherical-bomb method

This is very similar to the soap-bubble method, but the whole vessel is filled with the flammable mixture and instead of determining the expansion ratio from the final size of the bubble, simultaneous records are made of the size of the spherical shell of burnt gas and of the pressure in the vessel.

It can be shown (Manton, von Elbe & Lewis, 1953) that if a is the radius of the bomb and r_b and P are the radius of the shell of burnt gas and the pressure at time t, then

$$S_u = \frac{dr_b}{dt} - \frac{a^3 - r_b^3}{3P\,\gamma_u\,r_b^2} \cdot \frac{dP}{dt}$$

where γ_u is the ratio of specific heats of the unburnt mixture. This method is theoretically, very satisfactory, but in practice the burning velocity S_u comes out as the difference between two nearly equal quantities and cannot be measured with reasonable accuracy.

An alternative method of evaluating the burning velocity is to determine the mass fraction, n, of gas burnt at time t. Then

$$S_u = \frac{dn}{3dt}\frac{a^3}{r_b^2}\left(\frac{P_i}{P}\right)^{1/\gamma_u},$$

where P_i is the initial pressure. For small values of n,

$$n = (P - P_i)/(P_e - P_i),$$

where P_e is the pressure corresponding to combustion at constant volume and may be computed thermochemically. Unfortunately the method in this form assumes complete equilibrium behind the flame front and no heat losses. Any delay in attaining equilibrium in the rather large volume immediately behind the flame front will cause error, the burning velocity calculated using the above expression being less than the actual value. Any heat loss or quenching of combustion by the electrode leads will also affect the result. Results for ethylene/air and propane/air do in fact seem rather low. As the burnt volume increases there appears to be a slight increase in measured burning velocity, due perhaps to decreasing curvature of the flame surface or to the relatively smaller importance of losses near the spark gap. The rising pressure and consequent adiabatic heating will increase the burning velocity as the size of the burnt-gas shell increases.

The method is of special value for studying effects of pressure on burning velocity and for studying substances which are only available in small quantities. Gray and colleagues (e.g. Gray & Lee, 1959; Gray, Mackinven & Smith, 1967) have used the method frequently for work of this type; they employed a fairly large (20 cm) reaction vessel and followed the spark-ignited flame kernel by rotating-drum schlieren photography during the initial period when effects due to the rising pressure in the vessel could be neglected.

Bradley & Hundy (1971) have refined the method by using a hot-wire anemometer to measure the gas flow ahead of the flame front and a laser interferometer system, with high-speed cine recording, to study the flame front itself.

A variation on the usual bomb technique by Raezer & Olsen (1962) involves the study of two simultaneously ignited flame kernels. As the two flames approach each other the gas on the line joining their centres should be stationary and the flame velocity should fall to the burning velocity. This double kernel method seems to give reasonable results, but we distrust it because of the complexity of the gas movements.

The revised tube method

The determination of burning velocities in tubes is usually very unreliable. For tubes not much bigger than the quenching diameter a very uniform movement is obtained and it is therefore easy to measure the velocity. However, this cannot be a full-strength flame because of the excessive quenching. If tubes of larger diameter are used vibrations usually set in, and these either inhibit or accelerate the flame considerably. Comparative measurements for mixtures with different quenching distances are hardly possible because different tube diameters must be used for each mixture to keep in the region between excessive quenching and oscillatory burning.

Some of these limitations may be overcome by having partially closed tubes with small holes at each end to allow the pressure to adjust itself. The amount of gas pushed through these holes can be measured by following the movement of a soap bubble (Gerstein, Levine & Wong, 1951). The burning velocity is given by

$$S_u = (S_{sp} - S_g) A_t/A_f.$$

S_g is the velocity of the gas ahead of the flame front and can be measured by the displacement of the soap bubble; S_{sp} is the spatial velocity and can be measured by photocells, photography or ionisation detectors; A_t is the cross-sectional area of the tube, and A_f is the area of the flame front, which is usually hemispherical. The greatest inaccuracy is usually in the measurement of the flame area, as the surface is badly deformed near the walls. There is also often some difficulty in selecting suitable tube diameters which do not suffer from serious quenching effects and yet do not give vibrations. The method is certainly not a good absolute one, but may be useful for comparative measurements. One advantage, like that of the soap-bubble method, is that it can be used with small quantities of fuel.

Results by this method may be compared with those by other methods. Gerstein finds the maximum value of S_u for propane/air as 39 cm/sec, while Andersen & Fein (1949), by the nozzle method, get 45 cm/sec and Bartholomé (1949a), 47 cm/sec. This apparent consistency is, however, destroyed if we have a closer look at the variation of velocity with mixture strength, when appreciable differences appear. For ethylene/air the tube method gives a maximum S_u of 68·3 cm/sec compared with values of 78 by Bartholomé and 68 by Linnett, using other methods. From these results the method might appear as good as most, but we think that it cannot give results which are independent of tube diameter, and should only be used in special circumstances for comparison of similar mixtures under conditions when it has been shown to give results in agreement with those by other methods.

For small tubes wall quenching modifies the flame shape and limits its speed, and a discussion of the gas flow lines under these quenching conditions (Lewis & von Elbe, 1961, page 292) suggests that the flame speed might become equal to the burning velocity for the smallest tube which will just propagate a flame. Singer & von Elbe (1957) made measurements on the speed of downward propagation of flames in methane/air mixtures in small tubes of various sizes, and found that the speed was indeed close to the burning velocity for the smallest tube which would just allow a flame to propagate. Values of S_u determined in this way for various mixture strengths usually agreed with values determined by other methods to better than 10%. However, plots of flame speed against burner diameter show a rapid variation of speed with diameter, especially for the faster burning mixtures; a small error in determination of the minimum burner diameter would thus produce a big error in S_u. We do not feel convinced that for fast-burning mixtures a flame may not be able to propagate under quenching conditions at slightly less than the normal burning velocity. For the slow-burning near-limit mixtures selective diffusion effects may also be troublesome and modify the local mixture composition and hence the speed. For later work using this method see Fuller, Parks & Fletcher (1969) and the review by Andrews & Bradley (1972).

Comparison of burning velocity measurements

Measurements for propane/air have been made by many investigators, so that a comparison of results by various methods is possible. These are given in Table 4.1.

In this table we have given only those values for which a serious attempt has been made to measure S_u accurately. Variation in the initial room temperature may account for some of the scatter. The most probable value is about 43 cm/sec, since methods (2) and (10) seem most likely to be free from the various errors we have discussed. The early nozzle-slope method, as used by Mache & Hebra (3), may be too small because the boundary-layer is not considered, so that the assumed flow rate is in error. The nozzle-slope method as used by Bartholomé (1) appears, however, to be rather high. The total-area methods (4), (6), (7) and (8) all appear low, presumably because of quenching at the base, and because divergence of flow lines in

TABLE 4.1

Maximum burning velocity for propane/air

	Author	Method	Value cm/sec
(1)	Bartholomé (1949a)	Nozzle-slope	47
(2)	Andersen & Fein (1949)	Nozzle with particle-track and slope	44·5
(3)	Mache & Hebra (1941)	Nozzle-slope	36
(4)	Hahnemann & Ehret (1943)	Nozzle, total-area	41
(5)	Garner *et al*.	$r/R = 0.4$	48·5
(6)	Garner *et al*.	Burner, total-area	40
(7)	Harris, Grumer *et al*.	Burner, using total area from frustrum of visible cone	41·5
(8)	Culshaw & Garside	Burner, total-area	37
(9)	Gerstein, Levine & Wong	Revised tube	39
(10)	Gray, Linnett & Mellish (1952)	Burner, slope of schlieren cone	43
(11)	Manton, von Elbe & Lewis (1953)	Spherical-bomb	41
(12)	Strehlow & Stuart (1953)	Soap-bubble	40·4
(13)	Botha & Spalding (1954)	Flat flame with heat extraction	41·7
(14)	Raezer & Olsen	Bomb; double kernel	44
(15)	Fuller *et al*. (1969)	Revised tube	40·5
(16)	Edmondson & Heap (1970)	Flat flame	41

the preheating region causes the measured flame area to be larger than the area of the position of the first temperature rise. (5) seems rather high; this method is empirical rather than fundamental and we doubt if this $r/R = 0.4$ relationship necessarily holds for all flames. The spherical-bomb (11) and soap-bubble (12) methods are free from the errors peculiar to burners (e.g. quenching at burner rim, divergence of flow lines) but seem in practice a little low; this may be caused either by curvature of the flame front, or by quenching and incomplete combustion in the electrode region near the centre of the exploding volume. (13) is interesting because it avoids the difficulties inherent in locating the flame front and since it uses a flat flame cannot be affected by curvature of the flame surface; a short extrapolation back to zero heat extraction is necessary, but the inaccuracy thereby introduced should be small.

These comparisons show that a simple nozzle method, with correction for the boundary-layer, is best and most practical, and that the base of the flame should be excluded from the determination. The flat-flame method with heat extraction appears very promising, and for slow flames the Egerton-Powling flat flame is satisfactory. For relative measurements on mixtures of comparable burning velocity, the total-area or $r/R = 0.4$ methods may be useful, but are not to be recommended for absolute determinations.

A less complete comparison is also possible for ethylene/air, and this includes results by Linnett et al. by several methods. These are collected in Table 4.2. Methods (3), (4), (5) and (6) agree very well. The total area method (2) is as usual too low, and Bartholomé's nozzle slope value (1) again appears high. The spherical-bomb result (7) is also again lower than the probable value. Although four methods agree in a value of 68 cm/sec, we think the true value may be a little higher as these methods all seem more likely to give a low rather than a high value; for burner methods the back pressure of the flame may disturb the parabolic flow pattern a little, and for the soap bubble there may be quenching by the electrodes.

TABLE 4.2

Maximum burning velocity for ethylene/air

	Author	Method	Value cm/sec
(1)	Bartholomé (1949a)	Nozzle-slope	80
(2)	Culshaw & Garside (1949)	Burner, total-area	60
(3)	Linnett et al.	Burner, slope of schlieren cone	68
(4)	Linnett et al.	Soap-bubble	68
(5)	Linnett et al.	Burner, slope of shadow edge	68
(6)	Strehlow & Stuart (1953)	Soap-bubble	68.6
(7)	Manton et al. (1953)	Spherical-bomb	63
(8)	Raezer & Olsen (1962)	Bomb, double kernel	79

A similar comparison for methane has been made by Andrews & Bradley (1972). They list 26 values from 39 to 47 cm/sec and recommend 45 cm/sec.

The temperature dependence of burning velocity

Early observations on the effect of preheating the gas mixture on the burning velocity were made by Ubbelohde & Hofsaess and by Sachse who apparently expressed the velocity relative to the flow of cold gas before the preheating; these early results thus appeared to show only a modest increase in S_u with initial gas temperature. However, more recent work shows that when the burning velocity is referred to the actual gas, at the preheated temperature, the effect is quite marked.

Fig. 4.9 shows the effects on burning velocity of preheating methane, propane and ethylene air mixtures to over 600 K, from Dugger & Heimel (1952). Kuehl (1962) extended observations on propane/air to 811 K, at which temperature the burning velocity had risen to 250 cm/sec; he also studied the variation of burning velocity with mixture strength and his curves show that the point of maximum burning velocity moves to less rich mixtures when the gas is preheated. Flames of benzene, heptane and iso-octane were also studied by Heimel & Weast (1957) who again found a marked increase in S_u with initial temperature, T_i, and expressed their results in the form

$$S_u = B + C \cdot T_i^n$$

where B, C and n were empirical constants, the value of n lying between 2 and 3. They also used some theoretical relationships to connect the temperature variation of burning velocity with the activation energy of the chemical reaction. If gas mixtures are strongly preheated, to over 800 K preflame reactions may occur and these tend to lower the burning velocity; Dugger, Weast & Heimel (1955) found that the velocity then depended on the contact time for which the mixture was held at high temperature, long contact (i.e. several seconds) causing a fall in S_u.

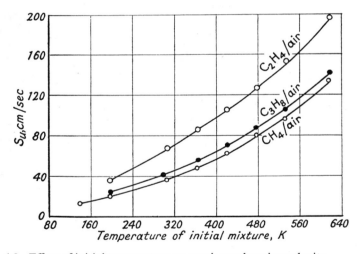

Fig. 4.9. Effect of initial temperature on maximum burning velocity.
[*From Dugger & Heimel* (1952)]

The pressure dependence of burning velocity

We have constantly stressed the need to use burners of large diameter for measurements of burning velocity. The burner diameter may be expressed as a multiple of the quenching diameter. Thus for propane/air the quenching diameter is 0·3 cm at atmospheric pressure. Serious cooling effects must therefore exist for over half this distance from the wall; a burner diameter of around 1 cm is therefore the absolute minimum for which we can expect to get even a part of the cone free from disturbances. To get to lower pressure we have to use bigger burners, and also keep the Reynolds numbers comparable. The conditions for flames at very low pressure have been investigated by Wolfhard (1943) who found that for all comparable measurements on flames the burner diameter must be kept proportional to 1/pressure. The plots of the stability regions show this clearly (*see also* page 17).

For CO/O_2 flames the pressure dependence has been studied by Kolodtsev & Khitrin (1936) up to 40 atm by the soap-bubble method. They found no change with pressure. Gilbert (1957) also made some measurements on CO/O_2 at reduced pressure and found little change in burning velocity over the range 150 to 760 torr; he used a burner-slope method, with a shadow of a grid to locate the flame and employed various burner diameters but did not keep the diameter strictly proportional to 1/pressure.

Wolfhard (1943) measured the velocity of acetylene/oxygen by the total-area method between 10 and 760 torr; below 10 torr the quenching distance becomes too big, compared with the burner diameter (about 2 cm) so that the burning velocity is affected. In the range for which the measurements are reliable S_u was found to be constant at about 900 cm/sec and independent of both pressure and mass flow. The value may perhaps be 10% to 20% low for reasons already brought out. Gilbert's measurements (1957) for C_2H_2/O_2 over the pressure range of about 50 to 380 torr also show S_u very constant. Gilbert's results for both C_2H_2/O_2 and propane/O_2 at lower pressure (down to 3 torr) are, as he shows, affected by the flame shape and depend on burner size, but do not indicate any appreciable variation of burning velocity with pressure for these flames.

For methane/oxygen, Smith & Agnew (1957) found a burning velocity of 564 cm/sec at 1 atm and 386 cm/sec at 0·1 atm, thus indicating a positive dependence of S_u on pressure; these values were obtained by a spherical-bomb method and do not agree too well with some other values in the literature. Strauss & Edse (1959) found positive pressure exponents for hydrogen and methane/oxygen flames between 1 and 90 atm. There is, however, a suspicion that the flame fronts become 'rough' at higher pressures and this leads to an apparent increase in burning velocity.

For hydrocarbon/air flames there is now fairly definite evidence of a general tendency for a negative dependence of burning velocity on pressure, values of S_u rising at reduced pressure. Gilbert's results (1957) for methane, ethylene, propane and isobutylene with air, using a burner method, all show some increase at low pressure, while Smith & Agnew's spherical-bomb experiments on methane (1957) show a definite negative dependence over the quite large range of 0·1 to 20 atm. Diederichsen & Wolfhard (1956), using a burner method, also found that the burning velocity of methane-air mixtures decreases to a value as low as 6 cm/sec at 40 atm.

For acetylene/air, which is rather hotter than other flames with air, there is little change in burning velocity with pressure at reduced pressures; Gaydon & Wolfhard (1950c) found a slight rise to a maximum around 200 torr and then a decrease; Gilbert (1957) found S_u practically constant for various mixture strengths to below 200 torr. While flames can be maintained to much lower pressures all measurements at very low pressures are limited by heat losses and inadequate burner size.

Thus it seems that for very hot flames with oxygen the burning velocity varies little with pressure or may have a slight positive dependence. For a fairly hot flame like acetylene/air it is independent of pressure or may have a slight negative dependence, increasing a little at reduced pressure, while for the cooler flames with air there is a fairly marked negative dependence. This dependence can frequently be expressed as a simple power law

$$S_{u(a)}/S_{u(b)} = (P_a/P_b)^n,$$

where $S_{u(a)}$ and $S_{u(b)}$ are the burning velocities at pressures P_a and P_b. Lewis (1954) has reported measurements of n for a large number of flames using the spherical-bomb technique, and his graph is reproduced in Fig. 4.10. This shows n negative for cooler flames and rising to a small positive value for the hottest flames. This relationship is very valuable as it compares many fuel mixtures under similar conditions, but we have already stressed that we distrust soap-bubble and spherical-bomb methods at reduced pressure. The general trend shown in the figure undoubtedly exists, but the absolute values of S_u and n are likely to be affected by heat losses, especially for the slower flames and lower pressures.

Hydrogen–oxygen–nitrogen flames show a pressure dependence that is markedly different from that of all other flames discussed so far. At low pressure the burning velocity becomes so low, even with large burners, that the flames are hard to stabilise below 10 torr. Although burning velocities have not been measured, as the flame is nearly invisible to the eye, there is no doubt about this effect. The reason for this decrease is the nature of the heat release mechanism in hydrogen flames. Recombination of atoms and radicals requires three-body collisions and this causes the hydrogen flame to reach its maximum temperature very late in the flame and this, in turn, influences the flow of heat in the ambient mixture thus reducing the burning velocity. For temperature measurements in hydrogen flames see Hinck et al. (1965) and, in particular the discussion remarks of Sugden.

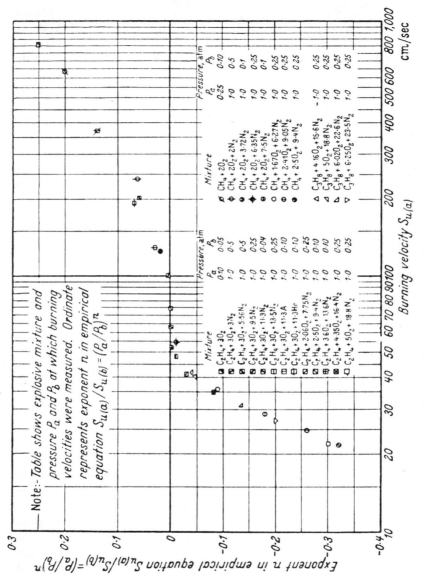

Fig. 4.10. Pressure exponent of burning velocity.

[*From Lewis* (1954)]

86   FLAMES

The influence of hydrocarbon structure on burning velocity

Table 4.3 gives a summary of burning velocities for various hydrocarbons with air (Gerstein, Levine & Wong, 1951). The data have been used by Hibbard & Pinkel (1951) to derive empirical formulae relating burning velocities with the number and types of chemical bond; the apparent relationship to bond strengths may be a disguised connection with the exothermicity of the reaction and thus with flame temperature. Simon (1951) has also used these data to interpret the burning velocities in terms of diffusion rates of OH, H and O atoms in the equilibrium flame gases. In both these treatments, ethylene has a much higher burning velocity than calculated. Acetylene/air also has a high S_u, variously given as 145 to 170 cm/sec, but is not included in these discussions.

TABLE 4.3

Maximum burning velocities for hydrocarbon/air flames

Fuel	S_u max cm/sec	% fuel at S_u max.	Fuel	S_u max. cm/sec	% fuel at S_u max.
*Methane	33·8	9·96	2, 2, 3 Tri-methylbutane	35·9	2·15
Ethane	40·1	6·28	*Ethylene	68·3	7·40
*Propane	39·0	4·54	Propylene	43·8	5·04
Butane	37·9	3·52	1 Butene	43·2	3·87
Pentane	38·5	2·92	1 Pentene	42·6	3·07
Hexane	38·5	2·51	1 Hexene	42·1	2·67
Heptane	38·6	2·26	2 Methyl-1-propene	37·5	3·83
2 Methyl-propane	34·9	3·48	2 Methyl-1-butene	39·0	3·12
2, 2 Dimethyl-propane	33·3	2·85	Propyne	69·9	5·86
2 Methyl-butane	36·6	2·89	1 Butyne	58·1	4·36
2, 2 Dimethyl-butane	35·7	2·43	1 Pentyne	52·9	3·51
2, 3 Dimethyl-butane	36·3	2·45	1 Hexyne	48·5	2·97
			Cyclohexane	38·7	2·65
			Benzene	40·7	3·34

* Better values are probably: methane 45, propane 43, ethylene 75. We also note acetylene/air 158, methane/oxygen 450 and acetylene/oxygen about 1140 cm/sec.

The influence of mixture strength, diluents and additives

Burning velocities usually reach their maximum values well on the rich side of stoichiometric. Figs. 4.11 and 4.12 show curves for hydrogen and for acetylene mixed with varying amounts of nitrogen; the 21% O_2 mixture is, of course, equivalent to air. In these curves the burning velocity, S_u, is plotted against the mixture strength λ, where λ is the ratio of the actual amount of oxygen to the stoichiometric amount required just to burn to CO_2 and H_2O; λ is the reciprocal of the equivalence ratio, ϕ, used by many authors. For acetylene, dilution with nitrogen moves the position of maximum S_u towards the stoichiometric point, but for hydrogen the maximum S_u shifts to still richer mixtures.

The burning velocity of mixed fuels does not obey a simple mixing rule, although Spalding (1956a) has developed a modified rule involving the burning velocities referred to the final temperature of the burnt gas mixture; this apparently holds for most fuel mixtures and fuel-diluent mixtures to within about 10%, but cannot be applied to cases where there are strong catalytic effects (e.g. of water or hydrogen on carbon monoxide flames) or inhibition (e.g. by hydrogen sulphide, Yumlu, 1968).

The most striking catalytic effects have been obtained with carbon monoxide flames. Pure dry CO with air or O_2 burns very slowly if at all, and quite small traces of water vapour or hydrogen have a marked influence in increasing the flame speed. Garner & Johnson (1928) studied CO/O_2 mixtures in a closed 'bomb' and found, for example, that 0·23% of water vapour raised the burning velocity from 100 cm/sec to 780 cm/sec. Hydro-

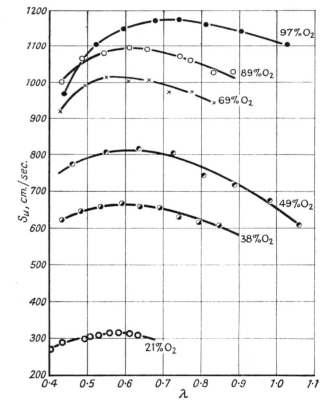

Fig. 4.11. Variation of burning velocity with mixture strength for hydrogen–oxygen–nitrogen mixtures; the percentages of O_2 in the O_2–N_2 mixture are indicated. [*From Bartholomé* (1950)]

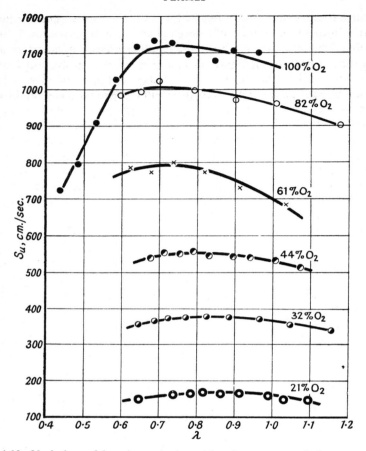

Fig. 4.12. Variation of burning velocity with mixture strength for acetylene–oxygen–nitrogen mixtures. [*From Bartholomé* (1950)]

gen, ethyl nitrate, ethyl iodide and chloroform had a similar effect, but carbon tetrachloride tended to reduce the flame speed; the actual burning velocities obtained may not, by modern standards, be very accurate, but the conclusion that these diluents do have a marked effect is convincing. Tanford & Pease (1947a) reproduced, from John, similar results for $CO/O_2/N_2$ mixtures with traces of water or H_2, and Friedman & Cyphers (1956) have made further measurements on $CO/N_2/O_2/H_2O$ by a burner method. It has also been shown (Simpson & Linnett, 1957) that small amounts of water have a strong accelerating effect on carbon monoxide/nitrous oxide flames.

The study of inhibitors is of great importance in fire prevention and in choice of fire extinguishers. However, such effects often occur in diffusion flames rather than premixed flames and are most important at the limits of flammability and are not always best studied in terms of variation of burning velocity. However, many organic halides, which are well known as inhibitors and fire extinguishers, do reduce the speed of flames. Methyl bromide* and methyl iodide are good inhibitors, and generally bromides and iodides have a greater effect on S_u than do chlorides. Simmons & Wolfhard (1955) found that for stoichiometric ethylene/air 2% of methyl bromide reduced S_u from about 66 to 25 cm/sec; for rich mixtures the effect was even greater, and for weak mixtures less but still real. For these organic inhibitors, especially the larger molecules, the effect on the change of mixture strength should not be confused with true inhibition as their addition makes the mixture richer. The effect of an inhibitor also varies to some extent with the nature of the fuel.

The effect of some eighty additives on the burning velocity of hydrogen/air flames was examined by Miller, Evers & Skinner (1963). Most hydrocarbons had a strong inhibitory action, and $3\frac{1}{2}\%$ of butane reduced S_u from 275 to about 15 cm/sec. Organic bromides were quite effective and some vapours of metallic compounds also reduced the flame speed, e.g. 0·5% of iron carbonyl, $Fe(CO)_5$, reduced the speed to 65 cm/sec.

For flames of organic fuels (methane, propane, hexane) Morrison & Scheller (1972) examined the effect of many additives, including halides, and vapours of compounds of metals and non-metals. They list 32 additives which reduce the burning velocity of hexane/air by at least 30%, and they also measured the effect of these and other inhibitors on the spontaneous ignition temperature (measured by a hot-wire method); they concluded that although a reduction in flame speed was sometimes accompanied by an increase in ignition temperature there was no systematic correlation. For hexane/air, metal compounds were particularly effective in reducing the burning velocity; only 0·014% by volume of lead ethyl, $Pb(C_2H_5)_4$, or 0·0165% of iron carbonyl being necessary to reduce S_u by 30%.

Rosser, Inami & Wise (1963) have examined the effect of powdered metallic salts on the burning velocities of methane/air. In many cases there is a marked reduction in S_u, e.g. 1×10^{-5} g cm^{-3} of Na_2CO_3 reduced S_u from 65 to 15 cm/sec. The decrease appeared to be initially a linear function of salt concentration, but levelled out to a limiting value, probably when the salt had reduced the free radical concentration to equilibrium. Although there was some effect of particle size they concluded that the main effects were explained by inhibition by metal atoms after dissociation of the salt.

* Methyl bromide was at one time used as a fire extinguisher, but it, and also carbon tetrachloride, have now been largely replaced by the less toxic $CBrClF_2$.

The burning velocity of turbulent flames

The subject of the structure and propagation of flames in turbulent gas flows is dealt with in the next chapter. Here we mainly wish to stress the difficulty of both defining and measuring the turbulent burning velocity, S_t. For laminar flow we defined S_u as the velocity of propagation of a plane flame front in the direction normal to this plane. Under conditions of turbulent flow the random gas movements distort, i.e. 'wrinkle', the flame front and may even fragment it, so that it is no longer plane and the local burning velocity varies from point to point. For flames on burners, the scale and intensity of turbulence varies with distance from the burner rim and it is even difficult to define the mean position of the flame front.

Nevertheless most practical flames are turbulent and there appears to be a need to understand how fast a flame propagates into a turbulent mixture. Three methods have been used (i) from area of the mean flame surface, using a burner, (ii) measurement of cone angle of a flame spreading from a flame holder into turbulent flowing gas, and (iii) study of growth of spark-ignited kernels in turbulent-flowing gas. In all methods, the turbulence may be the natural turbulence generated by fast flow (at $R_e > 2300$) in a tube, or it may be generated by passing through a grid of selected size. For flames on burners or flame holders the mean flame position may be deduced from ordinary photographs in the flame's own light, but usually instantaneous schlieren photography is used to determine the position of the flame and also to give information about the scale of turbulence. The new technique of laser-Doppler anemometry may be used to measure the local values of the turbulent fluctuation velocity, i.e. the intensity of turbulence.

For turbulent flames on burners the diffuse flame brush, shown in Plate 7a, may be used to locate an outer edge, whose area usually gives a velocity close to S_u, an inner edge, whose area was used in early work by Damköhler to define S_t, or a mean position (indicated by the broken line in the plate) used in later work by Bollinger & Williams (*see* page 143). The scale and intensity of turbulence vary with position in the flame and as S_t apparently depends on both these parameters, a flame-area method can at best give a mean value. Snyder (1962) has succeeded in modifying the Egerton–Powling flat-flame burner to produce turbulent flames (*see* Plate 7b).

The cone-angle method is probably better than burner-tube methods. Plate 21 shows flames, held on a flame holder, spreading out as a roughly conical flame front into turbulent gases, with different flow rates and with controlled grid-generated turbulence, from Lefebvre & Reid (1966); the burning velocity is usually calculated by the $S_t = U \sin \alpha$ method. The thermal expansion of the burnt gas presumably accelerates the gas downstream, but we have some doubt as to whether there may be some distortion of the flow lines by the expansion. The scale of the distortions of the flame front increases with distance from the flame holder in some cases and again this makes it difficult to make a unique choice of flame angle.

Successive schlieren photographs of an expanding flame kernel in a stream of turbulent mixture can give the speed of the turbulent flame front, but the thermal expansion of the gases within the kernel must be allowed for. Using the relationship $S_t = U\rho_b/\rho_u$, there is difficulty in determining the density ratio between burnt and unburnt gas. Possibly laser interferometric methods could be adapted to measure this ratio.

Turbulent burning velocities are greater than laminar ones, usually by between a factor of 2 and 12. Results are given in the next chapter when the theory of turbulence is discussed in more detail.

Chapter V

The Structure of the Reaction Zone and its Relation to Flame Propagation

Introduction

Since early editions of this book were written there have been major advances in the measurements of rates for specific chemical reactions, and in computer techniques for making detailed calculations. Nevertheless the detailed processes of flame propagation are very complex. They include chemical and physical processes such as reaction rates, activation energies, heat transfer and diffusion of active and stable species. Even an apparently simple reaction such as $2H_2 + O_2 = 2H_2O$ involves a very complex series of chain reactions with differing activation energies and many species with differing diffusion rates, so that in the strong temperature gradient which occurs at the flame front only the most detailed calculations can give a true picture of what is happening; some reaction-rate data are now available for about forty processes involving H, O, OH, HO_2 and H_2O_2 as well as the stable species, and Dixon-Lewis and colleagues have found that at least twelve of these processes are involved to a significant extent in actual flames. The reactions are so complex that it is seldom a satisfactory approximation to use an overall reaction rate or a single activation energy.

Thus, although good progress is now being made with provision of the basic data and in the computational handling of it, we shall still follow the previous pattern of discussing the physical principles involved in flame propagation and shall not go deeply into formal equations relating burning velocities and reaction processes. We shall use some simple formulations for the burning velocity only with the understanding that they are illustrative rather than quantitative. Although more refined formal equations may be developed they still involve approximations and there is seldom adequate information about the relevant parameters for them to give numerical results for the burning velocity. Our treatment will be based on experimental observations on flame structure and properties. Known facts such as spontaneous ignition temperatures, limits of flammability and burning velocity are certainly relevant, but what we really need to know is the temperature profile through the flame front, the rate of heat release at various points in this temperature field, the concentrations of active radicals and intermediary species through the flame front and, of course, the detailed reactions involved.

The experimental difficulties are considerable because of the small thickness of the flame front and the disturbances which are often introduced when sampling probes or thermometers are introduced. The use of flames at low pressure with thicker reaction zones, advances in mass spectrometric sampling, the use of optical absorption spectra, quantitative use of chemiluminescence processes, use of optical methods of temperature measurement and of thin non-catalytic thermocouples, have enabled the recent advances. A good treatment of techniques involving probe measurements is given in the book by Fristrom & Westenberg (1965). Several simple one-dimensional flames, such as those of hydrogen, methane, ammonia, hydrogen/bromine and hydrazine decomposition, have now been the subject of very detailed study, but general application of these methods to less simple flames is still impracticable.

Determinations of rate constants and activation energies for specific processes can be made by a variety of methods and combinations of methods. Study of discharge-tube processes, molecular beams, photochemical reactions and flash photolysis, combined with spectroscopic, mass-spectrometric, chemiluminescent (e.g. lithium addition to sample H atoms), electron spin resonance and gas chromatographic sampling have been employed. Some of the data for simple reactions have been critically summarised in the Leeds University reports on *High Temperature Reaction Rate Data* from 1968 onwards and by Jensen & Jones (1978). In the earliest work on applying reaction processes to flames it was assumed that the reactions observed in low-temperature slow combustion might be applied to hot flames, but this is not generally true as slow combustion usually involves surface reactions and large molecules such as peroxides and aldehydes, whereas in flames it is now known that the propagation is mainly determined by the production and diffusion of simpler species such as free hydrogen atoms, OH and CH_3 radicals.

The rate of propagation is obviously related to the structure of the reaction zone and certain general relationships may be deduced from the conservation laws. In this treatment we shall neglect the small change in pressure through the flame front.

The product of the density, ρ, and the velocity, v, will be constant,

$$\rho \cdot v = \rho_i \cdot v_i \tag{5.1}$$

where ρ_i and v_i are the initial values for the unburnt gases.

94   FLAMES

Conservation of energy requires that, for the one-dimensional case,

$$\frac{\partial}{\partial x}\left(k \cdot \frac{\partial T}{\partial x}\right) - \frac{\partial}{\partial x}(c_p \cdot T\rho v) + Q \cdot U = 0 \qquad (5.2)$$

where T is the absolute temperature, k the thermal conductivity (which depends on T), Q the heat of reaction, U the reaction velocity (mole cm^{-3} sec^{-1}) and c_p is the specific heat. In equation (5.2)

$$k \cdot \frac{\partial T}{\partial x} \text{ is the heat current}$$

$$c_p \cdot T\rho v \text{ is the heat carried by the flow}$$

and $Q \cdot U$ is the heat released by the chemical reaction.

From conservation of the number of atoms,

$$\frac{\partial}{\partial x}\left(D_j \cdot \frac{\partial n_j}{\partial x}\right) - \frac{\partial}{\partial x}(n_j \cdot v) - v_j \cdot U = 0 \qquad (5.3)$$

where n_j is the number of mole cm^{-3} of component j, D_j is the diffusion coefficient of j and v_j is the number of moles of component j which disappear in accordance with the stoichiometric reaction.

In equation (5.3), $D_j\,(\partial n_j/\partial x)$ is the diffusion current of the component j, and $n_j v$ is the number of moles of j carried by the flow, while $v_j U$ is the number of moles of component j which are removed by reaction.

If all the necessary data were available these equations would enable us to calculate v_i, which is, of course, the burning velocity. This is rarely the case, although, as already noted, some progress has now been made by numerical methods for a few simple fuels.

The preheating zone

We visualise the temperature course through the flame front as shown in Fig. 5.1. At a certain point on the x-axis, which for convenience we make the origin of co-ordinates, the temperature has risen to a value at which exothermic reactions just begin to be significant. The part of the curve represented by negative values of x thus corresponds to a region in which the temperature is rising by heat conduction alone. We shall refer to this as the preheating zone.

In this zone we can integrate equation (5.2) between $x = -\infty$ and $x = 0$ because in this region the term $Q \cdot U = 0$. We thus get

$$\bar{c}_p \cdot \rho_i \cdot v_i\,(T_0 - T_i) = k_0 \left(\frac{\partial T}{\partial x}\right)_0. \qquad (5.4)$$

where the index 0 denotes conditions at the point $x = 0$; this is sometimes called the ignition point.

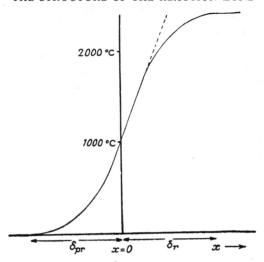

Fig. 5.1. Temperature course through flame front. The actual temperatures shown are only indicative and might apply to methane but not to other fuels.

In equation (5.4), $\bar{c}_p$ is the average value of the specific heat between T_i and T_0. If we assume the conductivity k to be independent of temperature, or if we can take a satisfactory average value, $\bar{k}$, then we can integrate again to get

$$T - T_i = (T_0 - T_i) \exp(\bar{c}_p \cdot \rho_i \cdot v_i \cdot x/\bar{k}). \qquad (5.5)$$

Thus the temperature rises exponentially. In practice there is a slight deviation from the exponential law because the conductivity varies rather rapidly with temperature. However, this formula enables us to estimate the thickness of the preheating zone in different flames. Unless we go to mixtures which are very rich in fuel, neither the density nor the thermal conductivity differ greatly between flames with oxygen and flames with air; the thickness of the preheating zone should therefore depend essentially on pressure and burning velocity. The following Table 5.1 gives the thickness of the preheating zone to be expected for various values of the burning velocity. For $\bar{k}$ and $\bar{c}_p$ we take the average values for air between 0°C and 1000°C. Theoretically the preheating zone will always be infinitely thick because of the first slow exponential temperature rise, but we restrict ourselves to the region between $T = T_0$ and $T = T_i + (T_0 - T_i)/100$. That is we restrict the effective preheating zone to the region between the ignition temperature and the position where the temperature has risen by 1% of the rise in the zone. This is done to give us a simple physical picture. Thus we have

$$(T - T_i)/(T_0 - T_i) = 1/100,$$

and so equation (5.5) reduces to

$$-2 = \frac{\bar{c}_p \cdot \rho_i \cdot v_i \, (-\delta_{pr})}{\bar{k}} \cdot \log_{10} e.$$

Here we have replaced x by $-\delta_{pr}$, i.e. by minus the thickness of the preheating zone as defined above. Replacing v_i by the burning velocity S_u to which it is always equal, we have

$$\delta_{pr} = 4 \cdot 6 \bar{k}/\bar{c}_p \, \rho_i \, S_u. \tag{5.6}$$

TABLE 5.1

Approximate values for effective thickness of preheating zone (for region of 99% of temperature rise in zone) for flames at 1 atm.

Burning velocity	δ_{pr}
10 m/sec	$2 \cdot 7 \times 10^{-3}$ cm
1 m/sec	$2 \cdot 7 \times 10^{-2}$ cm
10 cm/sec	0·27 cm
1 cm/sec	2·7 cm

The values calculated in Table 5.1 are, of course, very approximate. In actual flames the values of $\bar{c}_p$, $\bar{k}$ and ρ_i do vary quite a bit. Also the selection of the effective thickness as that covered by 99% of the temperature rise is quite arbitrary. However the figures serve to illustrate the order and to this extent are useful. If we were to take 90% instead of 99%, then the preheating zone would be only half as thick; 99·9% would mean a 50% increase in zone thickness.

Temperature measurements through a reaction zone have confirmed the existence of a zone in which the gases heat up solely by heat conduction and reaction rates are negligible. Klaukens & Wolfhard (1948) measured the temperature rise in the flame front of flat low-pressure flames with a thermocouple and found the preheating zone to extend to a temperature level of about 800°C. The exact point is of course difficult to locate as the chemical reaction will start gradually. Friedman (1953) and Friedman & Burke (1954) have made careful thermocouple temperature determinations in lean low-pressure propane/air flames. Above 400°C, slow reactions can be detected and 14% of the total heat release has occurred before the position of intense luminous light emission in the reaction zone. Burgoyne & Weinberg's studies (1954) of refractive-index gradients, using an inclined

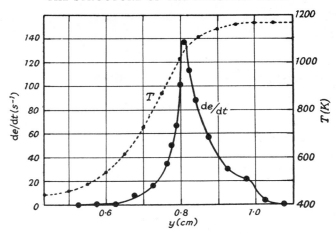

Fig. 5.2. Temperature course and rate of heat release through the reaction zone of a lean ethylene/air flame. [*From Burgoyne & Weinberg* (1954)]

slit (*see* page 312) have enabled the temperature course through lean near-limit flames to be plotted accurately. From this, the rate of heat release $d\varepsilon/dt$ can be deduced at any point. Fig. 5.2 shows their traces for ethylene/air; in this case exothermic reaction begins around 600 K and is greatest at about 1000 K. Measurements of this type have mostly been made on slow-burning mixtures with relatively thick reaction zones. Dixon-Lewis & Wilson (1951) did some work on a methane/air flame with a final temperature of 2150 K and found exothermic reaction beginning between 1100 and 1300 K; Fristrom, Prescott, Neumann & Avery (1953) studied propane/air flames with maximum temperatures of about 2000 K and 2200 K and their curves indicate that the inner edge of the luminous zone was around 1200 and 1400 to 1500 K respectively. For hydrogen, Dixon-Lewis, Sutton & Williams (1970) have made careful measurements with very fine thermocouples, protected against catalytic effects, and found exothermic reaction beginning at only 550 K. Their curves, at atmospheric pressure, shown in Fig. 5.3, indicate quite a sharp start to the heat release and it could well be that below 550 K there are even some endothermic reactions; for this flame the maximum heat release rate was at 900 K.

Although most recent work indicates exothermic reactions beginning at fairly low temperatures, it should be pointed out that many of the initial chemical reactions, especially the chain-branching ones which are so important, are endothermic, and in some cases (e.g. propane, *see* page 116 and hydrogen flames, *see* above) there may be indications of actual heat absorption in an early part of the preheating zone. Failure to observe appreciable heat release does not necessarily indicate absence of chemical processes.

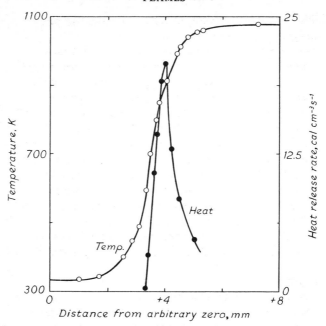

Fig. 5.3. Temperature profile and heat release rate for $H_2/O_2/N_2$; 18·83% H_2, 4·60% O_2, 76·57% N_2; initial temperature 336 K.

[*From Dixon-Lewis, Sutton & Williams* (1970)]

Further evidence on the existence of the preheating zone may be gained by comparison of direct and schlieren photographs of flame fronts. The flame begins to radiate close to the ignition point; the separation between the direct and schlieren images is thus a rough measure of the thickness of the preheating zone. For the schlieren (refractive index -1) is $\propto 1/T$ and a plot of the $1/T$ curve shows that it has its greatest gradient around $T = 200°C$, and this will be the position of the intensity maximum in the schlieren light. This is only true for flat flames; for conical flames there are complications which we have discussed in the preceding chapters. However, Linnett found experimentally that for cone-shaped flames the outer edge of the schlieren coincided with the first departure of the flow-lines from parallel flow. Thus qualitatively we may assume that the separation between the direct and schlieren images of the cone is a measure of the preheating zone. It is an experimental fact that the separation is greater for the slower flames, in agreement with equation (5.6). Again, Klaukens & Wolfhard's measurements were made for flames of acetylene with both air and oxygen, and the measured preheating zones confirmed expectations satisfactorily. We conclude therefore that flames generally possess preheating zones of the type predicted by theory.

The above observations apply to studies of actual flames. A very interesting and important comparison between flames and an artificial preheating zone has been made by Weinberg & Wilson (1970). They compared temperature gradients in an ethylene/air flame with those in an unignited ethylene/air mixture preheated either by flowing through a heated grid or impinging on a heated quartz plate. The temperature measurements were made by laser interferometry, and the heating grid was made catalytically inactive by coating it with silica. In the unignited mixture the heat release rate was much less than when the same mixture was ignited; thus at 700 K it was only about one twelfth of that for the flame. Also the apparent activation energy in the unignited mixture was constant at 37 kJ/mole (or 8·8 kcal), differing entirely from results in premixed flames. These results show the important effects caused by radical diffusion back from the ignited gas region into the preheating zone.

The ignition point

It was at one time thought that the ignition point in a flame might be related to the spontaneous-ignition temperature. In spontaneous ignition experiments the induction periods are, of course, much longer than the time available in the preheating zone, and it was therefore realised that the temperature at the ignition point might be higher than that found from conventional ignition experiments. This view presupposed that the unburnt gas itself had to generate reaction centres, and that ignition occurs when the chain reactions are self-accelerating, as in the ordinary change from slow combustion to ignition. Bartholomé (1949a; 1950) argued that in flames the available times will be so short that in practice reactions will only start at around a temperature of 2000 K. This would mean that, especially for flames with air, reaction would only occur in a very narrow temperature interval between 2000 K and the flame temperature. Other calculations, for methane/air, quoted by Dixon-Lewis & Wilson (1951) vary from 1450 K to 2100 K, but with a probable value of 1750 K to 1800 K. However, experimental evidence with acetylene, ethylene and propane flames suggests that the ignition point is below 800°C and may be as low as 400°C. For methane/air, we have Dixon-Lewis & Wilson's value of 1100 K to 1300 K. For hydrogen flames we have seen (Fig. 5.3) that exothermic reaction begins at only 550 K, which is actually well below the spontaneous ignition temperature of around 840 K.

Thus an important fact, now well established, is that there is no direct correlation between spontaneous ignition temperature and the ignition point. There is some evidence that those substances (mostly halogenated compounds) which inhibit spontaneous ignition do affect the temperature at which ignition occurs in a flame; Wilson (1965) concluded that his observations on inhibitors, e.g. CH_3Br on CH_4/O_2, support a mechanism in which a major effect of the inhibitor is to prolong the pre-ignition zone and shift the primary reaction to a higher temperature; the inhibition reactions have a lower activation energy, so that radicals which diffuse into the pre-ignition zone react preferentially with the inhibitor and so are not available to initiate chain reactions. However, Morrison & Scheller (1972) made a study of burning velocities and spontaneous ignition temperatures, measured by a hot-wire method, and concluded that there was no general relationship between the two. Similarly, catalysts and inhibitors which affect the spontaneous ignition have relatively little effect on the flammability limits; Egerton & Powling (1948) found that catalysts such as diethyl peroxide, ether, nitrogen peroxide and ozone, and inhibitors such as methyl iodide, hardly influenced the flammability limits apart from residual effects due to change in mixture strength and calorific value. The subject of inhibitors is again discussed later (*see* page 114).

We therefore conclude that, at the ignition point, reaction is not started by the unburnt gases producing their own active reaction centres, as these should be influenced by additives. It seems that the active centres come from contact with the flame, that is, by diffusion from either the burnt gas or from the reaction zone itself. This conclusion is probably not a general one, and may only hold for reactions which are of the chain-branching type. Reactions of many organic compounds with oxygen or air are of this type. For the decomposition flames of hydrazine, azomethane and nitric oxide, and for other flames like that between H_2 and NO, the starting of reaction centres by diffusion of active species from the flame may be less important and in these cases the ignition point may lie at a very high temperature, close to the flame temperature.

The reaction zone; radical concentrations

The techniques already mentioned have recently given useful information about the concentrations of some free radicals, as well as the stable molecular species, through the reaction zone. Most of the work has been done on flames at low pressure or on near-limit flames, as under these conditions the reaction zone is relatively thick and therefore easier to study.

The hydrogen flame. Hydrogen flames are rather different from those of organic fuels because they are practically non-luminous and the thin reaction zone is not obvious. The main radiation is in the near ultra-violet (3064 Å) from OH radicals, and for hot flames the strong thermal radiation from this radical dominates that produced by chemical processes so that even photographically the reaction zone is not at all conspicuous. For cooler flames, with excess nitrogen or other diluent, however, the OH emission is relatively much stronger in the reaction zone (Charton & Gaydon, 1958). Detailed probe sampling and thermocouple studies by Dixon-Lewis and colleagues and spectroscopic studies with added metal salts by Sugden and colleagues have given a clear picture of the structure of hydrogen flames. Thus for example, lithium in flame gases reacts according to

$$Li + H_2O \rightleftharpoons LiOH + H$$

and any increase in H-atom concentration shifts the equilibrium and increases the strength of Li radiation in the red (Bulewicz, James & Sugden, 1956), and so Li emission can be used to derive the H-atom concentrations. Alternatively the lithium concentration may be monitored by measuring the strength of the Li absorption line (Halstead & Jenkins, 1969). For sodium, the chemiluminescent reaction

$$H + H + Na = H_2 + Na^*$$

can give information about hydrogen atoms. Mass spectrometric studies with small amounts of added deuterium or heavy water also give information about H-atom concentrations (Dixon-Lewis & Williams, 1963).

In these hydrogen flames, chain-branching by the slightly endothermic bi-molecular reactions

$$O + H_2 = OH + H$$
$$H + O_2 = OH + O$$

and the propagation step

$$OH + H_2 = H_2O + H$$

lead to a rapid build up of an excess population of free atoms and radicals in the reaction zone. This is illustrated by the H-atom profile shown in Fig. 5.4.

Above the reaction zone rapid bi-molecular exchange reactions maintain a pseudo-equilibrium between H, O and OH, but the excess radical concentration only declines slowly through the ter-molecular reactions.

$$H + H + M = H_2 + M$$
$$O + O + M = O_2 + M$$
$$OH + H + M = H_2O + M.$$

For these recombination reactions the rate depends on the square of the radical concentration and, because of its ter-molecular nature, is pressure dependent. In flames at relatively low temperature, where equilibrium concentrations are low, the radicals recombine fairly fast immediately above the reaction zone, but there is very slow removal of the last excess

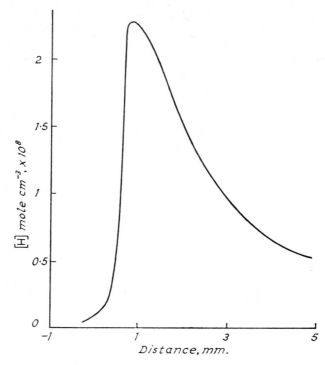

Fig. 5.4. H-atom concentration profile through a flame of 4·6% O_2, 18·8% H_2 and 76·6% N_2 at 1 atm on an Egerton-Powling flat-flame burner. This is built up from study of deuterium–hydrogen exchange reactions and sodium chemiluminescence.

[*Redrawn from Dixon-Lewis & Williams* (1963)

in the afterburning region, where there is a slow temperature rise. Radical excess is particularly likely to be observed in flames at reduced pressure. In hot oxy-hydrogen flames at 1 atm, where equilibrium radical concentrations may reach 10% these excess radical concentrations are much less important and less persistent and the radiation is almost entirely thermal.

Hydrocarbon flames. For hydrocarbon flames it seems that OH radical and free-atom concentrations rise to well above equilibrium values in the primary reaction zone where the hydrocarbon is being consumed. Beyond this primary zone the CO (and sometimes also the H_2) concentration reaches a peak, and then falls slowly as the CO_2 and H_2O concentrations rise to their equilibrium values. This second stage of burning from CO and H_2 to CO_2 and water vapour is relatively slow and is sometimes referred to as the afterburning region. There is some further temperature rise in this

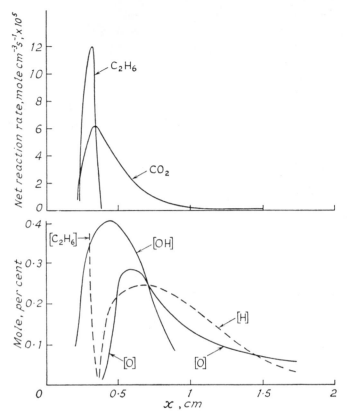

Fig. 5.5. Profiles of a 5% ethane/oxygen flame at 0·1 atm pressure. x is the height above the burner rim in centimetres. The upper part of the figure shows the rate of removal of ethane and of formation of CO_2. The lower figure shows the molar concentrations of ethane, OH radicals, hydrogen atoms and oxygen atoms.

[*Redrawn from Westenberg & Fristrom* (1965)]

afterburning region and this is indicated by the shelf in the heat-release rate curve in Fig. 5.2.

Early indications of the high OH-radical concentration were obtained by Gaydon & Wolfhard (1948); they found that the ultra-violet absorption band of OH could be observed down to the very base of the luminous reaction zone of low-pressure hydrocarbon flames, and that they perhaps extended even into the preheating zone. In a similar way Broida (1951) found that the OH concentration in methane flames was about $2\frac{1}{2}$ times higher in the reaction zone than in the burnt-gas region. Westenberg & Fristrom (1965) studied weak low-pressure ethane flames and deduced the

OH concentration from the rate of formation of CO_2. They also used electron-spin resonance spectroscopy to determine concentrations of free oxygen and hydrogen atoms. Their results are reproduced in Fig. 5.5. The upper figure shows the *rates* of formation of carbon dioxide and of removal of ethane, while the lower figure shows the concentrations of OH, O, H and ethane.

From these curves it appears that OH radicals occur well before the free atoms O and H. We also note that the H-atom peak is much broader than that of O atoms; this is probably because the H atoms diffuse faster away from the region where they are formed. In this weak mixture the equilibrium concentration of H atoms will be very low and it will be seen that it has fallen below the O-atom concentration towards the top of the reaction zone but is still presumably rather above equilibrium.

For methane Fristrom (1963) showed by mass spectrometric sampling of a flame at 1/20 atm pressure that OH, H and O had maximum concentrations near the reaction zone. Later work by Hastie (1973) has shown that CH_3 and H_2 also peak, but not quite at the same place, in the reaction zone, as shown in Fig. 5.6. Although the intensity estimates are in arbitrary units, Hastie says the radical concentrations probably all attain around 4×10^{-3} mole fraction at maximum. From a detailed treatment of the methane flame structure (*see* page 123) Smoot, Hecker & Williams (1976) conclude that the important initial steps for the methane flame are

$$CH_4 + OH = CH_3 + H_2O$$
and
$$CH_4 + H = CH_3 + H_2$$

followed by both

$$CH_3 + O_2 = CH_2O + OH$$
and
$$CH_3 + O = CH_2O + H$$

and then further reactions, also involving CHO, HO_2 and CO, at a later stage. It should be noted that this type of reaction chain requires the initial presence of OH radicals; in a flame these may diffuse from the reaction zone, whereas under spontaneous ignition conditions they would have to be formed, perhaps by formation and breakdown of a peroxidic compound or perhaps by surface reactions.

The light emission from the reaction zone of hydrocarbon flames is mainly from OH, CH and C_2 radicals. In their early studies on low-pressure flames the authors found that the relatively thick reaction zones were usually coloured more green at the base, due to C_2 Swan-band emission, and more blue-violet above, due to CH emission; this is well shown in the

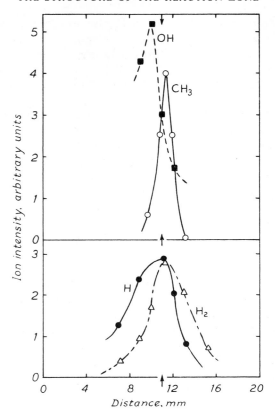

Fig. 5.6. Ion intensity profiles for H, OH, CH$_3$ and H$_2$ in the flame CH$_4$ (0·394) + O$_2$ (0·375) + N$_2$ (0·231) at about $T = 2400$ K. Measurements were made relative to the luminous reaction zone whose position is indicated by the arrow. [*From Hastie* (1973)

colour photograph, Plate 2. This separation is particularly clear when a small quantity of a hydrocarbon, such as acetylene, is added to a low-pressure oxy-hydrogen flame. Spectroscopic studies show that the C$_2$ and OH emission extends below that of CH. Studies of concentration profiles of electronically excited and ground state OH, CH and C$_2$ radicals, using emission and absorption spectroscopy, have been made for low-pressure methane and acetylene flames by Porter, Clark, Kaskan & Browne (1967); the radicals all have strong narrow concentration maxima in the reaction zone, and from a discussion of their observations they conclude that for acetylene flames the data indicate that C$_2$H$_2$ is consumed by reaction with OH, although some other mechanisms cannot be definitely excluded.

Among other detailed studies of concentration and temperature profiles, may be mentioned that of the ammonia/oxygen flame (Maclean & Wagner, 1967); using mass-spectrometric and optical-spectrometric techniques the species O_2, NH_3, N_2, H_2O, NO, N_2O, H_2, NH_2, NH, OH and several electronically excited radicals have been studied. For this flame, reactions do not start until fairly near the final flame temperature. NH, NH_2, N_2O and H_2 show concentration maxima but OH only appears later and does not seem to exceed the equilibrium value at any stage.

These studies of concentrations of various species through the flame fronts are very valuable in an understanding of propagation processes. Many of the studies have been done on flames with thick reaction zones either at low pressure or on near-limit mixtures, and the disequilibrium effects which occur under such conditions should not be unduly stressed. In hot flames equilibrium concentrations of free atoms and radicals are much higher and tend to dominate over minor anomalies, and also temperature and concentration gradients are very much higher so that heat transfer and diffusion occur much more rapidly.

The reaction zone; its thickness in relation to burning velocity

The rate of propagation of a deflagration flame is related to the rate of the chemical reactions and it is of interest to develop some general relationships between burning velocity, reaction rate and the thickness of the reaction zone. These relationships are free from all specific assumptions.

From continuity considerations we can set up the following equation

$$n_{fi} \cdot S_u = \int_{-\infty}^{\delta_r} U \cdot dx = \int_0^{\delta_r} U \cdot dx = \bar{U}\delta_r, \qquad (5.7)$$

where n_{fi} is the number of moles of fuel/cm³ in the unburnt gas, δ_r is the thickness of the reaction zone, and the equation defines $\bar{U}$ the average rate of reaction over the reaction zone. This equation only expresses the fact that the faster the flame and the thinner the reaction zone then the greater the average rate of reaction must be.

A further condition for a continuous change through the flame front must be that the temperature gradient at the end of the preheating zone is the same as at the beginning of the reaction zone. Following Damköhler (1940), we obtain

$$\left(\frac{\partial T}{\partial x}\right)_0 = \frac{T_e - T_0}{\delta_r} \cdot R, \qquad (5.8)$$

where T_e is the temperature at the end of the reaction zone and R a constant which must be close to 1.

Combining equations (5.4) and (5.8) and using
$v_i = S_u$

$$\delta_r = \frac{R \cdot k_0 (T_e - T_0)}{\bar{c}_p \cdot \rho_i \cdot S_u (T_0 - T_i)} \qquad (5.9)$$

or using (5.7)

$$S_u^2 = \frac{R \cdot k_0 \cdot \bar{U} (T_e - T_0)}{\bar{c}_p \cdot \rho_i \cdot n_{fi} (T_0 - T_i)}. \qquad (5.10)$$

No specific law for the reaction velocity has been introduced so equation (5.10) should be quite generally applicable. We thus find that the square of the burning velocity will be proportional to the average reaction rate in the reaction zone, $\bar{U}$.

Equation (5.9) shows that the thickness is inversely proportional to both the density (i.e. pressure) and the burning velocity. The thickness of low-pressure flames or slow-burning flames can be determined by probe sampling techniques, but for faster flames this is not practicable. However, for organic fuels the luminous region, in which C_2 and CH emission is strong, appears to coincide quite well with the reaction zone. The delay in emission of radiation after excitation of atoms or molecules is usually of the order 10^{-8} to 10^{-6} sec for lines or bands in the visible and near ultra-violet; this time is short compared with the time of passage through the reaction zone, and we shall not in general expect any displacement between the luminous zone and the reaction zone on this account; for very fast flames which have very thin reaction zones at or above atmospheric pressure the delay may just begin to be appreciable. Thus it is of interest to compare values of the thickness of the luminous zone, δ_l, for flames of different burning velocity S_u.

The general increase in the thickness of the reaction zone, from observed values of δ_l (Gaydon & Wolfhard, 1949b), with decreasing burning velocity is illustrated in the following table, which also gives roughly calculated values for the thickness of the preheating zone, δ_{pr}, using equation (5.6). We have stressed that optical effects make it difficult to measure δ_l for fast flames, and some of the values given in Table 5.2 are extrapolations from low-pressure observations.

The relation between burning velocity, temperature and radical concentration
The energy liberated in a flame is best represented by the adiabatic flame temperature. The temperature is roughly proportional to the heat of reaction up to about 2200 K, but at higher temperatures the dissociation of the products becomes important and the energy is shared between the enthalpy of the exhaust gases and the increasing amount of energy used in the dis-

TABLE 5.2

Approximate values of S_u, δ_l (observed) and δ_{pr} (calculated) for some flames (stoichiometric mixtures) at 1 atm.

Flame	S_u cm/sec	δ_l cm	δ_{pr} cm
C_2H_2/O_2	800	0·0021	0·0034
C_2H_2/air	150	0·0065	0·018
C_2H_2/argon-air	240	0·0045	0·011
C_4H_{10}/air	40	0·02	0·07

sociation. This effect in hot flames has to be considered if we try to relate S_u to the temperature.

Fig. 5.7 shows burning velocities plotted against flame temperature for H_2, C_2H_2, C_2H_4 burning with mixtures of oxygen and nitrogen. It shows that the points for the hydrocarbons lie fairly well on a single curve, but those for hydrogen lie on quite a separate curve. Points for alcohol, ethers

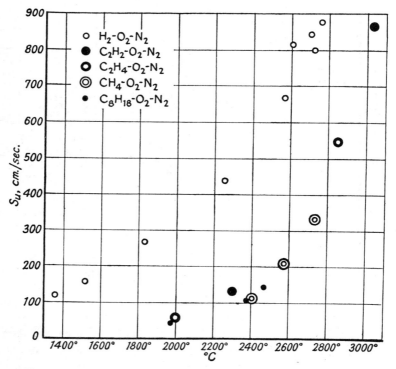

Fig. 5.7. Dependence of burning velocity on final flame temperature.

[*Partly from Bartholomé*

THE STRUCTURE OF THE REACTION ZONE   109

and even CS_2, lie on the same curve as the hydrocarbons, suggesting that the burning velocity does depend essentially on flame temperature. Only if the oxygen is replaced by some other gas which will support combustion, like N_2O, NO or NO_2 do the points lie on different curves. This is shown in Fig. 5.8, the data for which come mainly from Parker & Wolfhard (1953) and partly from Sachse & Bartholomé (1949). It seems that each curve may represent a certain type of propagation mechanism. The very low burning velocities for flames with NO may be noted. Flames with NO_2 present some special difficulties which will be discussed later. There are some exceptions to this general behaviour which we may discuss at once. The H_2/NO_2 flame behaves like a flame with O_2 and it is found experimentally that

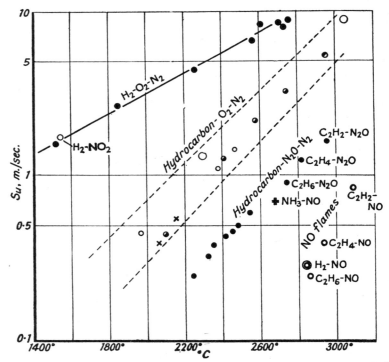

Fig. 5.8. The dependence of burning velocity on final flame temperature. Points for $H_2/O_2/N_2$ mixtures lie on a fairly straight line, and points for various hydrocarbons with O_2 and N_2 lie within a fairly narrow band; points for different hydrocarbons are marked by different types of points. Hydrocarbon/N_2O/N_2 mixtures also lie on a fairly good but separate straight line, and flames supported by NO on another line at lower burning velocity.

[*Results from Sachse & Bartholomé and from Wolfhard & Parker*

110   FLAMES

all the nitrogen in the burnt gases is found as NO which does not react but behaves as an inert gas, the experimental flame temperature being only 1550°C. Another exception is NH_3/NO.

The general and important conclusion is that over a wide range of temperature, between flames with O_2 and with air, the burning velocity against temperature curve is smooth without any apparent break.

We have seen in equation (5.10), that there is a close connection between burning velocity and mean reaction rate, and chemical reaction rates tend to rise, often exponentially, with temperature. The concentration of free radicals in a flame is also very sensitive to the temperature, and Tanford & Pease (1947a) were among the first to realise that the burning velocity may be related to the number of free radicals in the burnt gases. We now realise that the free radicals often peak in the reaction zone to an order or two higher than their equilibrium value in the burnt gases. However the comparisons we previously made between burning velocity, S_u, and equilibrium hydrogen atom concentration, [H], are still interesting.

Fig. 5.9 shows the variation of S_u with [H] as given by Bartholomé (1949b) with some additional values which we have available. For $H_2/O_2/N_2$ mixtures the results are fairly trustworthy as a fair range of mixture strengths is covered. Obviously [H] will tend to be greatest for hydrogen-rich mixtures, and S_u has its maximum on the rich side, too, so it seems that $S_u = f([H])$ may be better than $S_u = f(T)$. The curve of S_u against [H] is linear for an equilibrium value of [H] greater than 2%; this straight line either cuts the

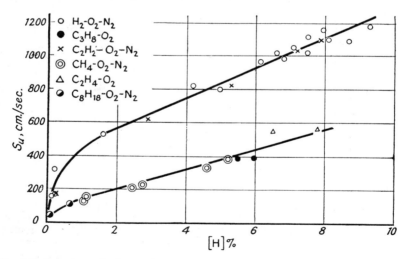

Fig. 5.9. Burning velocity against equilibrium concentration of H atoms. Values from Bartholomé, Jahn (adjusted by 20%) and ourselves.

axis at about $S_u = 4$ m/sec, or the curve changes to a more parabolic form on which the points for air flames may conveniently be included.

The lower curve is for hydrocarbons, other than acetylene. Again it is nearly a straight line which either bends a little to pass through the origin or cuts the axis at about $S_u = 50$ cm/sec. There is no doubt that the colder flames with a lot of N_2 fall on the curve, and if such a curve means anything at all it applies to both oxygen and air flames.

Rather strangely, points for C_2H_2 do not fall on the curve for other hydrocarbons but on the curve for H_2. We think, contrary to Bartholomé, that this is accidental, since, for equivalent temperature, acetylene flames have a lower equilibrium value of [H] than other hydrocarbons because of the higher carbon/hydrogen ratio. The deviation of the points for acetylene may therefore only indicate that S_u depends not only on the concentration of H atoms but on those of other radicals as well.

The real test of the dependence of S_u on [H] for hydrocarbons is to study the change with mixture strength. We have done this for C_2H_2 and C_2H_4 using S_u values from Bartholomé (1950) with a little extrapolation, and our own calculations of [H]. Fig. 5.10 shows the result, the mixture strengths being indicated. It is clear, especially for ethylene, that no simple relationship exists. There is no doubt that for hydrocarbons there are other factors besides [H] which are important. This is the same conclusion as that which can be drawn from a similar analysis by Pickering & Linnett (1951b). They

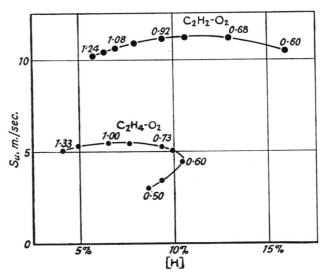

Fig. 5.10. Burning velocity against hydrogen atom concentration for acetylene/oxygen and ethylene/oxygen. The mixture strength, λ, is indicated.

tried to vary not only the amount of diluents (N_2 and CO_2) but also the pressure. Their results, plotted in Fig. 5.11, for 9% C_2H_4 + x% O_2 + remainder N_2 in which N_2 is progressively replaced by O_2 are especially interesting. Starting with the rich mixture, the temperature increases as we replace N_2 by O_2, but as we get to leaner and leaner mixtures it begins to drop again. The burning velocity increases all the way, whereas [H] increases steeply at first and then drops. [OH] and [O] are the only

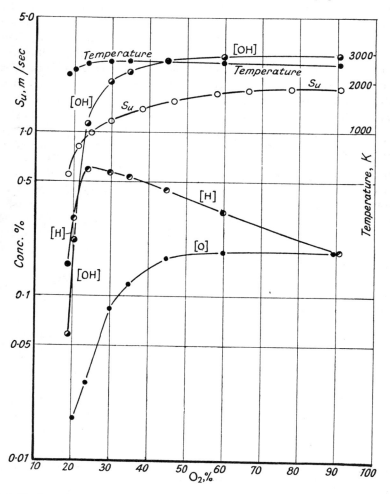

Fig. 5.11. Equilibrium concentrations of OH, H and O, the burning velocity and temperature against mixture strength for a 9% C_2H_4 + xO_2 + $(91 - x)N_2$ flame. [*After Pickering & Linnett*]

components which increase all the way, though slowly, up to the 91% O_2 mixture. This certainly reveals a complicated picture.

The burning velocity of flames of CO with O_2 is very sensitive to the presence of traces of moisture or hydrogen. These small additions do not appreciably affect the equilibrium flame temperature, but do alter [H] and [OH]. The early studies by Tanford & Pease (1947a) did suggest good correlation between S_u and [H], but not with [OH] or [O]. Later, Friedman & Cyphers (1956) remeasured the burning velocities and did not find such satisfactory correlation.

The experimental observations show that while there is a general tendency for burning velocities to rise with final flame temperature, and also to rise with equilibrium values of [H], there is no specific dependence on either alone. We shall see later that both temperature and the concentrations of free radicals in the reaction zone, rather than in the burnt gas, are important in flame propagation.

The effect of additives on burning velocity, flammability limits and spontaneous thermal ignition

Often it has been assumed that near the limits of flammability the spontaneous thermal ignition temperature becomes higher and the flame temperature falls, and the limit is determined by the condition at which these two temperatures coincide. This is not, however, the case; spontaneous-ignition temperatures vary very little with mixture strength and may be measured even far beyond the limits of flammability of mixtures initially at room temperature.

Sachse & Bartholomé (1949) compared the burning velocities of fuels which behaved very differently in spontaneous-ignition studies. In Table 5.3 the burning velocities for fuels with very different octane numbers are

TABLE 5.3

Burning velocities and octane numbers

Fuel	Octane No.	S_u, cm/sec
n-heptane	0	36·4
Special fuel (I.G. 702)	45	35·0
Synthetic petrol (VT 702)	70	37·2
Iso-octane	100	31·6
Triptane	120	33·2

compared. There seems to be no connection at all between S_u and octane number. Table 5.4 shows also that the well-known anti-knocks and pro-knocks have practically no influence on burning velocity. Thus it seems either that the point in the flame front at which reactions commence is unimportant in determining S_u or, more likely, that the ignition mechanisms in steady propagation and in the initial stages of spontaneous ignition are different.

TABLE 5.4

Influence of anti-knocks and pro-knock on S_u

Fuel mixture	S_u, cm/sec
Iso-octane	31·6
+ 0·36% lead tetra-ethyl	32·7
Synthetic petrol (VT 702)	37·2
+ 0·12% lead tetra-ethyl	33·6
+ 0·36% lead tetra-ethyl	35·7
+ 5% ethyl nitrate	34·6

Although the anti-knock and pro-knock catalysts have little influence on flame speed, some substances, especially halogenated compounds, do have a definite inhibitory action, both flame speed and limits of flammability being affected. For a methane flame a reduction in flame speed is accompanied by a *narrowing* of the luminous reaction zone, an increase in the thickness of the preheating zone and a displacement of the ignition point to higher temperature. Studies by Wilson (1965) of inhibition of methane flames by HBr and CH_3Br show that in the uninhibited flame the OH concentration has a maximum, well above its equilibrium value, but that on addition of HBr or CH_3Br the maximum is eliminated. The net reaction rate without inhibitor shows a rather broad maximum, whereas with inhibitor the reaction is confined to a narrower region (Fig. 5.12) of more

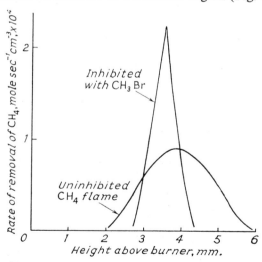

Fig. 5.12. Net reaction rate for methane, uninhibited and inhibited by CH_3Br.
[*Redrawn from* Wilson (1965)]

intense reaction at higher temperature. Similar observations by Levy et al. (1962) led them to conclude that HBr inhibited in the first stage of methane oxidation, but was inoperative in the second stage.

For hydrogen flames ($H_2/O_2/N_2$ mixtures) hydrogen bromide reduces the flame speed and maximum heat-release rate (Day et al., 1971), and computed values of [H] in the reaction zone show that its rise is delayed, its maximum is reduced and its overall width is again reduced. Here the mechanism of inhibition is attributed to chain removal of H atoms by

$$H + HBr = H_2 + Br$$
$$Br + H + M = HBr + M.$$

It had been suggested (Mills, 1968) that inhibition by halogens was due to formation of negative ions, but this is now believed not to be the case (Spence & McHale, 1975).

For propane, and probably other higher hydrocarbons, a more complex picture emerges (Pownall & Simmons, 1971). Here the heat-release profile (Fig. 5.13) is broader when hydrogen bromide is added. This is attributed to additional reaction between the bromine atoms and propane

$$C_3H_8 + Br = C_3H_7 + HBr.$$

The initial, slightly negative, heat-release rate may be noted.

These inhibitory effects of halogenated compounds support the view that free radicals, especially hydrogen atoms, are important in flame propagation, their removal reducing the flame speed and altering the structure of the reaction zone, although the final flame temperature is little affected.

Early thermal and diffusion theories of flame propagation

From the discussion in the preceding sections it is clear that both heat transfer *and* radical diffusion are important in flame propagation. At the time the first edition of this book was written there was still some controversy over the rival merits of heat transfer *or* diffusion.

In the thermal theory the basic assumption was that normal heat transfer processes raised the reactants to the spontaneous ignition temperature and that this propagated the flame; allowance might be made for the ignition temperature being higher, because of the shorter time involved, than the value from normal ignition studies. This would lead to an ignition point in the flame which was always above, perhaps well above, the normal spontaneous ignition temperature and to a high activation-energy requirement similar to that obtained from ignition studies.

Early attempts to put the thermal theory of flame propagation into quantitative form were made by Zeldovich & Frank-Kamenetsky (1938a;

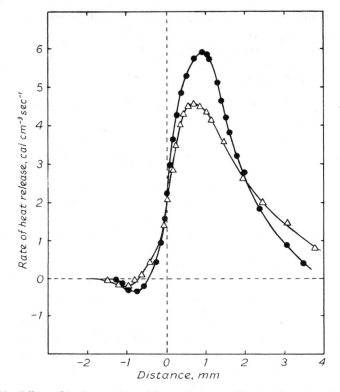

Fig. 5.13. Effect of hydrogen bromide on the rate of heat release in a flat flame of propane (1·23%) with oxygen and argon. ●, No HBr; △, 0·22% HBr.
[*From Pownall & Simmons*

1938*b*; 1938*c*), Zeldovich (1947), Zeldovich & Semenoff (1940) and Semenoff (1940). They assumed that reaction only started at a point near that for the final flame temperature, the temperature at the ignition being below the final temperature only by an amount not exceeding RT_e^2/E, where R is the gas constant, T_e the final temperature and E the activation energy. A second assumption was that the reactions could be represented by equations derived from classical kinetic theory, taking the form

$$U = K \cdot a \cdot e^{-E/RT} \text{ for a first-order reaction,} \qquad (5.11)$$

or $$U = K \cdot a^2 \cdot e^{-E/RT} \text{ for a second-order reaction} \qquad (5.12)$$

where U is the reaction rate, a is the concentration of combustible and K is a frequency factor. With certain further assumptions the general equations (5.2) and (5.3) may be integrated to obtain for first order

$$S_u^2 = \frac{2k\,c_p\,K}{\rho_i L^2}\left(\frac{T_i}{T_e}\right)\left(\frac{k}{c_p\,\rho D}\right)\frac{n_i}{n_e}\left(\frac{RT_e^2}{E}\right)^2 e^{-E/RT_e} \quad (5.13)$$

and for second-order reactions

$$S_u^2 = \frac{2kc_p{}^2 Ka_0}{\rho_i L^3}\left(\frac{T_i}{T_e}\right)^2\left(\frac{k}{c_p\,\rho D}\right)^2\left(\frac{n_i}{n_e}\right)^2\left(\frac{RT_e^2}{E}\right)^3 e^{-E/RT_e} \quad (5.14)$$

where k, c_p and ρ are the thermal conductivity, heat capacity and density at the final flame temperature, a_0 is the concentration of combustible, L is the calorific value of the mixture, D is the diffusion coefficient at the final temperature and n_i and n_e are the initial and final numbers of molecules. Frank-Kamenetsky & Zeldovich thought that these equations were hardly applicable to flames of hydrogen and hydrocarbons but might be used for carbon monoxide. Murray & Hall (1951) observed the decomposition flame of hydrazine and found that the burning velocity agreed with that predicted by this purely thermal theory. The value of this work was that the thermal theory had been rigorously applied with such assumptions that a numerical evaluation of the burning velocity was possible.

Bartholomé et al. (1950) developed a thermal theory along similar lines and attempted to apply it to hydrocarbon flames, again with limited success, but again it may have uses for some decomposition flames and perhaps for flames like that of H_2 burning with NO whose reaction rate depends on an initial thermal decomposition process, in this example of the NO. A later development of the thermal theory by Boys & Corner (1949) did not make such dangerous assumptions as those made by the Russian school or by Bartholomé but involved various constants such as activation energies which were not usually available for use with their explicit equations and so made the theory difficult to test.

That thermal conduction does play an important role in flame propagation is supported experimentally by the marked effect of preheating the mixture, which increases the burning velocity, and by the observations of Spalding using heat-abstraction to a porous-plate burner to reduce the burning velocity. However, we have seen that the ignition point appears to lie at much lower temperatures than predicted by purely thermal theories, and the absence of any marked effect of pro-knock and anti-knock additives is not readily explained on a thermal theory.

The diffusion theories start with the basic idea that the propagation rate depends on the speed of diffusion and on the concentration of active radicals (mostly H atoms), and we have already seen that for some flames, such as that of moist CO, there is indeed an apparent relationship between S_u and [H]. The theory developed by Tanford (Tanford, 1947; Tanford &

Pease, 1947b) assumed that H atoms diffuse from the burnt gas into the reaction zone where the reaction rate is everywhere proportional to [H]. Tanford showed that the concentration of H built up by diffusion is many orders higher than that due to thermal dissociation in the preheating zone. His calculations are not of high accuracy, because he used an average value for the diffusion coefficient, whereas really D varies roughly as $T^{1.75}$ and the diffusion of H atoms will be practically arrested at medium temperatures by the fall in D and would be an order or two lower than his values; however, he assumed diffusion from the equilibrium values in the burnt gas, whereas we have seen that in the reaction zone actual values for the cooler flames are an order or two higher in the reaction zone than in the equilibrium region. His theory must therefore be rather approximate, but may be useful for those flames dominated by diffusion. Tanford's general formula is

$$S_u^2 = \sum_i \frac{K_i \cdot C \cdot p_i \cdot D_i}{Q \cdot B_i},$$

where the sum is taken over all the active species, i; usually it is sufficient to consider only the one species, H atoms. K_i is the rate constant, C is the concentration of combustible, D_i is the diffusion coefficient, Q is the mole fraction of combustion products, p_i is the mole fraction of active species i in the burnt gases and B_i is a constant which allows for loss of the active species due to reaction.

Although there is good evidence that for many flames, diffusion of active species, such as free atoms or radicals, is important, it is fairly obvious that flame propagation cannot be explained entirely by radical diffusion without also considering heat transfer. If the presence of a free radical alone was sufficient to initiate reaction then the fuel/oxidant mixture would be self-igniting. A number of self-igniting (hypergolic) fuels are known; presumably, for these, free radicals produced by chance effects such as ionising radiation (cosmic rays, etc.) are sufficient to start reaction, usually after a short delay. Normally combustion occurs by a chain mechanism which involves a highly endothermic chain-initiation step, then less strongly endothermic chain-branching processes and then exothermic or thermoneutral chain-propagation. Thermal ignition will normally be limited by the chain-initiation step, and will only occur, for short induction times, at fairly high temperatures. The role of diffusion may be to overcome this initiation step, so that heating to an appreciably lower temperature may suffice to provide enough energy for the branching step and so cause reaction.

In considering these thermal and diffusion theories of flame propagation, we must remember that heat transfer and molecular diffusion tend to obey similar laws and behave in the same way, so that it is not always easy to distinguish between them. If we have a propagation mechanism which depends on reactions being started by radical diffusion, but with these same reactions also requiring an activation energy, then the flame propagation will depend on *both* heat transfer and radical diffusion. In such a case, the less efficient process would tend to be rate-determining; if, for example, there is an ample supply of radicals, then the heat transfer required to supply the activation energy will be limiting and factors affecting this heat transfer will have a more important influence than those affecting the supply of radicals. But the fact that heat transfer appears to be important should not then be interpreted as meaning that radical diffusion does not occur or is not also important.

The complex joint roles of heat transfer and radical diffusion are well illustrated by the effects of isotopic substitution (e.g. deuterium for hydrogen) on S_u and other flame properties. Watermeier (1957) measured the burning velocities of CO/O_2 with added H_2 and D_2 in a bomb experiment, and a few of his values are reproduced in Table 5.5. Effects on thermal conductivity and final flame temperature should be negligible, so a strong effect of the faster diffusion of H atoms is indicated. However, Watermeier concluded that a pure diffusion theory could still not explain the results quantitatively. In an attempt to sort out the relative importance of diffusion and heat transfer Gray & Smith (1967) compared flame speeds of H_2 and D_2 and at reduced pressure obtained maximum values of S_u of 4180 and 3030 cm/sec. The average ratio of S_u for H_2 and D_2 was 1·4:1.

TABLE 5.5

Values of burning velocity of stoichiometric CO/O_2 with added H_2 or D_2

H_2, %	S_u, cm/sec	D_2, %	S_u, cm/sec
0·1	35·8	0·1	28·1
0·2	42·2	0·2	31·7
0·3	55	0·3	36·1

Estimates of diffusion rates, conductivity and change in activation energy for breaking the H–H bond could lead to a ratio of S_u of 1·6:1, so it seems that the complexity of the system prevents any simple interpretation. Similarly Gray & Holland (1970) compared the burning velocities of N_2H_4 ($S_u = 1330$ cm/sec) and N_2D_4 (683 cm/sec) and apportioned the change due to different thermal conductivity as about 1·1, to change in adiabatic flame temperature as 1·3 to 1·5 and to kinetic isotope effect about 1·5; they also found an effective activation energy of only 35 kcal/mole compared with an expected 58 kcal/mole to break the NH_2–NH_2 bond.

Combined thermal and diffusion theories

Another interesting form of the thermal theory has been developed by Bechert (1949; 1950). This is essentially a thermal theory but diffusion effects are included by making the activation energy a parameter. This accounts for possible changes in reaction mechanism due to diffusion of radicals. This is useful because the activation energy E is also rather difficult to define for flame gases for other reasons as well. If we want to retain an expression for the reaction velocity of the form $e^{-ER/T}$ it seems that we must determine E from relevant experiments and that it may not have the usually accepted meaning. Bechert gives first a dimensional analysis which is useful for interpretation of experimental material. We cannot discuss his treatment at length but quote two relations for a flame involving reaction between two molecules, e.g. a hydrocarbon/air flame. He finds

$$S_u^2 \propto \frac{D \cdot Z}{n_i},$$

where n_i is the number of fuel molecules/cm³ in the unburnt gas. For the thickness of the flame front he gives

$$\delta_r \propto \frac{k}{\rho \bar{c}_p \cdot S_u} = \frac{D}{S_u},$$

where Z is the number of successful collisions leading to the formation of the final product, and D is the diffusion coefficient.

It is of interest briefly to consider the significance of the activation energy. The proportion of particles with energy in excess of that required to overcome the energy barrier (i.e. with energy exceeding the activation energy E) can only be calculated if we have a Maxwell-Boltzmann distribution of energy. This will generally be true for slow reactions but this distribution may not be attained in fast reactions. It is indeed astonishing that the expression $e^{-E/RT}$ is used so freely without reservations. We think that for fast reactions there will be a disturbance of the equilibrium distribution leading to an excess of fast particles above the predicted number. We may thus expect the reaction to be accelerated, not because of any specific energy chains but because of the general departure from equilibrium. This is supported by our spectroscopic studies (Chapter X); apart from a fairly definite chemiluminescence in some cases, we also find a general disturbance of the Maxwell-Boltzmann distribution of energy with a marked excess of excitation to high electronic levels (*see* the results for Fe atoms, etc., page 288). The effective temperature for electronically excited levels depends on the energy of the levels. We must therefore either abandon the use of the expression $e^{-E/RT}$ in flame propagation or give E a new significance to relate it to the actual reaction velocity in the flame front. This explains the success of Bechert's theory in which E is a parameter; the values of E required to give the correct burning velocities do, in fact, come out much lower than the values obtained from slow combustion reactions. This is probably mainly because when radicals are being supplied by diffusion it is only necessary to provide activation energy for moderately exothermic branching reactions, rather than for highly exothermic initiation steps (*see* page 118); however, in very fast reactions, departures from thermal as well as chemical equilibrium may also be of importance.

It is very difficult, without over simplification, to give a useful mathematical treatment of flame propagation because the relevant quantities such as thermal conductivity, diffusion coefficients and reaction rates all depend in a complex way on both temperature and gas composition. Thus it is not possible to write down explicit algebraic equations. Usually it is necessary to use reiteration methods, feeding in such limited experimental data as are available for reaction rates, conductivities and diffusion coefficients under particular conditions. Modern computer techniques are, of course, a powerful tool for this work. Among important theoretical papers are those by Hirschfelder, Curtiss & Campbell (1953), Von Karman & Penner (1954), Spalding (1956b) and Klein (1957). Hirschfelder (1963) has given a clear account of the physical principles underlying the mathematical formulations. Spalding, Stephenson & Taylor (1971) set out the basic differential equations, using unsteady-state instead of steady-state assumptions, and simplified the equations by a change of variable; some assumptions are still, of course, necessary, e.g. no convection, no viscous effects, constant pressure, unit Lewis number (*see* page 125) and no radiative heat transfer. Application of the method to the relatively simple case of a hydrazine decomposition flame, using experimental rate constants and using three previously accepted reaction schemes all led to values of S_u which are low by a factor of around two when compared with experimental data.

The deepest understanding of flame propagation has come not from generalised treatments but from careful experimental study of individual flames, backed by numerical computations using established rate constants for the chain reactions and using realistic assumptions about diffusion and heat transfer rates.

The hydrogen flame. Flames of $H_2/O_2/N_2$ mixtures have been the subject of a major attack by Dixon-Lewis and colleagues, the main results being published in a series of, to date, nine papers in Proceedings of the Royal Society from 1967 to 1975, with additional symposium papers. Experimental observations, some of which have already been referred to, include detailed temperature and heat-release profiles and study of both stable species and radicals through the reaction zone. The computational method consists of setting up the time-dependent heat conduction and diffusion equations with reaction rates, assuming an arbitrary set of temperature and composition profiles, and then following by standard finite-difference methods the transient process which converges to the steady state. Besides the burning velocities, temperature and heat-release profiles and radical concentration profiles have been compared with experimental data for various, mostly rather slow burning, flames. Initial work (Dixon-Lewis, 1970) using the three basic chain-propagating steps and the three exothermic chain-terminating steps was not very satisfactory, and a more complete system using six additional reactions involving HO_2 became necessary (Day, Dixon-Lewis & Thompson, 1972; Dixon-Lewis, Isles & Walmsley, 1973). This led to good agreement, at least for slow rich-mixture flames, for burning velocities and other flame properties. The basic reactions, as numbered by Dixon-Lewis *et al.*, with rates (in cm mole sec units) are,

chain-propagating/branching
(1) $OH + H_2 = H_2O + H$ $k_1 = 3.3 \times 10^{13} \exp(-2700/T)$
(2) $H + O_2 = OH + O$ $k_2 = 2.05 \times 10^{14} \exp(-8250/T)$
(3) $O + H_2 = OH + H$ $k_3 = 1.8 \times 10^{13} \exp(-4700/T)$

exothermic chain-terminating
(15) $H + H + M = H_2 + M$ $k_{15, \text{all } M} = 4.5 \times 10^{15}$
(16) $H + OH + M = H_2O + M$ $k_{16, H_2} = k_{16, N_2} = k_{16, O_2} = 2.0 \times 10^{16}$
                                $k_{16, H_2O} = 24 \times 10^{16}$
(17) $H + O + M = OH + M$ $k_{17, M} = 0.25 \, k_{16, M}$

subsidiary reactions
(4) $H + O_2 + M = HO_2 + M$
(7) $H + HO_2 = OH + OH$
(7a) $H + HO_2 = H_2O + O$
(12) $H + HO_2 = H_2 + O_2$
(13) $OH + HO_2 = H_2O + O_2$
(14) $O + HO_2 = OH + O_2$

For these subsidiary reactions only the ratios of reaction rates are used,

$2 \, k_2/k_{4,H_2} = 0.091 \exp(-9000/T)$; $k_{4, O_2} = 0.35 \, k_{4, H_2}$;
$k_{4, N_2} = 0.44 \, k_{4, H_2}$; $k_{4, H_2O} = 6.5 \, k_{4, H_2}$; $k_7/k_{12} = 6.03$;
$k_{7a}/k_{12} = 0.67$; $k_{13}/k_{12} = 0.3$; $k_{14}/k_{12} = 2.5$.

With these values, agreement is generally good, S_u being predicted accurately, the heat release starting correctly around 550 K, and with the overshoot in concentrations of H and OH in the reaction zone well shown. There is a minor discrepancy in the temperature profile in the later stages of combustion and Dixon-Lewis, Isles & Walmsley discuss the possibility of the use of a lower value for k_{15}, possibly down to 1.5×10^{15} with a consequent adjustment in k_7/k_{12}.

Methane. Detailed computations for methane/air flames have been made by Smoot, Hecker & Williams (1976) using modified equations from Spalding, Stephenson & Taylor (1971) and rate data from various sources including Jensen & Jones (1971). Some 28 reactions were considered, of which five were not significant and three were of questionable importance. The remainder, with their numbering, are listed, with rates, in Table 5.6; values are in cal/gmol, sec and K; R is the gas constant, 1·987.

TABLE 5.6

Methane/oxygen reaction mechanism

No.	Reaction	Rate constant
A1	$CH_4 + OH = CH_3 + H_2O$	$3 \times 10^{13} \exp(-5000/RT)$
A2	$CH_4 + H = CH_3 + H_2$	$2 \times 10^{14} \exp(-11\,900/RT)$
A3	$CH_4 + O = CH_3 + OH$	$2 \times 10^{13} \exp(-6900/RT)$
B1	$CH_3 + O = CH_2O + H$	$3.5 \times 10^{13} \exp(-3300/RT)$
B2	$CH_3 + O_2 = CH_2O + OH$	$1 \times 10^{12} \exp(-15\,000/RT)$
C1	$CH_2O + M = CO + H_2 + M$	$2 \times 10^{16} \exp(-35\,000/RT)$
C2	$CH_2O + OH = CHO + H_2O$	$2.5 \times 10^{13} \exp(-1000/RT)$
C3	$CH_2O + O = CHO + OH$	3×10^{13}
D1	$CHO + O_2 = CO + HO_2$	3×10^{13}
D2	$CHO + OH = CO + H_2O$	1×10^{14}
E1	$CO + OH = CO_2 + H$	$5.5 \times 10^{11} \exp(-1080/RT)$
F2	$HO_2 + OH = O_2 + H_2O$	2.5×10^{13}
F3	$HO_2 + H = OH + OH$	$2 \times 10^{14} \exp(-2000/RT)$
F4	$HO_2 + H = O_2 + H_2$	$6 \times 10^{13} \exp(-2000/RT)$
F5	$H + O_2 + M = HO_2 + M$	$1.4 \times 10^{16} \exp(-1000/RT)$
G1	$H + O_2 = OH + O$	$2.2 \times 10^{14} \exp(-16\,800/RT)$
G2	$O + H_2 = OH + H$	$1.7 \times 10^{13} \exp(-9460/RT)$
G3	$OH + H_2 = H_2O + H$	$2.2 \times 10^{13} \exp(-5200/RT)$
H1	$H + OH + M = H_2O + M$	$7 \times 10^{19} T^{-1}$
H3	$H + H + M = H_2 + M$	$2 \times 10^{19} T^{-1}$

Some of the reactions included in Table 5.6 are, of course, the same as those used by Dixon-Lewis for the hydrogen flame, and it may be noted that the rate constants differ slightly, especially H1 and H3 which take a different form. The rate constants listed in the table lead to values of the burning velocity for methane which are too low. For stoichiometric methane/air the predicted value is 27·9 cm/sec compared with an experimental value of over 40. Smoot *et al.* point out that raising *all* reaction rates by a factor of 4 would bring S_u up to 47·9 but would halve the thickness of the reaction zone, the calculations for which already come rather small. If the values for the physical properties (thermal conductivity and diffusion coefficients) were, alternatively, increased four times, then S_u would be 53 cm/sec and the reaction zone thickness would be doubled, which would be rather better perhaps. They have also considered using higher values for the reaction rates for B1 and B2 recommended by P. M. Becker; these would give better values for S_u for the stoichiometric mixture, but would give values which are much too high for rich mixtures, as indeed would the other alterations considered. In one approximation the usual assumption of unit Lewis number $= 1$ is not used, and this may be an improvement. The treatment brings out well the importance of H and OH diffusion in the propagation and gives reasonable agreement with most properties of the reaction zone, but clearly is not yet quite perfect.

Other hydrocarbon flames. No other flames have yet been treated in such detail as those of hydrogen and methane, but Peeters & Mahnen (1973) have established values for many of the rate constants involved in the ethylene/oxygen flame and have studied the structure of the reaction zone in detail. Earlier valuable work on this flame was done by Fenimore & Jones (1963).

For the acetylene flame Browne *et al.* (1969) have determined the main rate constants and Eberius, Hoyermann & Wagner (1973) have made a mass spectrometric study of weak-mixture flames and shown that 40% of the temperature rise in the preheating zone is due to heat conduction and 50% is due to diffusion of radicals, mainly H atoms, into this zone.

In considering the mechanism of flame propagation we have concentrated on the role of heat transfer by thermal conduction and of radical diffusion. Another possibility which has frequently been discussed is the role of radiation from the flame. Schorpin (1950) has given a mathematical and semi-quantitative treatment of flame propagation by radiation. Generally, however, the unburnt gases are much too transparent to the flame radiation for the light to assist the propagation of small flames. The ultra-violet light, which might produce dissociation and initiate reaction centres, is relatively weak and consists mainly of OH radiation and some CO-flame spectrum in the near ultra-violet and this is not absorbed at all by normal flame gases. The infra-red radiation is rather more in amount and in some gases might be absorbed to some extent, but the most this could do would be to increase slightly the heat transfer. For some flames, e.g. H_2, there will not be any absorption even of infra-red radiation.

In flames of oil sprays, especially on a large industrial scale, preheating by radiation may become significant, and in coal-dust flames radiation may be the most important term in the heat transfer (*see* page 173).

Heat loss by radiation may have some influence in arresting flame propagation, especially for near-limit flames. For flames of very rich mixtures, which are highly luminous due to carbon formation, heat losses may be very important, reducing the final flame temperature and affecting the flame speed and the mixture limit for propagation.

The concept of excess enthalpy

The success of Lewis & von Elbe's work on minimum ignition energies for spark ignition, and the connection between these and the quenching distance (page 25) led them to speculate on an extension of this relationship to flame propagation. A simple picture is to consider that ahead of the reaction zone the gases in the preheating zone receive energy by thermal conduction from the burnt gas region, while still retaining their chemical energy. Thus in the preheating zone there is an enthalpy excess, or energy hump, which travels with the flame front. If we equate the excess enthalpy per unit area, h, to the minimum energy for spark ignition, I, assuming a minimum volume of diameter d or area πd^2, where d is the quenching distance, we obtain

$$I = \pi d^2 h.$$

Also this excess enthalpy will be equal to the heat flux; if k is the thermal conductivity and T_u and T_b are the initial and final temperatures, then

$$h = k(T_b - T_u)/S_u.$$

These equations lead to

$$I = \pi d^2 k(T_b - T_u)/S_u. \tag{5.15}$$

Measurements of quenching distance, minimum ignition energy and burning velocity are available for a number of fuels, and comparison tends to support the approximate validity of equation (5.15).

However, this simplified treatment will not stand up to more quantitative examination (Spalding, 1955). Diffusion of combustion products, especially water vapour, into the preheating zone also occurs, and this lowers the chemical potential, and thus the enthalpy of the mixture. Whether or not there is an enthalpy excess or deficit thus depends on the relative importance of heat transfer and product-diffusion (*see* von Elbe & Lewis, 1959), and it has now become practice to derive the Lewis number, which is usually taken as

$$L_e = k/\rho C_p D_{eff}$$

where ρ is the density, C_p the specific heat of the mixture and D_{eff} an effective mean diffusion coefficient for the combustion products diffusing into the combustible mixture. A Lewis number of 1 then indicates that conduction and diffusion just cancel so that there is no net enthalpy excess. It seems that for many flames the Lewis number is indeed around unity, although in particular flames there may be appreciable enthalpy excess and in others an enthalpy deficit.

The simple picture of the enthalpy excess is thus untenable, and Spalding (1955) has shown that a relationship of the type

$$I = \text{constant} \cdot kd^2 (T_b - T_u)/S_u$$

may be deduced from more general considerations and does not indicate that approximate validity of equation (5.15) means that the excess enthalpy concept is necessarily true. However, there does seem to be some similarity between spark ignition, which provides a sufficient quantity of thermal energy and a sufficient concentration of radicals to produce a propagating flame, and actual propagation in a steady flame. Perhaps it is sufficient to retain the thermal conduction contribution, which brings the mixture up to the state where chemical reactions become important, and to ignore the diffusion term, because dilution with a small amount of inactive combustion products does little to suppress reaction. The excess enthalpy concept is, of course, just another way of looking at the problem, and its correctness or otherwise does not invalidate the more detailed thermal and diffusion mechanisms dealt with in previous sections.

The concept of flame stretch

It was noticed by Karlovitz *et al.* (1953), initially for turbulent flames on burners, that near the rim the luminous reaction zone got thinner when the gas flow was increased, and that at still higher flows holes appeared in the flame surface at the rim prior to lifting of the flame. It seems that under the influence of a strong velocity gradient, which occurs in the boundary layer, the shearing forces stretch the flame so that its surface area is always increasing. This was linked with the concept of limiting flame enthalpy. A flame is stable and will propagate provided that heat release (and free-radical production) within the flame volume exceeds losses. Because of the flame stretch the heat balance is adversely affected so that the reaction rate and burning velocity are reduced and the flame may even be extinguished.

The concept was applied by Karlovitz *et al.* (1953) and Karlovitz (1959) to turbulent flames, but Reed (1967; 1971) and Edmondson & Heap (1969a; 1970) have considered its application to blow-off in laminar flames (*see* Fig. 5.14). They defined the Karlovitz number as

$$K = (n_0/U)(dU/dy)$$

where U is the flow velocity, dU/dy is the velocity gradient across the flame and n_0 is an effective flame thickness given by $n_0 = k/\rho c_p S_u$, with k the thermal conductivity, ρ the gas density, c_p the specific heat and S_u the

THE STRUCTURE OF THE REACTION ZONE

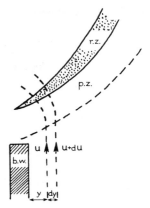

Fig. 5.14. Flame stretch near burner rim. r.z. = reaction zone. p.z. = preheating zone. b.w. = burner wall.

burning velocity. The critical velocity gradient at the boundary-layer at blow-off g_b should then be given by

$$K = g_b n_0 / S_u$$

and they have used this relationship to determine K at blow-off. The velocity gradient in the boundary-layer is assumed to be linear and it is presumably assumed that g_b can be determined from the flow rate assuming parabolic laminar flow distribution. A logarithmic plot of g_b against S_u by Reed, covering nearly three orders of magnitude, for weak flames of ethylene/air and several other hydrocarbons shows a good straight line of slope 2, indicating that g_b is proportional to S_u^2, with Karlovitz number 0·23. For flames of rich mixtures the outer cone supports the primary reaction zone and the Karlovitz number for blow-off is higher. For other blow-off conditions, rather different values of the Karlovitz number are obtained; for inverted wire-stabilised flames K varies from 1·3 to 2·0. Edmondson & Heap basically confirmed Reed's observations, using methane inhibited with CH_3Br and using inverted flames stabilised on thin plates, but conclude that for ordinary flames on cylindrical burners changes in S_u due to interdiffusion with air at the rim interfere with the observations. The stability of Bunsen-type flames has already been discussed (page 8) and the effects of gas flow through the dead space at the rim have been noted (page 56). Although wall quenching is probably the main cause of the fall in S_u for slow flames, flame stretch probably contributes to the causes of blow-off in fast flows. This concept of flame stretch has recently been examined by Strehlow & Savage (1978). They find that there is appreciable stretch at the tip of the inner cone of premixed flames, but in some parts of the flame surface there may even be negative stretch, i.e. compression.

Reaction rates

Under laminar flame conditions the average rate of the chemical reactions can, of course, be determined directly from measurement of the thickness of the reaction zone and the burning velocity. In practical combustion systems the rate at which fuel can be burnt depends mainly on the rate of spread of flame into the mixture. The overall combustion rate and heat release rate may be increased by design of burners to increase the area of flame surface and by increasing the effective flame area by turbulence (*see* page 135). It is of considerable interest to know at what stage the chemical reaction rate itself, rather than the propagating mechanism, becomes limiting. The fully stirred homogeneous reactor, sometimes referred to as the Longwell reactor (Longwell & Weiss, 1955) has been designed for this type of study.

The homogeneous reactor consists of a thermally well-insulated spherical vessel into which premixed gases are injected from the centre with such force that they mix almost instantaneously with the hot burnt gases and an attempt is made to maintain a homogenous reaction throughout the vessel. The contents of the vessel are therefore at uniform temperature and composition. The aim is to measure the maximum reaction rate by making the mixing time small compared with the chemical reaction time. Burnt gas is withdrawn from the vessel through one or more openings in the shell of the vessel. Various designs have been tried so as to obtain the most rapid and uniform mixing. Fig. 5.15 shows schematically a form of burner used by Kydd & Foss (1965). Maximum reaction rates are normally determined by increasing the mass flow until the flame suddenly extinguishes. Alter-

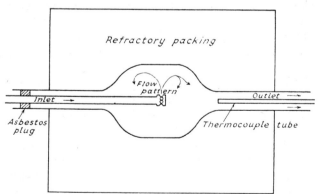

Fig. 5.15. Diagram of homogeneous fully-stirred reactor.

natively extinction may be produced by lowering the temperature, altering the mixture strength or adding inert diluents.

A number of determinations of maximum heat release rates, pressure dependence of reaction rate and activation energy have been made using the Longwell-type reactor. Results are seldom strictly comparable, but agreement is not always good and there is no doubt great difficulty in ensuring an adequate mixing rate, especially for fast-burning mixtures. For fuels consisting mainly of propane, Longwell & Weiss (1955) found a pressure exponent of 1·8 and a value of 1·8 to 2 was also obtained by Clarke, Odgers & Ryan (1962) and Clarke, Odgers, Stringer & Harrison (1965); a value of 2 implies a bi-molecular reaction. Hottel, Williams & Baker, however, find a mean pressure exponent of only 1·3; this value may be applicable to lean mixtures and the higher value to stoichiometric mixtures. Activation energies are of the same magnitude as those determined chemically, but again rather variable; Clarke *et al.* give a value around 26·6 kcal/mole for propane wheras Kydd & Foss (1965) give 61 kcal/mole near the lean limit of combustion; Hottel, Williams & Baker find around 30 kcal for stoichiometric mixtures and around 50 for rich mixtures. Clarke *et al.* find a maximum reaction rate for propane/air around 40 mole sec^{-1} litre^{-1} atm^{-2} (this assumes a pressure exponent of 2). Combustion rates beyond the normal limits of flammability for propane have been measured in a homogeneous reactor by Kydd & Foss, while Clarke, Odgers, Stringer & Harrison have studied mixtures of propane, oxygen and nitrogen and made predictions of the maximum combustion rate for pure propane/oxygen mixtures.

In the homogeneous reactor, sampling with a probe (Hottel, Williams & Baker, 1957) shows that the initial breakdown of the hydrocarbons is very rapid but that complete combustion of hydrogen and carbon monoxide is relatively slow. Similar slow completion of combustion under laminar flame conditions has already been referred to in the section on radical concentrations in the reaction zone (page 102 and is referred to again in the chapter on temperature measurements (page 284). These afterburning effects are usually most noticeable in near-limit mixtures and are often associated with excess populations of free hydrogen or oxygen atoms which are only removed slowly by ter-molecular reactions.

With complex fuels or mixtures of fuels it may happen that reaction rates for two or more components are different and this may lead to the reaction zone having a visibly complex structure. Thus we have found (Gaydon & Wolfhard, 1948) that when iron carbonyl is introduced into a hydrocarbon flame at low pressure, the flame develops a blue-green fore zone. This is presumably due to the decomposition of the carbonyl and occurs at relatively low temperature in the preheating zone. Somewhat similar effects are observed with flames of hydrocarbons supported by NO_2 as oxidiser (Parker & Wolfhard, 1953); these show a yellowish fore zone due to thermal decomposition of NO_2 below the combustion zone.

Other interesting structures are sometimes observed with fuels burning with oxides of nitrogen. H_2 and NO only burn apparently under conditions in which the final flame temperature exceeds 2800°C; in mixtures of H_2, N_2O and NO Parker & Wolfhard (1953) have observed a double flame front, the lower front due to combustion of H_2 with N_2O and the upper to burning with NO. Similar effects occur with H_2, NH_3 and NO mixtures and in lean mixtures of hydrocarbons with NO. NO itself possesses decomposition flame, but only if it is preheated to 900°C so that the final flame temperature exceeds 2700°C (Parker & Wolfhard, 1953).

In some cases reaction rates vary in a curious way with temperature, and many hydrocarbons, ethers, etc. can give the well-known cool flames at quite low temperature, little reaction at rather higher temperature and then normal ignition at still higher temperature. This two-stage combustion has been fully discussed by Spence & Townend (1949), and in some propagating flames a cool-flame region may precede the main ignition zone. This has been briefly pointed out in Chapter II, and Plate 3a shows a cool flame situated well below the main flame front. Under these two-stage conditions the flame front has quite an unusual structure and temperature profile.

The formation of oxides of nitrogen

Pollution from nitric oxide and nitrogen dioxide, usually referred to collectively as NO_x, has become an increasing problem in recent years and has stimulated a lot of research into the mechanism of NO formation. Pollution mostly arises from car exhausts and high temperature flames. Ordinary Bunsen-type flames form very little NO because the reactions under such conditions are too slow for the NO to reach anything like its equilibrium concentration, which would be of the order 0·2% in a natural-gas/air flame and 1% in acetylene/air. However in internal-combustion engines the higher pressure in the cylinders, and the rather higher temperature as well, accelerate the NO formation processes. Other high-temperature flames, such as some used in experimental work in attempts at power generation by MHD, also produce undesirably high concentrations of oxides of nitrogen.

A full discussion of the chemical kinetics of NO_x reactions would require a lot of space; here we give only a brief summary. For further information the reader is referred to the numerous recent original papers on the subject, including some fourteen in the *14th International Symposium on Combustion* (1973) and later ones by Fenimore (1975), Ay & Sichel (1976), Morley (1976) and Hayhurst & Vince (1977).

A little NO_x is formed quickly in the thin reaction zone of Bunsen-type flames, and this has been called 'prompt' NO_x, but the main formation in the post-reaction zone is very slow. This main reaction can be explained by the Zeldovich mechanism

$$O + N_2 = NO + N \quad k = 1\cdot 4 \times 10^{14} \exp(-75\,000/RT) \text{ cm}^3 \text{ mole}^{-1} \text{ sec}^{-1}$$
$$N + O_2 = NO + O \quad k = 6\cdot 4 \times 10^9 \, T \exp(-6250/RT) \text{ cm}^3 \text{ mole}^{-1} \text{ sec}^{-1}.$$

The large activation energy for the first reaction is due to the high dissociation energy of N_2, 225 kcal/mole, and this accounts for the slowness of the main reaction at the temperature of ordinary flames with air at atmospheric pressure. In pure oxygen/nitrogen mixtures the atomic oxygen would have to be formed by thermal dissociation of O_2, and the overall activation energy for NO formation would be the heat of formation of O (59 kcal/mole) plus that of the first reaction above, i.e. a total of about 134 kcal/mole. In flames the atomic oxygen will be produced by chain-branching reactions which are only slightly endothermic, of the type discussed on page 122, such as

$$H + O_2 = OH + O \quad -16\cdot 5 \text{ kcal/mole}.$$

Two other reactions are also involved to a lesser extent in the overall scheme for NO formation by the main process

$$N + OH = NO + H \quad k = 4\cdot 0 \times 10^{13} \text{ cm}^3 \text{ mole}^{-1} \text{ sec}^{-1}$$
and
$$NO + O = O_2 + N \quad k = 1\cdot 55 \times 10^9 \, T \exp(-38\,500/RT)$$
$$\text{cm}^3 \text{ mole}^{-1} \text{ sec}^{-1}.$$

The former is relatively important in flames of rich mixtures and the latter in weaker mixtures.

These reactions explain the formation of NO_x in flames of hydrogen and of carbon monoxide quite well, but for organic fuels rates may be higher, and the contribution of the 'prompt' NO_x formed in the reaction zone may be significant. Thus with lean propane/air in a stirred reactor the NO formation was found by Engelman *et al.* (1973) to be some four times that expected, and in rich mixtures might be up to ten times.

We have seen (page 101) that in lean flames of hydrogen and of carbon monoxide there is a considerable excess concentration of oxygen atoms in the reaction zone, due to their rapid formation by fast bimolecular chain reactions and slower removal in three-body processes. This excess [O] will lead to a higher rate of NO formation by the main reaction processes. This overshoot in [O] may also be responsible for the prompt NO in organic flames, or at least part of it. However, an alternative mechanism, favoured by Fenimore, involves the carbon radicals known spectroscopically to be present in the reaction zones of premixed organic flames,

$$C_2 + N_2 = 2\,CN$$
or
$$CH + N_2 = HCN + N.$$

The second of these reactions is supported by Hayhurst & Vince (1977) who have shown that prompt NO depends on the number of C atoms rather than on C_2. However, the subject of prompt formation of NO_x is still somewhat controversial.

In flames of organic fuels containing bound nitrogen the NO formation rate is very much higher, occurring in the reaction zone at about the same rate as the combustion processes.

Flame propagation in turbulent gases

For premixed flames, the differences in laminar and turbulent flow have been discussed briefly in Chapter II, page 14, and the problem of defining and measuring the flame speed under turbulent conditions has been mentioned in Chapter IV, page 90. Here we shall consider in rather more detail the nature of turbulence and its effect on flame propagation.

The relatively random movements of turbulence vortices, and the interactions between them, are very complex, but to simplify the understanding of the processes and to facilitate mathematical treatment it has been usual to define various quantities. The turbulence balls may be visualised as rotating volumes of gas which have a similar vector of velocity, the random fluctuation velocity u, which has a mean value $u' = \sqrt{\bar{u}^2}$. The intensity of turbulence may be defined as the ratio of this fluctuation velocity to the average local gas flow velocity, $I_t = u'/v$. Isotropic turbulence denotes the case in which the r.m.s. fluctuation velocity is equal in all three dimensions. The turbulence balls may be visualised as travelling an average distance l, the mixing length, before colliding with each other; this mixing length may be roughly equal to the diameter of the balls, and the balls may sometimes merge or lose their identity on collision. As previously mentioned, an important characteristic of flowing gas which affects the onset of turbulence is the dimensionless Reynolds number $R_e = vL/v$ where L is a characteristic length (page 14); L is equal to the pipe diameter d for flow through a pipe, but its meaning in some other cases is less obvious.

Turbulence may be generated for flow through a long pipe (or burner tube) by the shearing forces in the boundary-layer close to the wall; turbulence then usually sets in around $R_e > 2300$. It may also be generated, at lower flows, by inserting a grid or screen in the pipe; turbulence generated in this way will have a scale (or mixing length) related to the spacings of the grid; it will gradually die out well above the grid. Turbulence may also be generated in flow over a surface such as an aerofoil, and, especially important in combustion, at the boundary between gases flowing at different speeds, e.g. close to the mouth of a burner where the moving combustible gas mixture exerts a shearing force on the stationary surrounding air. Temperature and density differences across the mixing layer are additional factors besides the relative-velocity effect.

One of the earliest and most valuable treatments of turbulence in Bunsen-type flames was by Damköhler (1940), who showed that the scale of turbulence increased towards the centre of the tube. Fig. 5.16 shows, for steady flow through a tube, the variation of the dimensionless quantity l/R (where R is the tube radius) with fractional distance across the tube, r/R (r being the distance from the centre). It will be seen that l depends mainly on the size of the tube and the distance from the centre and hardly at all on the Reynolds number. It is useful to keep in mind the fact that the mixing length is about 6% of the tube diameter near the centre and becomes much smaller towards the wall. This is the reason for the 'brush' of turbulent flames being thicker in the centre of the flame.

This picture of more or less spherical turbulence 'balls' of comparable size bumping into each other is probably a major over simplification.

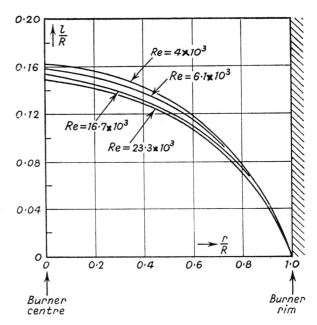

Fig. 5.16. The dimensionless quantity l/R as a function of fractional radius for various Reynolds numbers.   [After Damköhler

Probably the turbules approximate more nearly to rotating cylinders, and the larger ones in the same region probably all rotate in the same sense. Roshko (1976) has also stressed that it is necessary to separate the random and the non-random processes; large coherent structures which are related to the flow characteristics usually exist. Small turbulence eddies decay quickly; they must obviously disappear when they become comparable in size with the molecular mean free path. Larger turbulence balls appear, from short duration photographic exposures, to persist for an appreciable time, see Plate 10b. Between the larger turbulence balls, so prominent in instantaneous schlieren photographs, there are probably smaller scale turbulent threads which contribute to energy and material transport between the larger structures and are involved in the dissipation of turbulence energy. In general, the shearing forces at the boundary-layer take energy from the gas flow to maintain the turbulence, and then this is broken down to molecular (heat) energy by the smaller eddies.

A good visual picture of the structure of a turbulent flow is given by Roshko's study (1976) of a plane turbulent mixing layer. Fig. 5.17 shows vortices (redrawn from successive schlieren photographs) developing at

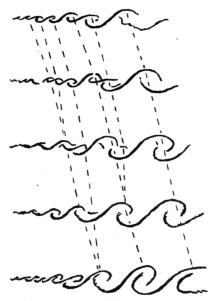

Fig. 5.17. Mixing layer between flows of nitrogen (upper) and He/Ar mixture (lower) of same density. $R_e = 3 \times 10^6$. Redrawn from schlieren photographs; frames 1, 3, 9, 13 and 19 of a sequence. The broken lines connecting the vortices show how they merge.

[*Redrawn from Roshko*

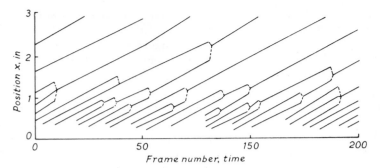

Fig. 5.18. x–t diagram of eddy trajectories in mixing layer between helium and nitrogen streams. '$R_e = 1\cdot 6 \times 10^4$ per inch'. [*From Roshko*

the mixing layer between a high-speed flow of nitrogen and a slower flow of a helium/argon mixture of the same density. The initial turbulence in the flow is of quite small scale and the visible vortices are much larger coherent structures. It can be seen that, as the vortices travel along, two or sometimes three of them merge to form a larger vortex. Fig. 5.18, also from Roshko, shows in an x–t diagram how this occurs.

There are three main effects of turbulence on flames, depending on the scale and intensity of turbulence.

Small-scale turbulence may affect the reaction zone itself by increasing the transfer of free radicals and of thermal energy. Larger scale turbulence distorts or wrinkles the flame front, thus increasing the effective flame area and combustion rate; this concept of the wrinkled flame front has proved the most profitable. Very intense turbulence may disrupt the flame front so that there is no longer a continuous flame surface and it may quench the flame by the 'flame stretch' principle.

The wrinkled flame concept. The wrinkled flame front is responsible for the formation of the typical flame brush. Instantaneous schlieren photographs show the distortion of the flame very well, e.g. Grover, Fales & Scurlock (1963) and Fox & Weinberg (1962). Richmond *et al.* (1957) using pipe flow with R_e up to 160 000 showed that the wrinkled flame was continuous within the flame brush. They focused a small section of the flame on to a photomultiplier tube so that the flame could be scanned throughout the turbulent brush or a particular section could be studied with very high time resolution. Interference filters allowed selection of different wavelengths. If the flame broke up into separate flamelets one would expect that the level of light intensity would occasionally go to zero, and occasionally show high peaks if there was overlapping of flame fronts or if the photomultiplier saw the flame end-on. It was found, however, that the light level never went below the level of a laminar flame viewed normal to the flame front. On the other hand Parker & Guillon (1971), who studied stoichiometric methane/air flames with grid-generated turbulence, found unburnt pockets of mixture up to 1 cm downstream of the visible flame; they used a fine iridium resistance thermometer with special electrical circuitry to give high time resolution. The concept of a continuous wrinkled flame-front surface is probably valid under many conditions, but it should not be assumed that it always occurs; especially there are differences between tube-flow turbulence and that generated artificially with a screen.

With the simple model of the wrinkled flame front we might expect (Damköhler, 1940) that the front would propagate at a speed S_t equal to the sum of the laminar burning velocity and the turbulent perturbation

$$S_t = S_l + u'$$

where u' is the root mean square fluctuation velocity. A slightly more detailed treatment by Shchelkin (1943) assuming a continuous conically wrinkled front gave

$$S_t = S_l \sqrt{[1 + B(u'^2/S_l^2)]}$$

where B is a constant near unity.

A number of other variations of this type of formula have been proposed, and Andrews, Bradley & Lwakabamba (1975), in a valuable review, have listed thirteen papers giving treatments of the wrinkled flame, sometimes assuming that the front is partly split into separate eddies or that further turbulence is generated within the flame. The effect of flame-generated turbulence was especially considered by Karlovitz, Denniston & Wells (1951) who deduced an expression

$$S_t = S_l + (2S_l u'')^{\frac{1}{2}}$$

where
$$u'' = \frac{S_l}{\sqrt{3}}\left(\frac{\rho_u}{\rho_b} - 1\right).$$

Although there is good experimental evidence (e.g. Durst, Melling & Whitelaw, 1972) of an increase of turbulence in the flame front, most recent work has tended to doubt the role of flame-generated turbulence (Lefebvre & Reid, 1966; Dashchuk, 1971). Even though such turbulence is generated it is difficult to see how it could diffuse back into the unburnt gas to affect the burning velocity.

However, many of these hypotheses do not appear to take sufficient account of recent experimental evidence on the actual structure of the wrinkled flame. A feature is that it is not wrinkled smoothly but shows surfaces which are convex towards the unburnt mixture with cusp-shaped lines connecting these convex parts. This is shown in the photograph by Fox & Weinberg (1962) and in Plate 7c. The process of formation of this structure may be understood from consideration of Fig. 5.19. If we imagine that turbulence produces initially a large-scale sinusoidal perturbation of the thin flame front, we may then suppose that flame propagation proceeds at the laminar burning velocity from each element of flame surface without further perturbation (rather like Huygens principle in optics); thus the envelope of the flame develops as in (b), (c) and (d). Thus the formation of the cusps is accounted for and we see that the retarded part of the flame front catches up by the spreading of flame across this region, while the advanced part continues to propagate at the laminar speed. In practice turbulence will continue to produce further perturbations and a complex structure of the type shown in Plate 7c is obtained. In the cusp regions of the flame we have a situation rather similar to that at the tip of a Bunsen flame, the effective value of the burning velocity being greatly increased by the three-dimensional nature of the heat flow in this region and by divergence of flow lines. This cusp effect is considerable so that the effective burning velocity averaged over the surface of the wrinkled flame is appreciably greater than the laminar value for a plane flame front.

A very detailed study of a stoichiometric propane/air flame has been made by Fox & Weinberg (1962) using a variety of optical techniques

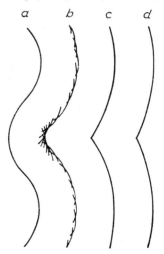

Fig. 5.19. This illustrates the propagation of a flame front with an initially sinusoidal wrinkle and shows how the cusp formations develop. The flame propagates from left to right. (*a*) is the initial form, and (*b*), (*c*) and (*d*) show successive shapes.

(instantaneous schlieren photography, ray deflection, particle-track photography) and comparing laminar and grid-generated turbulent flames. From a detailed study of the results they were able to deduce the instantaneous flame contours and Fig. 5.20 shows some redrawn contours superimposed on each other. It may be recalled (page 90) that in determining

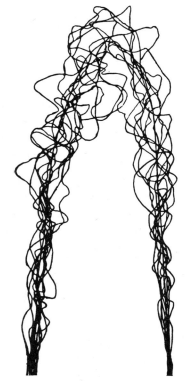

Fig. 5.20. Superimposed instantaneous flame contours. [*From Fox & Weinberg*

turbulent burning velocity from a flame brush, either the mean position of the flame or its inner edge may be used. Fig. 5.21 shows the local turbulent burning velocity, S_t, from the cone angle of the brightest part of the flame brush and the approach flow obtained by particle-track photography (curve 2). This flame has a laminar burning velocity, S_l, of 40·4 cm/sec, but the local value varies across the flame and is plotted in curve 1; the marked increase at the centre caused by curvature effects at the flame tip, where the apparent value of S_l reaches 250 cm/sec, is shown. Fox & Weinberg have also determined the local value of the 'laminar burning velocity of the turbulent flame', S_l', which is the value the laminar burning velocity would have to have in the wrinkled flame to give the observed instantaneous contours. They studied flames with two flow rates, 367 and 247 cm/sec on the centre line which, with the grid in place, gave turbulence mean fluctuation velocities of 18 and 12 cm/sec, i.e. about 5%. The turbulence scale

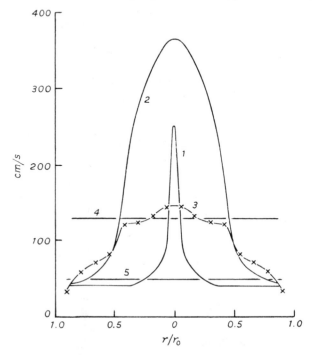

Fig. 5.21. Burning velocities of a laminar and grid-induced turbulent propane/air flame. Curve 1, laminar flame, local value of S_l. Curve 2, S_t based on line defined by maximum concentration of instantaneous contours (i.e. brightest part of the flame brush). Curve 3, 'laminar burning velocity of turbulent flame', S_l'. Line 4, mean S_t based on surface area defined by inner locus of superimposed contours, i.e. inner edge of flame brush. Line 4, mean value of S, based on area defined by line in curve 2. [*From Fox & Weinberg*]

was in each case about 1·8 mm, close to the grid spacing. For these flames the mean S_l' were 145 and 125 cm/sec, some 3·6 and 3·1 times the true S_l. Fox & Weinberg's results indicate that about half of this effect could be accounted for by increase of burning velocity due to curvature of the wrinkled flames in the cusp lines, but that the remainder must be explained as being due to increased heat and radical transport by turbulence at the small-scale end of the turbulence spectrum. These results apply to grid-generated turbulence in relatively slow flows. They give a very clear insight into the structure of the wrinkled surface of such flames; it is not clear to what extent such conclusions would apply to more intense tube-flow turbulence at much higher Reynolds numbers.

Effects of small-scale turbulence. The stirring action of turbulence eddies which are smaller than, or at least comparable with, the flame thickness will cause increased heat and radical transfer. For laminar flow the burning velocity depends on thermal conductivity, heat capacity and gas density (*see* page 94)

$$S_l \propto (k/c_p\rho)^{\frac{1}{2}}.$$

From the kinetic theory it can be shown that $k/c_p\rho$ is nearly identical with the kinematic viscosity v. Turbulence produces an additional transfer process which is governed by the turbulence exchange quantity ε, which is the product of the turbulence scale and the mean fluctuation velocity $\varepsilon = lu'$. Thus

$$S_t/S_l = [(\varepsilon + v)/v]^{\frac{1}{2}} \approx (\varepsilon/v)^{\frac{1}{2}}.$$

Fox & Weinberg point out that this relationship is more satisfying theoretically than useful in practice. Although l defines the scale of turbulence, in reality there is need to know more about the spectrum of the turbulence scale. Even when the scale greatly exceeds the flame thickness, $l \gg \delta$, the smaller eddies which serve to dissipate the main turbulence vortices may increase the heat and radical transfer and so increase the effective burning velocity, so that the above definition of ε is insufficient. Even with fairly large-scale grid-induced turbulence, we have seen that the small-scale turbulence makes an appreciable contribution to the increase in burning velocity of the wrinkled flame.

Flame disruption and flame stretch. As the turbulence becomes more intense the flame may become more broken, and the next stage may be like that illustrated in Fig. 5.22, where the front is still continuous but pockets of burning gas break off and are consumed in the hot product gases downstream from the main front.

Summerfield and co-workers (1955) considered the flame brush to be a rather deep zone of chemical reaction, and found that the light emission from CH and H_2O showed a larger shift of the position of their intensity maxima in the flame than could easily be explained by the movement of a fluctuating reaction zone of the wrinkled-flame type. Such measurements have to be evaluated with care, however, because the processes of light emission by CH and by H_2O are entirely different. The temperature distribution through the brush showed that maximum CH light emission coincided with an average attainment of 85% of maximum temperature, which is again difficult to reconcile with a wrinkled-flame model. Although

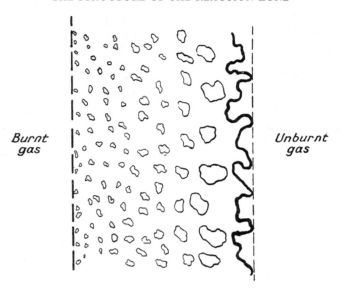

Fig. 5.22. Combustion zone in a highly turbulent gas.
[*After* Shchelkin (1943)]

Summerfield's experiments were open to some criticism they provided a great stimulus to further detailed study of turbulent flame structures.

If, as Summerfield assumed, large changes in flame structure occur under turbulent conditions, changes in reaction mechanism and light emission would also be expected. Gaydon & Wolfhard (1954) found that the light yield of C_2, CH and OH radiation was only slightly reduced by rather mild grid-generated turbulence. Similar experiments were made by John & Summerfield (1957) with open and enclosed burners, the latter to avoid changes of mixture strength by entrainment of air. In both cases the total intensity of C_2, CH and CO-flame radiation decreased slightly with increasing turbulence, but not enough to make a large change in reaction mechanism plausible. A thickened reaction zone, as in Fig. 5.21, would increase the depth of the region of light emission but should not affect the total of chemiluminescent radiation.

Kovasznay (1956) derived a criterion for the dissolution of a turbulent flame front by equating the velocity gradient in the turbulent approach flow with the velocity gradient in a laminar flame front:

$$\alpha = u'\delta/lS_l$$

where l is again the scale of turbulence and δ is the thickness of the laminar flame front. For $\alpha > 1$ dissolution is predicted. Shetinkov (1959) has attempted to deal with the theory of flames when there is no longer a coherent front at all. Andrews, Bradley & Lwakabamba (1975) in their review have listed a number of papers in which the effect of disruptive turbulence is taken into account in deriving formulae for the turbulent burning velocity. Most of these still involve the laminar value, S_l.

One effect of extreme turbulence may be to extinguish a flame. If a thin reaction zone is so extended by turbulence that it is diluted with cold unburnt gas at a rate greater than reaction can maintain the temperature, then the flame will be extinguished. This concept of 'flame stretch' has already been discussed briefly (page 126) and its application to turbulence has been expounded by Karlovitz (1959) and Lewis (1973). It is likely that the dilution and cooling of the reaction zone by flame stretch is responsible for the observed slight fall in chemiluminescent radiation. It also contributes to a narrowing of the flammability limits under turbulent conditions and (*see* below) an increase in the ignition energy.

Minimum ignition energy in turbulent flows. The increased heat transfer due to the stirring action of turbulence increases both the minimum spark energy for ignition and the quenching distance. A technique similar to that used by Lewis & von Elbe for stagnant conditions (*see* page 25) has been used for gases flowing in a duct with grid-generated turbulence by de Soete (1971) and by Ballal & Lefebvre (1977*a*; 1977*b*). Such studies are obviously of technical importance for re-ignition of aero engines under flight conditions. There are some complications in making absolute measurements because results are to some extent affected by the electrical circuitry, which determines the spark duration and character, and by electrode design, which has to be fairly robust for fast flow conditions and may itself affect the gas flow.

Results indicate that there is a steady increase in the minimum critical electrical energy for spark ignition with increasing mean fluctuation velocity u', and that the optimum electrode separation also increases with u'. There is, however, little influence of the turbulence scale, l, or of the mainstream flow velocity. Ballal & Lefebvre (1977*b*) find that the minimum ignition energy is proportional to the cube of the quenching distance and that the quenching distance under turbulent flow conditions is related to the thermal conductivity, heat capacity, density and laminar burning velocity by the relationship

$$d_{qt} = A(k/c_p\rho)/(S_l - 0\cdot 16u')$$

where A is a constant related to the Peclet number.

It should be realised that although turbulence, by increasing heat and mass transfer, increases flame speeds, it does at the same time make ignition more difficult and narrows the flammability limits.

Values of turbulent burning velocity. Difficulties in defining S_t and measuring it have been mentioned in Chapter IV (page 90). Using the locus of the brightest part of the flame brush of a turbulent Bunsen-type flame, with natural pipe turbulence, to determine the flame area gives a mean value of S_t which increases with the approach flow Reynolds number to rather over twice the laminar value S_l; some old results by Bollinger & Williams are plotted in Fig. 5.23.

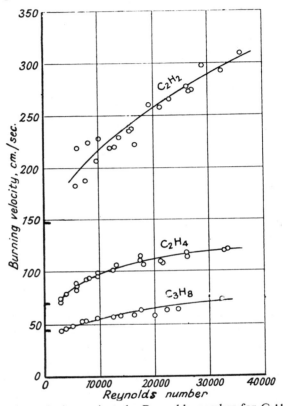

Fig. 5.23. Burning velocity against the Reynolds number for C_2H_2, C_2H_4 and C_3H_8/air mixtures of maximum burning velocity. The laminar burning velocity, S_l is marked on the ordinate in each case.

[*From Bollinger & Williams* (1949)]

When the flame spread method is used the apparent increase in burning velocity seems to be much greater. Thus Lefebvre & Reid (1966) used screens or grids to generate turbulence of up to 14% in a propane/air mixture in a duct, and their results, as replotted by Mizutani (1972), are shown in Fig. 5.24. The experimental points at various intensities of turbulence are shown, with theoretical curves from Mizutani based on a rather complex wrinkled-flame model with some allowance for flame-generated turbulence. These results, and also similar ones given later by Ballal & Lefebvre (1975), indicate that the turbulent burning velocity varies with intensity of turbulence from around twice to over twelve times the laminar value.

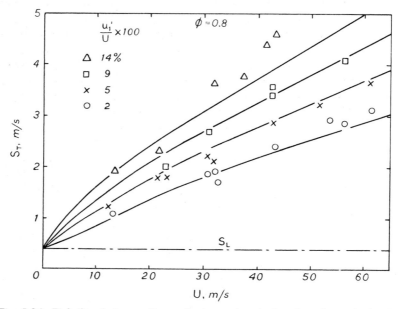

Fig. 5.24. Relation between flow velocity and turbulent burning velocity from flame spread in a duct for propane/air with equivalence ratio 0·8.
[*From Mizutani with experimental points by Lefebvre & Reid*]

Studies of the rate of expansion of a spark-ignited flame kernel in turbulent gas by Palm-Leis & Strehlow (1969) show variations in dR/dt with radius R which are difficult to interpret, but seem for grid-generated turbulence in propane/air to indicate relatively low values of S_t/S_l, mostly around 1·4, but increasing with turbulence intensity and with radius R up to 2·2.

For natural pipe-flow turbulence the meaning of the Reynolds number is fairly clear, although the scale and intensity of turbulence vary across the flame so that we use a mean value of S_t, as in Fig. 5.22. For grid-generated turbulence the scale of turbulence can be measured from schlieren photographs and the intensity from anemometry measurements of u', but the meaning of Reynolds number is not so clear. The subject has been discussed by Andrews, Bradley & Lwakabamba (1975) in their review; they discuss the use of the Taylor microscale which uses a characteristic length based on the turbulence scale and frequency, from u', and also the Kolmogorov scale based on the rate of turbulent energy dissipation and the kinematic viscosity. With this complex phenomenon of turbulence it seems that the old adage holds that the deeper you go the muddier you come up! Many papers on both the theory and technical implications of turbulence in flames are given in the various Combustion Symposia, including the 1973 European Combustion Symposium. There are valuable recent papers be Ballal (1979).

Chapter VI

Diffusion Flames

Diffusion flames initially received less attention than premixed flames in fundamental research, despite the fact that diffusion flames are more frequently used industrially. Thus the burning of producer gas with air in steel furnaces is in the form of a turbulent diffusion flame. Similarly the combustion of coarse oil sprays, as in diesel engines and gas turbines, is largely controlled by diffusion-flame type processes, although finer droplets may evaporate and burn more nearly as premixed gases. The industrial combustion of pulverised fuel (i.e. coal dust) occurs mainly by a diffusion controlled process around each grain. The difficulty with diffusion flames, unlike the premixed flames, is that there is no fundamental characteristic like burning velocity which can readily be measured; even the mixture strength has no clearly defined meaning.

It is usually assumed that for diffusion flames the chemical reaction rate is not the rate-determining process. Spectroscopic investigations have shown that this is in the main true, but only for the hottest part of the reaction zone; the preheating zones on both sides of the reaction zone show interesting sequences of chemical reactions and in some cases these zones are nearly ideal for the observation of reaction rates.

Chemists have, in the past, been little interested in diffusion flames because it was believed that they did not allow the study of chemical reactions. The physicist was similarly unable to measure any fundamental constants. Research on diffusion flames has therefore been concerned with the more formal study of the flames. Calculations and measurements on the shape and height have been made, and the lifting-off has been recorded. The turbulent combustion of non-premixed gases has, of course, received a good deal of attention. We shall concentrate on the more fundamental processes rather than the details, as the mathematical treatment of turbulent diffusion does not seem to cover all the essential features.

In this chapter we shall first discuss the earlier work on small diffusion flames and then go on to the more recent work on larger flames, including turbulent flames. In the later sections we shall deal with some of the more applied aspects of diffusion flames such as the combustion of droplets and solid particles, and also the spread of flame over liquid surfaces.

The height and shape of small flames

Burke & Schumann (1928) have given a formal theory of circular and flat diffusion flames. They employed either a set of one tube inside another or two rectangular ducts. In the first case (Fig. 6.1) the fuel flows up the inside tube of diameter B, and comes into contact with air flowing up the outer tube of diameter A. The mean velocity of the fuel and the air are kept the same by using mass flows proportional to B^2 and $(A^2 - B^2)$ respectively.

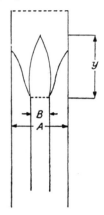

Fig. 6.1. Shape of circular diffusion flame with either air excess (flame closes over axis of burner) or fuel excess (flame bends towards the walls).

A different set of tubes has to be used if the mixture strength is to be changed. Similarly for the rectangular ducts, the size has to be selected to give the same velocity in the two parts.

It can be shown that for the circular flame, the task is to solve the differential equation

$$\frac{\partial C_{ry}}{\partial t} = D \left\{ \frac{\partial^2 C_{ry}}{\partial r^2} + \frac{1}{r} \cdot \frac{\partial C_{ry}}{\partial r} \right\},$$

where C_{ry} is the concentration of gases at the co-ordinates r and y, r is the radius, y the height and D the diffusion coefficient.

The solution of this equation leads to Bessel functions. Burke & Schumann had to make some simplifying assumptions which cannot be wholly justified, but they succeeded in giving a formal picture of the diffusion flame as far as height and shape are concerned. They assumed that the reaction takes place in an infinitesimally thin zone which forms a surface around the issuing jet of fuel. This surface will close over the top of the fuel supply jet if air is in excess, but will approach the wall of the outer tube if fuel is in excess (Fig. 6.1). In both cases the height of the flame can be defined. They locate the reaction zone as the position where air and fuel are in stoichiometric proportions. This is certainly a definition and is necessary for mathematical treatment of the problem, but it has no physical meaning as fuel and air will certainly react not only where they are in stoichiometric proportions but also wherever the mixture is within the limits of flammability, and at high temperature in practically all proportions. A further assumption is that the rate of diffusion alone is the rate-determining process, that therefore it is sufficient to calculate the amounts of fuel and air diffusing into the reaction zone to obtain the amount of gas reacting. This assumption, as will be shown below, is doubtless correct, but the difficulty is that to obtain a mathematical solution it is necessary to use a value for the diffusion coefficient which is independent of temperature and gas composition. Their calculations have been refined by later workers because of these limitations; as the gases heat up, the diffusion coefficient increases, roughly as $T^{1.75}$, and this is only partially compensated for by an increase in the velocity of the gas stream, which only increases proportionately to T. However, this is not the main objection. The reaction zone starts directly on top of the inner tube, and gas reacting here will attain a temperature close to the theoretical maximum. The radial diffusion into the reaction zone thus starts with a coefficient having the value as for room temperature and rising to a value which for flames with oxygen may be greater by a factor of 30 at the hottest part. It is the change of diffusion coefficient in a horizontal direction rather than in the vertical direction which may lead to an incorrect picture. However, Burke & Schumann were able to calculate not only the height of the flame but also the shape of the reaction-zone surface. For their calculations they had to choose an average value for the diffusion coefficient, rather than the value at room temperature, in order to get agreement with experiment. It may be noted that this full mathematical treatment of the diffusion problem for a circular burner is a radial problem and is capable of explaining the rounding of the flame towards the top.

It has been pointed out by Jost (1946) that some of Burke & Schumann's results can be derived without the need of solving the differential equation, and as it is the main task of this type of theory to predict the height of flames under various conditions of pressure, gas velocity, dilution, etc., we may briefly recall Jost's discussion. For a flame with excess air, which closes over the inner burner, the height will be given by the point at which air just diffuses to the axis of the inner tube. The distance covered by diffusion can be calculated by the equation

$$\bar{x}^2 = 2D \cdot t,$$

where x is the distance, D is the diffusion coefficient and t is the time. With some loss of accuracy we may write this

$$x^2 = 2D \cdot t.$$

Now the time for the gases to travel to the required height y will be

$$t = y/v,$$

where v is the gas velocity. Thus by substitution

$$x^2 = 2Dy/v \text{ or } y = x^2 v/2D.$$

Noting that the volume flow $V = \pi x^2 v$, we may rewrite this as

$$y = V/2\pi D. \tag{6.1}$$

This relation is rather approximate, because the expression $x^2 = 2Dt$ only gives the average diffusion rate, whereas actually we require to reach a stoichiometric concentration on the axis of the flame; also, the flame is circular, not flat, so that we have a two-dimensional problem. Hence, equation (6.1) may be replaced by the more general form

$$y = \theta V/\pi D, \tag{6.2}$$

where θ is a numerical constant replacing the $\frac{1}{2}$ in (6.1). This shows that: (i) The flame height will remain the same if the mass flow is kept constant but the burner dimensions are changed (keeping the ratio of the two burners the same, of course). (ii) The flame height should be inversely proportional to D. Actually the heights for CO and H_2 flames are in the ratio 1:2·5, whereas at room temperature the values of D are in the ratio 1:4; this suggests that the theory does not include all the factors which determine the flame height. (iii) The theory predicts that the height is independent of pressure if the mass flow is kept constant; this is true as long as there is no carbon formation (Parker & Wolfhard, 1950).

A fairly simple treatment of this type can be extended to a flame in stationary air. Making various approximations and assuming that the diffusion constant, D, varies nearly as T^2, Guyomard (1952) developed an expression of the form

$$y = 2V_0 T_0 / 5 \cdot 76 \pi D_0 T_m \tag{6.3}$$

where V_0, T_0 and D_0 are the volume flow, temperature and diffusion constant at room temperature and T_m is a mean flame temperature. This is of similar form to (6.2), but with some allowance for variation of diffusion coefficient with temperature (*see also* Roper, 1977; 1978).

For flames in which carbon formation occurs (e.g. most hydrocarbons) the flame height has no definite meaning as we cannot distinguish between the end of the combustion of the fuel vapour and the completion of the combustion of the carbon particles. In some cases this latter stage may not be complete, the unburnt carbon escaping as soot.

Several advances on Burke & Schumann's theory have been made. Barr & Mullins (1949) studied so-called 'vitiated' flames; in these the fuel burns in an atmosphere which is diluted with combustion products (CO_2, H_2O and extra N_2); these flames are of interest because of the use of secondary combustion in turbines, furnaces and industrial appliances. They have extended Burke & Schumann's theory to include different velocities of fuel and air; it is difficult to make reliable calculations because to get a rational mathematical solution it is necessary to assume that the fuel and air keep their separate velocities, i.e. that no transfer of momentum takes place. This additional assumption is obviously not exact, but as the error so introduced may not be more than in the other assumptions already made in Burke & Schumann's treatment, the results may still be of value if it is realised that they do not claim to give more than a dimensional analysis. Barr & Mullins have been able, by using average values of D from several experiments, to express flame heights in terms of flow rates, air/fuel ratio and amount of vitiation, all on one curve. Plate 8b shows diffusion flames in slightly vitiated air for various increasing flow velocities; Plate 8a shows the constancy of height with pressure.

We should like to draw attention to the complex structure of the flames of hydrocarbons in vitiated air studied by Barr & Mullins and shown in Plate 8b. There are two distinct zones, as may be seen especially well in the flame for a gas velocity of 1·53 cm/sec. In this the true diffusion flame can be seen as a thin reaction zone extending vertically from the burner to the luminous zone. In this part the gaseous reactions take place and diffusion is certainly the rate-determining factor. This region does not close towards the centre, but ends in the luminous part of the flame. This luminous part is of quite different character, and is where the combustion of the solid carbon particles takes place, and the laws governing the rate of reaction must be of an entirely different nature. This can easily be seen from Fig. 6.2, which shows that the flame height is only independent of pressure for very small flames. For larger flames carbon or soot formation causes the height to increase suddenly manyfold. However Jones & Rosenfeld (1972) have shown that even when carbon formation occurs the Burke & Schumann treatment can be useful, the fraction of carbon liberated being determined by physical rather than chemical conditions so that the sooting can be predicted.

The study of diffusion flames has been further extended by Hottel & Hawthorne (1949), Wohl, Gazley & Kapp (1949), Barr (1954) and Fay (1954). The treatments are somewhat similar and try to extend Burke & Schumann's theory to long flames, either laminar or turbulent, of fuels

DIFFUSION FLAMES 151

burning in an unlimited air supply. The structure of a flame of this type is best illustrated by Fig. 6.3 and by Fig. 6.4, which gives actual measurements of composition at different heights for a laminar hydrogen flame. Within the accuracy of measurement there is no oxygen inside the flame

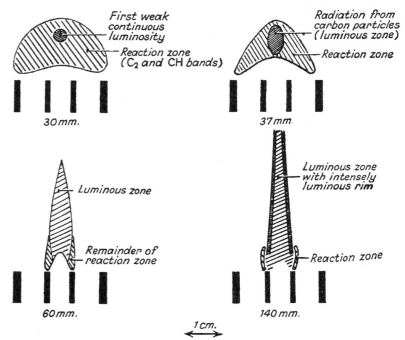

Fig. 6.2. Structure of C_2H_2/air diffusion flame at various pressures.

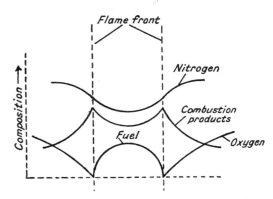

Fig. 6.3. Idealised diagram of diffusion flame; the flame front is considered infinitesimally thin.

front and no fuel outside it. The products of combustion have their maximum there and diffuse towards both sides, both fuel and air being more and more diluted as we approach the flame front. The flame height may be located at the point where the fuel concentration falls to zero; here, provided there is no carbon formation, the reaction zones close in to the axis of the burner.

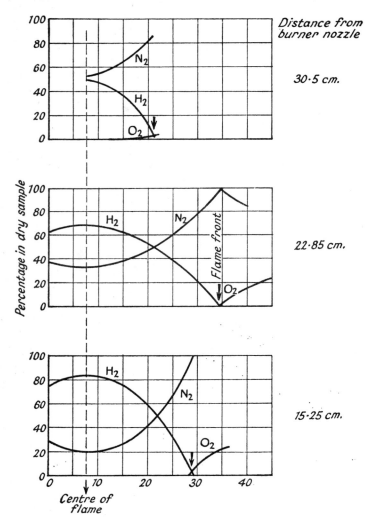

Fig. 6.4. Gas composition in an H_2/air diffusion flame at different heights above burner nozzle. [*Redrawn from Hottel & Hawthorne* (1949)

The shape of an actual diffusion flame can be seen from Plate 9a and b. The first shows the cylindrical flame widening out as it rises, and the latter is a shadow photograph which gives a good impression of the fuel being slowly used up, this being shown by the narrowing of the inside schliere.* The outside boundary of the flame does not give a good shadow as the heat

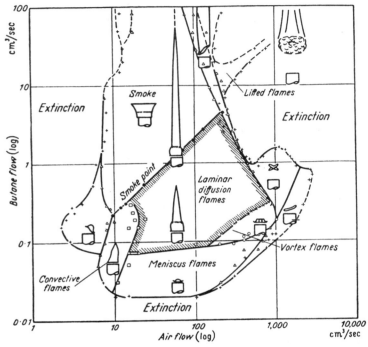

Fig. 6.5. Types of diffusion flame obtained with various flow rates.
[*From Barr*

spreads some distance from the reaction zone, and the shadow or schliere, due to change in temperature, occurs where $1/T$ is changing most rapidly, which usually lies at rather low temperature, often about 200°C. The reaction zone itself is not visible in shadow or schlieren photographs as there is no sharp composition or temperature gradient in this hot low density region. The upper parts of the flames shown in Plage 9a and b already show some turbulence, and if the flow rate is increased this turbulence increases, and leads us into the study of turbulent diffusion flames.

While these laminar diffusion flames are the type most commonly en-

* This schliere is essentially due to change of composition, the change of refractive index being due to the change from fuel to air and combustion products; it allows us to follow the disappearance of the fuel.

countered and have been the subject of most study, the flames may take very different forms. With very low fuel flows, curious meniscus-shaped flames sometimes occur; flames which are cap-shaped, but lift so that they only hold to the burner at one point, are also observed. Changes of flow pattern, due to differing densities of fuel and air and to thermal convection, are important. With faster flows, and especially with a flame running in a fast air stream, lifted flames and curious vortex-shaped flames occur. These are illustrated for butane in air, for various flow rates, in Fig. 6.5 (from Barr, 1954).

Although the structure of an established diffusion flame does not depend greatly on reaction rates, the limiting conditions, at which the flame extinguishes, obviously do. Some progress with the mathematical treatment of the extinction condition has been made by Zeldovich (1949), Spalding (1954), Spalding & Jain (1962) and Schmitz (1967).

An important influence, especially on larger laminar diffusion flames, is that of buoyancy, which causes the hot product gases near the stoichiometric surface to rise rapidly, entraining more air at the base of the flame. This air entrainment has been studied quantitatively by Robson & Wilson (1969). The buoyancy effects depend, of course, on gravity and some interesting experiments have been done under simulated zero-gravity conditions using a freely falling combustion system ('drop tower'). Under these conditions the hot product gases accumulate round the flame which enlarges and becomes less stable (*see*, for example, Edelman *et al.*, 1973).

The candle flame

An ordinary candle provides an interesting example of a diffusion flame, and it was the subject of a considerable amount of attention in the last century. Here we shall discuss a few later measurements; further discussion also appears in the chapter on carbon formation, and we shall refer again to combustion on wicks later in this chapter.

Plate 12*a* shows a schlieren photograph from work by Parker & Wolfhard. Two schlieren can be seen, one due predominantly to temperature well outside the reaction zone, and a concentration schliere of the fuel vapour leaving the wick. The luminous carbon zone commences inside the reaction zone, becomes denser as we go higher and closes at the top. For very small candle flames (or other hydrocarbon diffusion flames) the blue reaction zone can be seen outside the carbon zone, and this reaches right to the top of the flame. For larger flames, however, this blue reaction zone disappears about half-way up the flame, leaving the carbon zone exposed to the outside air; for very large flames the centre of the zone may cool down to a temperature at which oxidation of the carbon particles becomes slow, and then the particles escape and the flame is observed to

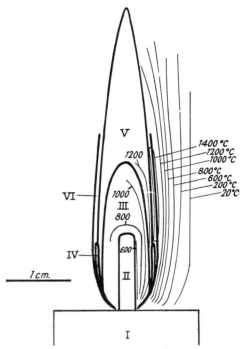

Fig. 6.6. Relative temperatures in a candle flame from thermocouple measurements: I, Body of candle. II, Wick. III, Dark zone. IV, C_2 and CH zone. V, Luminous zone. VI, Main reaction zone.

smoke. The height of a flame before sooting commences is often taken as a measure of the tendency of a particular fuel to form carbon and soot. The temperature field of a candle flame is shown in Fig. 6.6. The temperature at the surface of the wick is quite low, and the temperature gradient near the wick enables us to calculate the heat transfer to the wick; it can be shown that this is equal to the amount of heat required to heat up and vaporise the paraffin. The hottest region is, of course, the blue reaction zone. Fig. 6.6 serves to illustrate the temperature distribution and explain the schlieren picture; the measurements were made with a thermocouple and are uncorrected for radiation losses and so are not accurate.

The Wolfhard–Parker flat diffusion flame

A detailed study of the structure of the reaction zone of a diffusion flame has been made with the aid of measurements on the emission and absorption spectra of the zone. To develop a more rigorous theory of diffusion flames we think that knowledge of the structure of the reaction zone is necessary and the results are therefore given in detail.

All theoretical investigations so far have assumed that the reaction zone is infinitesimally thin. This is far from the truth; for flames with air the visible zone is normally a few millimetres thick, and for flames with oxygen it may be up to a centimetre. It is interesting to note that the thickness of oxygen diffusion flames does not depend on pressure; flames with air do, however, become thicker at reduced pressure.

Flames from circular tubes are unsuitable for detailed examination of the reaction zone, and a special rectangular burner (Fig. 6.7) has been developed to give a flat flame (Wolfhard & Parker, 1949b). End views of flames of ammonia with oxygen and ethylene with oxygen are shown in Plate 13c and d. For both of these there is a yellow zone on the fuel side;

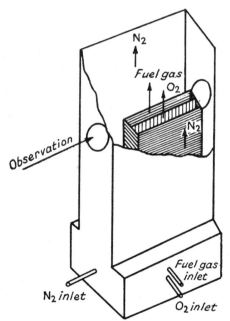

Fig. 6.7. Burner for flat diffusion flame.

for the ammonia flame it is due to emission of the ammonia α band (NH_2), and for ethylene it is due to carbon particles. As we pass across the flame towards the oxygen side there is a dark zone before the main blue reaction zone.

The flame on this type of burner can be made to burn very steadily, without flickering or sideways movement, and this is therefore an ideal source for spectroscopic examination. An image of the flame is thrown on to the slit of the spectrograph, the spectrograph being turned so that its slit is horizontal (alternatively a special optical device may be used to rotate the image of the flame through 90°); it is thus possible to study the variation of the spectrum in a horizontal direction across the reaction zone. Measurements have usually been made at a height of 1 cm above the burner top. The flame has a horizontal length of 5 cm, so that it is necessary to use a very small optical aperture for examining it, otherwise there will be a mixing of light from different parts of the reaction zone when the image is formed.

In a diffusion flame of this type it can readily be verified that diffusion is the only rate-determining process. For example, in a premixed ethylene/oxygen flame the average rate of consumption of oxygen in the reaction zone is about 4 mole cm^{-3} sec^{-1}, whereas for the diffusion flame measurements of the flow, height and thickness of reaction zone (about 1 cm) give an oxygen consumption of only 6×10^{-5} mole cm^{-3} sec^{-1}. It may also be noted that in the diffusion flame the average temperature of the reaction zone is near the theoretical maximum, whereas in a premixed flame the reaction starts at quite a low temperature. Comparison of these very different rates of oxygen consumption shows that in the diffusion flame the chemical reactions cannot be the rate-determining factor, and the rate must be governed by diffusion towards the reaction zone.

To get a simple mental picture of the reaction zone let us visualise the fuel and oxygen being initially separated by a diaphragm which is suddenly removed and the gases are instantaneously ignited along the boundary. Initially the temperature along the boundary will rise, and combustion products will be formed. Thus there will be a flow of heat away from the zone on both sides and also a diffusion of products. The oxygen and fuel concentration in the zone will fall to a very low value, and so they will diffuse towards it. The diffusion of oxygen and fuel towards the zone will tend to increase their concentrations above the equilibrium value and further reaction will occur almost instantly to keep the concentrations down to the equilibrium values. Owing to the high reaction velocity and slow rate of diffusion there will never be any serious departure from equilibrium in the reaction zone. A steadily increasing thickness of combustion products will build up between oxygen and fuel. In an actual flame, this will take the form of a wedge of products. A flame of this type, in which there is never any appreciable departure from chemical equilibrium may be regarded as a sort of ideal diffusion flame.

Under some conditions diffusion flames may depart from this ideal condition. Thus at reduced pressure the rate of diffusion increases, while that of reaction may decrease. Thus at low pressure there will be a tendency for fuel and oxygen to mix to some extent in proportions greater than for equilibrium, and we tend towards a sort of premixed condition. We get a similar trend with dilution with an inert gas, although in this case the diluent only decreases the reaction velocity without increasing the diffusion. Air flames have a much thinner reaction zone than oxygen ones, and diffusion gradients are greater, while the reaction velocity will be reduced by the lower temperature and dilution. Although flames with air have visually a similar appearance to those with oxygen, it is not certain to what extent they may depart quantitatively from the ideal condition. There is still a lack of experimental data for air flames.

Spectroscopic study of structure

Using the flat diffusion flame full spectroscopic examination has been made of flames of ammonia and ethylene with oxygen. We shall discuss the ammonia flame first (see Wolfhard & Parker, 1952). Plate 11a shows the spectrum, taken across the reaction zone. The hair line along the spectrum serves as a reference mark and this corresponds to a fixed position in the flame. Passing across the flame from the ammonia side towards the oxygen side, we first find the ammonia α bands of NH_2 and these are closely followed by bands of NH. There is then a gap in emission, corresponding to the dark central region of the reaction zone, and then we come to OH emission, which is strong, with a little continuum in the visible region; this continuum appears to be the same as that for the blue radiation from the burnt gas of premixed flames. A little farther across towards the O_2 side we find the Schumann-Runge bands of O_2.

The absorption spectrum shows bands of NH_3, NH, OH and O_2; the positions are slightly different from those in emission because in absorption the strength depends mainly on concentration, while in emission it is influenced by temperature as well. The temperature has been measured by the spectrum-line reversal method using both Na and OH, the results being in agreement. The results are shown in Fig. 6.8, with NH_3 to the left and O_2 to the right. For NH_3, because of the influence of temperature on the vibrational populations and hence on the absorption spectrum, the only definite information we get is the point where all the NH_3 has disappeared; this is marked A in the figure. Just before this, NH_2 emission can be seen, then NH absorption followed by NH emission. The temperature reaches its maximum about 2 mm farther to the right, and this is obviously the region where most heat is being liberated by formation of H_2O.

The OH absorption has been used to measure quantitatively the OH partial pressure. OH reaches its maximum partial pressure beyond the temperature maximum; other things being equal a mixture of O_2 with H_2O will obviously give a higher partial pressure of OH than will a mixture of H_2 with H_2O. The OH emission is strongest between the positions of maximum temperature and maximum OH concentration. Oxygen can

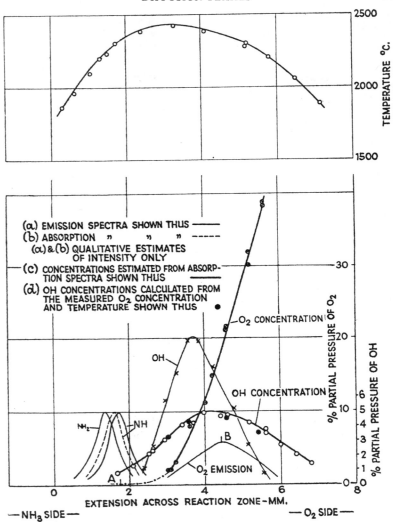

Fig. 6.8. Change in absorption and emission spectra across an ammonia/oxygen flat diffusion flame, 1 cm above the burner.

similarly be measured by its absorption and it will be seen that its concentration falls rapidly towards the centre of the reaction zone. At some positions we know O_2 and OH concentrations and temperature; calculations on the equilibrium show that the values are consistent with the existence of chemical equilibrium in this part of the flame.

Bands of OH and NH and also bands of O_2 around 2200 Å can all be reversed and all give about the same temperature by the reversal method. Thus even the O_2 emission is probably thermal and not due to chemiluminescence as has sometimes been assumed (Griffing & Laidler, 1949). The radiation from diffusion flames does not show any obvious sign of departure from thermal equilibrium.

Thus the main reaction zone (i.e. where the temperature is highest) of an ammonia/oxygen flame is a hot zone in chemical and thermal equilibrium, the oxygen and hydrogen arriving at this zone by diffusion at a rate which is slow compared with the time required to restore the equilibrium. On the NH_3 side there is a preheating zone into which the ammonia diffuses and is decomposed to H_2 and N_2. The further diffusion of the H_2 towards the main zone forms a barrier to oxygen which prevents it from reaching the ammonia.

Thus the ammonia appears to decompose without the presence of oxygen. Ammonia does not decompose readily at low temperature, and some of it penetrates to a region where the temperature is around 2000°C. Obviously, in the region where there is a high concentration of ammonia we cannot be in true chemical equilibrium, and it seems that departures from equilibrium may exist in the preheating zone.

The ethylene/oxygen diffusion flame behaves very similarly. The emission spectrum is shown in Plate 11b. In emission, we have, from the fuel side, first a continuum from luminous carbon, and close to this a region of C_2 and CH emission. After a gap, we come to strong OH bands and then, right over towards the oxygen side, bands of O_2 may be observed. In absorption the OH and O_2 bands are again used to determine the concentration of these substances, and the results are collected in Fig. 6.9. The temperature was again determined from OH and Na reversal. It will be seen that the O_2 concentration falls to zero before the carbon zone is reached. This result is only confirmed for diffusion flames with O_2 (i.e. not necessarily for air).

The carbon zone is therefore a region where the hydrocarbon is decomposed thermally in the absence of oxygen. OH radicals do, however, just reach this zone. Hydrogen produced by the decomposition diffuses again towards the main reaction zone where its reactions liberate heat, as shown by the existence of maximum temperature here. In the reaction zone we again appear to be in thermal and chemical equilibrium and there is no abnormally high electronic excitation as there is in the reaction zones of premixed flames. The radiation from carbon particles occurs in a region which is well below the maximum temperature. Between the carbon zone and the reaction zone there is a dark region which may be attributed to the absence of suitable emitters; the blue radiation from the reaction zone may be partially O_2 Schumann-Runge and partially CO + O radiation, both

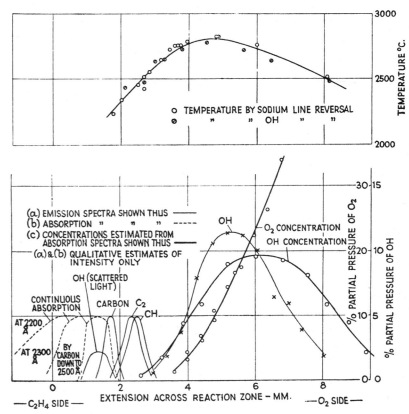

Fig. 6.9. Change in absorption and emission spectra across an ethylene/oxygen flat diffusion flame, 1 cm above the burner.

types being very dependent on the existence of high temperature. The carbon particles persist for a long while because there is no oxygen to burn them and they can only react with water and CO_2 to form CO and H_2.

The C_2 and CH radiation is strongest near the base of the flame and dies out before the top. Some C_2 could be formed by evaporation of carbon particles on the hotter (O_2) side, and this is actually the region where C_2 emission is found. Both C_2 and CH concentrations should, however, be extremely small in equilibrium (*see* e.g., Jessen & Gaydon, 1969). We suspect that the C_2 and CH emission is not of thermal origin, but arises from chemical processes due to slight premixing of fuel and oxygen by interdiffusion in the cool dead space at the base of the flame.

In diffusion flames at low pressure the C_2 and CH radiation is stronger due to faster diffusion and slower attainment of equilibrium. The C_2 and CH regions actually merge with the main reaction zone at low pressure. Even at atmospheric pressure in an air flame the C_2 and CH is relatively stronger than in a flame with oxygen.

We cannot follow the decomposition of most hydrocarbons spectroscopically because they do not possess suitable absorption bands, but if we introduce a little benzene with the fuel we see (Plate 11c) that the C_6H_6 bands disappear well before the carbon zone is reached. It is now known that the spectrum of benzene changes from banded to continuous at high temperature, but there does not appear to be any replacement of the bands by continuum. There must be some transition stage between benzene and carbon, and this is marked by the occurrence of a continuous absorption spectrum; this 'pyrolysis' continuum is discussed in Chapter VIII. The appreciable temperature gradient in the preheating zone may to some extent cause optical distortion and make quantitative interpretation of the continuous spectrum uncertain.

Sampling studies, using a small quartz probe and a mass-spectrometer for analysis, of methane/air diffusion flames (Smith & Gordon, 1956), support the conclusions from optical spectra that the methane is pyrolysed before coming into contact with oxygen. Various pyrolysis products are detected (*see* page 224), but apart from a rather uncertain trace of formaldehyde, there is no sign of intermediate oxidation products. The flow pattern and velocity distribution in flat diffusion flames have been studied in detail by James & Green (1971) using laser-Doppler anemometry.

Counter-flow diffusion flames

The flat diffusion flame developed by Wolfhard & Parker has given valuable information about the structure of diffusion flames and has the advantage that in many ways it is a two-dimensional idealisation of a normal flame of a fuel gas burning on an open burner. However, for the study of certain fundamental processes it has the limitations that the combustion is influenced by the small region of premixed gases at the base, which serves as a holding flame, and by the wedge of combustion products which grows thicker as we go higher above the burner rim. Pandya & Weinberg (1963; 1964) have developed a simple opposed-jet flame into a more sophisticated counter-flow diffusion flame which is almost flat and has some especially useful features.

A simplified diagram of the burner is shown in Fig. 6.10. The fuel gas is supplied from below through a wide (6 cm diameter) tube filled with glass beads and fitted with a matrix of wound crinkled and smooth layers of metal foil (as in the Egerton-Powling burner) to produce a uniform velocity in the issuing gas; a porous metal plate may be used as an alternative to the matrix. The supporting gas (oxygen + nitrogen) is supplied from above through a similar tube, which is water cooled. A flat disc-shaped flame is produced in the position indicated; this tends to curl up at the edges and this tendency is reduced by protective metal flanges which extend the region of horizontal outward flow from the burner axis. The most stable flame is obtained when the volume flows of fuel and supporting gas are equal, and this can be achieved by distributing the diluent nitrogen (or other inert gas)

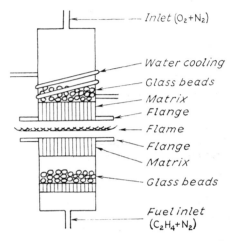

Fig. 6.10. Burner for counter-flow diffusion flame.

between the fuel/nitrogen and oxygen/nitrogen in suitable proportions. Most work has been done with the fuel/oxygen ratio in stoichiometric proportions, but excess of either fuel or oxidant may be used. Two variables are the overall flow rate and the spacing between the ends of the burner tubes; these control the duration in the flame region and the temperature gradient across the flame.

In these flames, effects at the burner rim are, of course, eliminated. The velocity of outward flow of gases increases with distance from the burner axis, the velocity being in first approximation proportional to distance from the axis. Thus the layer containing product gases does not form a widening wedge but is of constant thickness. Pandya & Weinberg (1964) and Pandya & Srivastava (1975) have applied a variety of techniques to the study of these flames, including particle-track photography, interferometry, deflection mapping and temperature measurement with thermocouples and by sodium-line reversal. Fig. 6.11 shows the temperature contours for an ethylene/nitrogen/oxygen flame; this shows that away from the burner edge the isotherms are parallel to the burner surfaces and the maximum

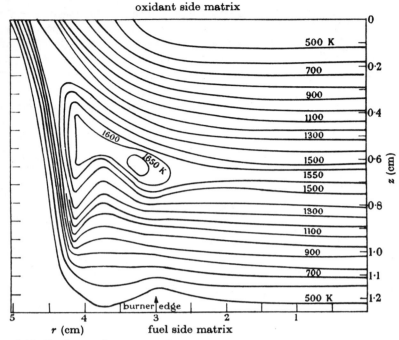

Fig. 6.11. Isotherms for counter-flow diffusion flame. Gas flows, fuel side 5·7 cm^3 s^{-1} C$_2$H$_4$ + 75 cm^3 s^{-1} N$_2$, oxidant side 22·4 cm^3 s^{-1} O$_2$ + 60 cm^3 s^{-1} N$_2$.

temperature is around 1550 K, compared with an adiabatic maximum temperature of 1850 K.

With moderate flow rates the appearance of these flames is similar to that of the flat Wolfhard-Parker flames, with a yellow (carbon particle) region on the fuel side and a bluish zone on the oxidant side. C$_2$, CH and OH emission again occurs (Ibiricu & Gaydon, 1964). With fuel excess sooting occurs, especially in the region of the flanges, and this type of flame has been found especially useful for studying the electrical control of the formation and deposition of soot (Place & Weinberg, 1966; Jones, Becker & Heinsohn, 1972). With higher flow rates the yellow carbon zone becomes weaker and with fast flows the flame becomes blue and then at a certain critical flow develops a hole at the burner axis prior to blowing off.

The extinction of diffusion flames may be studied by using the condition for occurrence of this hole in the flame as a criterion. A simple opposed-jet flame was used for studies of this type (Potter, Heimel & Butler, 1962; Anagostou & Potter, 1963) and the mass flow rate per unit area when the hole appeared was taken as a measurement of flame strength.

Tsuji & Yamaoka (1967) have used a porous cylinder as burner; the fuel gas is supplied through the porous sides of the cylinder and the flame is maintained in an upward flow of air. These opposed-jet and counter-flow flames have also been used to study flame inhibition by alkali metals (Friedman & Levy, 1963) and halogenated compounds (Ibiricu & Gaydon, 1964).

Measurements of temperature in these opposed-jet flames show that at very slow flows the temperature falls and extinction at slow flows is due to heat loss to the burner. Over most of the flow range the temperature does not vary much with flow rate and the sudden blow out at high flow is due to limitations on the chemical reaction rate, not to a fall in temperature (Tsuji & Yamaoka, 1971).

Comparison of diffusion with premixed flames

The gases of premixed flames start with a high chemical potential, which breaks down suddenly in the reaction zone. For ideal diffusion flames we have no such high chemical potential (there may be a small potential on the fuel side due to a delay in its breaking up to its elements). For diffusion flames there are preheating zones on each side of the main reaction zone and here the fuel and oxygen carrier may be broken up, and if this breaking up requires a high activation energy a small chemical potential may exist in the preheating zones.

The C_2, CH and OH radiation from the reaction zones of premixed flames is not of thermal origin (Gaydon & Wolfhard, 1948; 1949c; 1950a; 1950b). In diffusion flames with O_2, the OH bands show a normal rotational intensity distribution and have a reversal temperature in agreement with the flame temperature. In the diffusion flame, C_2 and CH emission is relatively very weak and comes from a different region of the flame from that of OH. We have not definitely established whether or not the C_2 and CH emission from diffusion flames is purely of thermal origin, but it may not be.

We have shown (Gaydon & Wolfhard, 1951a) that there is a marked abnormally high electronic excitation in the reaction zones of premixed flames. This causes excessively high reversal temperatures for lines in the ultra-violet and abnormal intensity distribution in the line spectra of introduced metals such as Fe or Pb. Diffusion flames with O_2 at atmospheric pressure do not show this phenomenon, although at low pressure, when the C_2 and CH bands also become strong, the excitation of Fe atoms again becomes abnormal and the OH bands again show high effective rotational temperatures. Diffusion flames with air also show some slight anomaly.

We believe that the non-thermal excitation in the reaction zones of premixed flames results from interaction of oxygen with organic fuels, and may be associated with the break-up of unstable intermediaries. In diffusion flames such interaction is not possible because the fuel is decomposed before it reaches the oxygen; there may be slight interaction of residual amounts of fuel and O_2 in the region where the C_2 and CH bands occur. At pressures above 1 atm the O_2 should be used up even farther from the fuel side, and we should expect C_2 and CH emission to be still weaker.

In conical premixed flames of the Bunsen type, we have seen (*see* page 34) that preferential diffusion of hydrogen from the centre may lead to a concentration of carbonaceous compounds towards the tip of the flame. For diffusion flames of mixed fuels (such as coal-gas, consisting of hydrogen and hydrocarbons) a similar effect may occur, the lighter constituents, such as hydrogen, diffusing more rapidly into the reaction zone. This is probably the explanation of Landolt's observation (1856) that for coal-gas diffusion flames the proportion of hydrogen decreased with height up the flame; this observation was one of the main arguments for the theory of the preferential combustion of hydrogen which held the field for a large part of the last century.

Turbulent flames

Plate 9c and d shows direct and shadow photographs of turbulent flames of city gas. The flame appears as an inverted cone and the shadow photograph shows that the fuel stream disappears at a lower point than previously. The outer boundary of the buoyant hot gas stream does not show any of the regular waviness as in the case of the laminar flame (Plate 9a). In these flames the turbulence seems to be generated in the flame itself and is probably due to the relatively high velocity gradient between the flame gases and the surrounding still air. These photographs do not indicate any turbulence in the issuing fuel jet itself. Butane behaves somewhat differently as the kinematic viscosity is lower than for city gas and so the turbulence sets in at rather lower velocities. As the gases emerge into the flame the kinematic viscosity increases and the Reynolds number decreases. Butane flames do not therefore have the turbulent boundary-layer as seen in Plate 9 for city-gas flames. Plate 10a and b shows shadow photographs with very short and rather longer exposure times on a butane flame (for which turbulence is already present in the supply tube). With the longer exposure the flame appears streaky, and from this we conclude that the turbulence balls retain their identity while they travel a considerable distance. High-speed photography shows in addition that the sideways motion of the turbulence balls is quite small.

Wohl, Gazley & Kapp have measured the height of city-gas flames for various volume flows (Fig. 6.12) and burner diameters. For very low velo-

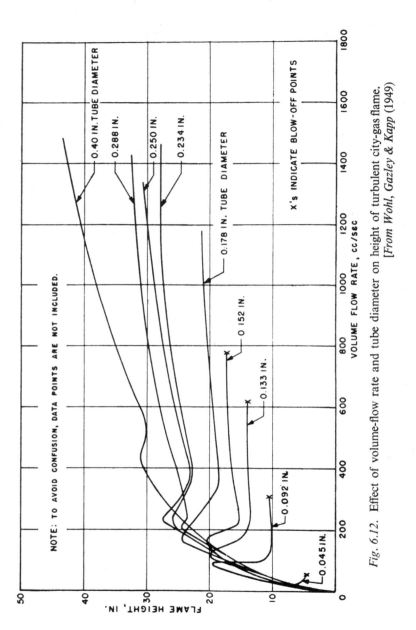

Fig. 6.12. Effect of volume-flow rate and tube diameter on height of turbulent city-gas flame. [*From Wohl, Gazley & Kapp* (1949)]

cities the height is independent of burner diameter but not a linear function of gas flow as predicted by Burke & Schumann for enclosed flames. The flame reaches its maximum height shortly before the onset of turbulence; this depends, of course, on burner diameter. The flame shortens when turbulence sets in, and thereafter the height only increases very slowly with increasing volume flow, presumably because at higher velocity the turbulence is greater and keeps the length down. Butane flames do not show the decrease in height when turbulence sets in. The reason is probably connected with the high kinematic viscosity.

It is clear that we have to distinguish between two types of turbulence in diffusion flames. (i) The turbulent motion may have already set in before the gas leaves the orifice due to the high Reynolds number in the tube. For a tubular jet we may expect turbulence when the Reynolds number exceeds about 2300, but for nozzle-like orifices or short tubes turbulence may not set in until much higher values of Reynolds number (*see also* page 42). (ii) Turbulence may commence above the orifice. Scholefield & Garside (1949) have studied the turbulence in unignited ethylene jets. It can be seen from Plate 13*b* that in these jets the turbulence at first starts quite a distance above the orifice, but the point of onset falls as the Reynolds number is increased. This turbulence in the jet begins long before the stream leaving the orifice is itself turbulent and must be due to friction at the boundary layer between fuel and air. Thus the 'height to turbulence' depends on the Reynolds number and on burner diameter and length. The height to turbulence in an ethylene flame is greater than in the unignited jet; this is presumably due to the change in viscosity and density in the hot flame gases.

We have noted that turbulence may start at the orifice, even when the approach stream is not turbulent. In experiments with gas flows from two concentric tubes, one gas jet being hot, the other cold, Onsager & Watson (1939) found that turbulence began with approach-flow Reynolds numbers as low as 150 to 200. Even with a turbulent approach flow, effects at the orifice rim may still be important, and we have already (page 134) discussed the work of Roshko (1976) in which large non-random wavy coherent structures are developed at the boundary-layer between two fast flows of different density.

For ignited ethylene jets, two types of flame appear to be possible: (i) a flame starting on the burner rim, and (ii) a lifted flame which seems to be suspended some distance above the burner. In the latter case there will be appreciable entrainment of air below the base of the flame, and the flame loses some of the characteristics of a diffusion flame and for hydrocarbons becomes bluer in colour, more like a premixed flame. The reasons for lifting and blow-off in diffusion flames are not understood, but some observations have been made by Scholefield & Garside (1949).

DIFFUSION FLAMES

The flickering of a flame occurs when the height to turbulence descends to within the visible flame; the hot gases may commence to become turbulent above the tip of the visible flame without causing the flame itself to flicker. Instabilities connected with the onset of turbulence are discussed in the following chapter.

Thin flames

Very thin, flat diffusion flames are sometimes used in the laboratory and industry. The so-called 'batswing' burner is an example. The rate of combustion in ordinary diffusion flames tends to be low, but it may be increased either by having a turbulent flame, or by having a very thin flame with a big surface area. It is doubtful whether the only rate-determining process in these thin flames is the rate of diffusion. We are inclined to think that for such flames some premixing will in fact occur, although visually the flame still retains the separate carbon and main reaction zones. Thin diffusion flames of this type have an advantage over premixed flames for industrial heating in that they cannot strike back. Very thin jets of fuel gases can easily be obtained by allowing two jets to impinge on each other. A description of experimental work on commercial flat flames has been given by Minchin (1949).

The Coanda burner

Another way to improve combustion in a diffusion flame is to make use of the Coanda effect, using a burner shaped as in Fig. 6.13; the flow of combustible gas follows an aerofoil surface for some distance, entraining air as it does so. The high-pressure gas emerges through a slot at sonic

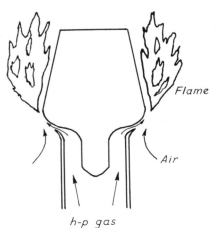

Fig. 6.13. Coanda burner, showing gas flow and surrounding flame.

velocity, creating a low-pressure region on both sides of the jet; the low pressure on the wall side causes the flow to cling to the wall, while that on the outside causes large air entrainment, up to twenty times the fuel flow (Wilkins et al., 1977). Burners of this type have been demonstrated by D. H. Desty at British Petroleum and are in commercial use, but so far little has been published in the scientific literature; photographs of flames have been shown in B.P. literature and reproduced by Weinberg (1975). Several modifications of design have been produced and they have proved valuable for very large scale combustion systems, especially for burning off waste oil-well gases. Flares on these burners, because of the high air entrainment are much less smoky than normal diffusion flames.

Furnace flames

Diffusion flames are used in some types of furnace, especially for melting iron in the open-hearth furnace. The combustion in these occurs under turbulent conditions, the fuel stream breaking up into small turbulence balls which may behave as small diffusion flames. The distances over which diffusion must occur is small, so that the rate of diffusion is not the limiting factor, and the rate of combustion is determined mainly by aerodynamic conditions. In fact for the study of combustion in furnaces of this type, experiments on the turbulent flow of water in model furnaces can be helpful and also computer simulation (Patankar & Spalding, 1973). A full discussion of this subject would be too technical for inclusion here and the reader may be referred to the books *The Science of Flames and Furnaces* by Thring (1962), and *Combustion Technology* by Palmer & Beer (1974), especially the chapters in the latter by R. H. Essenhigh.

Combustion of droplets and dusts

The combustion of fuel sprays is of considerable importance industrially and a large amount of work has been done recently on the burning of single droplets and fine sprays. Single large droplets each support their own individual diffusion flame, very fine sprays burn like a premixed flame, and there is an interesting region of intermediate droplet size. The subject has been reviewed by Williams (1973).

For single droplets, each spherical droplet tends to support a spherical diffusion flame, but that this flame is deformed by the air flow due to convection and falling of the droplet. There is an outward radial flow of vapour and inward diffusion of oxygen. The life, t_l, of a single stationary drop is proportional to the square of the initial drop diameter, d_0

$$t_l = d_0^2 \lambda,$$

where λ is an evaporation constant. This evaporation constant appears to be practically independent of the vapour pressure of the droplet, but does vary with the latent heat of vaporisation. There is a minimum size, around 5×10^{-4} cm, below which a single droplet cannot support its own individual diffusion flame, even in stagnant air. For larger droplets, the surrounding diffusion flame becomes increasingly deformed at higher airflow rates; beyond a certain critical flow, the surrounding flame suddenly breaks away, leaving a burning 'tail' behind the droplet, and then this too extinguishes. This critical flow velocity, U_c, is proportional to the droplet diameter, d; the ratio U_c/d is smaller for fuels with a high latent heat. The mass burning rate of a droplet in oxygen is similar to that in air, but the extinction velocity is higher. If m_0 is the mass burning rate of a droplet in in the absence of convection (forced or natural), then the actual mass burning rate, m, in an air stream, increases with flow, and with certain assumptions it can be shown that

$$\frac{m - m_0}{m_0}$$

should be proportional to the square root of the Reynolds number; experimentally, this relation is found to hold quite well for ethyl and butyl alcohols. Stationary droplets, e.g. on the end of a quartz fibre, burn about 25% faster than free droplets because of the greater heat transfer by convective flow round the stationary droplets. In making quantitative calculations it is also necessary to take into account the fact that for burning droplets the combustion products tend to fill in the wake behind the droplet and reduce its drag. Induced turbulence has a negligible effect on the mass burning rate, but does reduce the velocity at which extinction occurs; this extinguishing effect is particularly marked for small-scale turbulence. Inhibitors, such as methyl bromide, do not appreciably alter the mass burning rate, but again do markedly influence the extinction velocity (Wood, Wise & Rosser, 1957). There are recent papers on single droplet combustion by Nuruzzaman & Beer (1971), Krier & Wronkiewicz (1972) and Gollahalli & Brzustowski (1975).

Law (1978) has shown that in general there are two regimes; in a relatively low-temperature environment, the ignition is controlled by the chemical kinetics, while in a hot environment the rate of heating of the droplet is the dominant factor.

Studies under simulated zero-gravity conditions, using a free falling combustion chamber, by Kumagai, Sakai & Okajima (1971) show that for single droplets the relationship between d_0^2 and t_l still holds, but the size of the surrounding flame increases with time, presumably because of accumulation of product gases, although it may fall slightly during the final stages of burning.

Photographic studies of the burning of single streams of monodisperse droplets (Nuruzzaman, Hedley & Beer, 1971) have shown that the linear relationship between t_l and d_0^2 still holds, but that the slope of the line is only about a half of that for single droplets due to oxygen depletion by surrounding droplets.

Plate 12*b* shows the flame from a fine spray in still air. It resembles the flame front of a premixed flame. For this flame the fuel, kerosene, has a flash-point well above room temperature so that vaporisation cannot give a rich enough mixture for a premixed flame before we reach the combustion zone. It can also be shown that radiation from the flame is quite inadequate to heat the droplets to give the required vaporisation. It seems that the droplets are vaporised in the preheating zone. For the higher hydrocarbons with air the normal flame velocity is low, and so we have a relatively thick reaction zone and a still thicker preheating zone. The heat transfer to very small droplets is very rapid, and it can be shown that there is ample time for vaporisation in the preheating zone and even for mixing of the fuel vapour with air by diffusion. If a dense mist of fine droplets in air is passed through a vertical tube and ignited at the top, the mixture burns just like an ordinary premixed flame, and it can be seen visually that the mist clears, due to vaporisation of the droplets, about 1 mm below the luminous inner cone.

Burgoyne & Cohen (1954) show excellent photographs of a fine tetralin aerosol burning just like a Bunsen flame with a well-defined inner cone. With flames of this type, they have been able to measure burning velocities, using the total-area method. The combustion of these very small droplets is thus essentially similar in character to that for premixed gases. Measurements of the velocity of flame spread for droplets of kerosene in air by Polymeropoulos & Das (1975) have shown that the velocity increases with improved atomisation to rather over the burning velocity S_u for premixed gases, and then falls to this value for very fine sprays. Explosions in very fine mists, like those in premixed gases, may therefore be very powerful and destructive (Burgoyne & Newitt, 1955) and were at one time the cause of a number of serious crank-case explosions in ships.

The vaporisation of the droplet depends not only on droplet size, but also on the time of passage through the reaction zone. If we introduce relatively small kerosene droplets into an acetylene/air premixed flame, the drops may pass right through the flame, because, with acetylene, the flame speed is much greater and the reaction zone is thinner, so giving less time for evaporation. In Burgoyne & Cohen's work on aerosols, it is also seen that for the slightly larger droplets a Bunsen-like inner cone is still produced, but that some of the droplets pass right through, giving luminous streaks.

Burgoyne & Cohen succeeded in producing flames of aerosols of droplets of selected uniform size, by controlled condensation of tetralin vapour. For this fuel, the critical droplet size is between 10 and 40 μm. Below 10 μm, the flame is blue and has a coherent front. As the size is increased the flame rapidly becomes more luminous and the burning velocity tends to increase. Above 20 μm, the direct photographs clearly show the streaks of the individual burning droplets and that the coherent flame front has disappeared. The lower limit of mass concentration is considerably extended for the larger droplets (46 mg/l for 10 μm; 18 mg/l above 45 μm), and the equilibrium final flame temperature at the limit falls from around 1200°C for small droplets (10 μm) to below 600°C for large drops (40 μm). For these larger droplets, the burning velocity is higher (by at least a factor of two) than that with fine sprays of similar mass concentration (around 60 mg/l).

The burning of coal dust shows some resemblance to that of droplets. It is industrially important because of the extensive use of pulverised fuel in power stations. Single particles of relatively large size initially burn like oil drops, with a surrounding diffusion flame of burning volatile products which are liberated at a temperature of around 400°C; Kallend & Nettleton (1966) have shown that the temperature at the surface rises quickly to 400°C and then only slowly, while that at the centre rises slowly to 400°C and remains at this temperature until the volatile compounds have been evolved. For smaller particles, surface reaction is more important. Howard & Essenhigh (1967) have indicated that for a short initial period (around 80 msec) surface combustion of carbonaceous material occurs, then up to about 150 msec volatiles burn, either in a diffusion flame or close to the surface, and then the carbonaceous residue burns. For particles above 65 μm the volatile emission is sufficient to form a surrounding diffusion flame, but surface combustion is more important for smaller particles and below a critical size of about 15 μm surface combustion is always dominant.

Howard & Essenhigh speak of an ignition temperature of 1100°C, but later shock-tube studies of the ignition of coal dust in oxygen (Nettleton & Stirling, 1967) indicate that the ignition temperature depends on particle size, coal type and oxygen partial pressure but may be as low as 700 K to 800 K (i.e. around 500°C). Their observations also show that heat transfer to the particle is mainly by conduction and convection, rather than by radiation. Radiation may be more important in larger flames, and the theory of radiative heat transfer and coal ignition has been treated by Essenhigh & Csaba (1963). The rate of combustion of size-graded chars is again limited at high temperature by the rate of diffusion to the surface rather than by the chemical reaction rate (Field, 1969), and it has been shown that the chars burn internally as well as externally.

The burning of lycopodium dust is of interest because the particles are all the same size (around 30 μm) and nearly spherical. Mason & Wilson (1967) have been able to stabilise flat flames on a nozzle burner and have measured burning velocities which are up to 25 cm/sec for flames with air. It is only possible to stabilise flames for rich mixtures. The flames are rather similar to those of ordinary gaseous fuels, but some particles can be observed to pass right through the main reaction zone, and the dead space (related to the quenching distance) is relatively very thick. It has again been shown that for these small flames radiative heat transfer to the particles is not important.

Some further comments on flames of metallic dusts are included in Chapter XII, page 395.

Flames on wicks; mixed fuels

A flame on a wick is rather similar to that round a large droplet; the combustion may be quite different from that at a free surface. Thus for a mixture of a flammable and a non-flammable liquid we can see that at a free surface the more volatile component would dominate the gas phase (according to Raoult's law) and, according to which component was the more volatile, we should or should not have a flame. If, however, a similar mixture is burnt on a wick, the two components will vaporise in proportion to their concentrations in the mixture and not in proportion to their vapour pressures, and the composition of the remaining fuel will not change (Burgoyne & Richardson, 1949; 1950). We think, however, that there must be some separation of the fuels along the wick, the more volatile component being vaporised near the cooler base of the wick, and the less volatile at a higher temperature farther up the wick. Although the different components of the fuel may vaporise into different parts of the flame the subsequent diffusion and gas flow by convection may prevent any discernible separation of the combustion of the different components. Because of this difference in combustion characteristics it is possible to have a fuel of liquid hydrocarbon mixed with an inert diluent, such as carbon tetrachloride, which will burn on a wick, in, say, an oil-lamp, but will not burn at an open surface if spilt.

For fine mists we have seen that the rate of combustion does not depend on vapour pressure so that for mixed fuels we should not expect any separation. For large drops of mixed fuel, however, there may be preferential combustion of the more volatile constituents. This is important in Diesel engines, as heavy fuel-oils are mixtures of components with a wide range of properties; the tarry residues in such engines are evidence of this preferential vaporisation and combustion.

Combustion of liquids at a free surface

The burning of an open pool of liquid depends very much on the vapour pressure. When it is sufficiently high at ambient temperature for a flammable vapour/air mixture to be formed above the surface then application of any ordinary source of ignition will cause a flame to spread rapidly across the surface at a speed which is related to the burning velocity. The *flash-point* is defined as the lowest temperature at which vapour is given off in sufficient quantity to enable ignition; this temperature depends to some extent on the conditions of measurement (e.g. vessel shape and size and the position and character of the ignition source) and specified conditions of measurement are required for safety standards (Brame & King, 1955; Steere, 1967). For less volatile fuels ignition can only be achieved by a strong heat source applied for sufficient time to heat up and volatilise some of the fuel prior to ignition; further spread of the flame then depends on heat transfer to the fuel from the hot flame gases, and its movement across the surface is relatively slow.

The rate of burning of the fuel depends on heat transfer to the surface from the hot flame gases. This appears to be mainly by conduction from the hot gases for small flames and for flames which are relatively transparent, such as that of alcohol, but for larger flames (roughly those on burners of over 10 cm diameter) of higher emissivity (e.g. benzene, kerosene) the heat transfer is mainly by radiation (Rasbash, Rogowski & Stark, 1956). The buoyant rising of the hot gases and consequent air entrainment plays an important part in determining the flame shape and hence the heat transfer; Akita & Yumoto (1965) have used particle-track photography to study the gas movements and have reproduced an interesting photograph showing the air entrainment and vortex formation. Under some conditions, especially for alcohols, flame pulsations may be observed (Akita, 1973). The burning rate and rate of flame spread across the surface also depend on heat transfer within the liquid; the temperature distribution below the liquid surface and the convection currents have been studied by Burgoyne & Katan (1947) and Burgoyne, Roberts & Quinton (1969). For liquids of fairly high flash-point it may be possible to extinguish a fire by stirring the liquid from below so that cool oil of low vapour pressure is brought to the surface. In a steadily burning oil the surface usually heats up to just below the boiling point. In fuel oils of complex composition selective evaporation of the more volatile constituents often occurs, so that the residue becomes less volatile with time.

The effect of the diameter of the burning pool of liquid on the combustion rate has been studied experimentally by Blinov & Khudiakov (1957) over a very big range of diameters, and some theoretical relationships have been given by Spalding (1955) and Hottel (1959), while Agoston (1962) and Fons (1961) have made further studies with a moderate range of burner sizes. For small burning pools the burning rate, expressed either as rate of fall of the liquid surface or as fuel consumption rate per unit area, decreases with increasing diameter, reaching a minimum value, often at around a diameter of about 10 cm, and then increases again to a constant value for large pools. The effect may be explained by the onset of turbulence, at around the minimum burning rate (Fons, 1961); this shortens the flame and increases the heat transfer. Atallah (1965) has studied the variation of flame height with pool size. For large pools, any motion in the surrounding air may set up rapid rotation of the hot gases, developing into a 'fire whirl' (Emmons & Ying, 1967).

When the temperature of the fuel is above the flash point, or more correctly when it is above a temperature T_s which is specific for a particular fuel and is rather above the flash point, then flame speed is independent of fuel temperature (Burgoyne & Roberts, 1968; Nakakuki, 1973). This is because there is always some layer above the surface where the fuel/air ratio is stoichiometric. For ethanol and hexane the speed of flame spread has been observed to be about $1\frac{1}{2}$ m/sec and independent of fuel temperature.

Chapter VII

Flame Noise and Flame Oscillations

The noise from large flames, especially turbo-jet aero-engines, has become of increasing importance in recent years; in the first two editions of this book we had to lament the lack of fundamental knowledge about flame noise, but now work, especially by Lighthill, Bragg & A. Thomas, has greatly improved the position, and we have been able largely to recast this chapter.

The production of noise by flames and the causes of flame oscillations and the effect of these oscillations on flame noise and stability are discussed first, while the older work on the effect of sound waves on small flames is covered in later sections.

Flame noise

A steadily burning laminar flame makes practically no noise, whereas explosions, even of quite small bubbles of gas mixtures, produce strong sound or shock waves. Flames also become noisy when they are turbulent. The view, expressed in our earlier editions, that noise is associated with pressure pulses due to irregularities in direction and speed of movement of the flame front, has been confirmed and put on a quantitative basis by recent work.

The amplitude of a sound wave is proportional to the pressure variation ΔP, and the sound intensity, I, measured as the energy flux, is given by

$$I = \overline{\Delta P^2}/\rho c \qquad (7.1)$$

where $\overline{\Delta P^2}$ is the mean square of the pressure fluctuation, c is the velocity of sound and ρ is the density. For a flame, the expansion of the gas as a result of the heat release during combustion tends to produce a rise in pressure round the flame and an outward flow (*see* page 55 for a calculation of the pressure difference across a flame front), but fluctuations in pressure only occur when the rate of combustion changes, i.e. when there is some *change* in the flame.

The simplest source of sound is known as a monopole and may be visualised as a pulsating balloon. Thomas & Williams (1966) have shown that a spherically expanding explosion behaves as a monopole source. They studied spark-ignited explosions in spherical soap bubbles, using streak-schlieren photography to follow the movement of the flame front and a condenser microphone to study the pressure pulse. The output from the microphone was displayed on an oscilloscope, and was calibrated using a known sound-pressure level from a pistonphone.

Now for a monopole at a distance d from the microphone it can be shown that the pressure fluctuation ΔP is given by

$$\Delta P = \frac{\rho}{4\pi d} \cdot \frac{d}{dt}\left(\frac{dV}{dt}\right)$$

where dV/dt is the rate of change of the source volume. For a spherical flame front of radius r and volumetric expansion ratio E (i.e. E is the ratio of the volume of burnt gas to the volume before it was burnt)

$$\frac{dV}{dt} = 4\pi r^2 \cdot \frac{dr}{dt}\left(\frac{E-1}{E}\right).$$

Thus

$$\Delta P = \frac{\rho}{4\pi d} \cdot \frac{d}{dt}\left\{4\pi r^2 \cdot \frac{dr}{dt} \cdot \left(\frac{E-1}{E}\right)\right\}$$
$$= \frac{\rho(E-1)}{dE}\left\{2r\left(\frac{dr}{dt}\right)^2 + r^2 \cdot \frac{d^2r}{dt^2}\right\}. \quad (7.2)$$

During the steady burning period the flame travels at a uniform speed $dr/dt = S_u E$, where S_u is the burning velocity and E the volumetric expansion ratio; for this uniform speed $d^2r/dt^2 = 0$. Thus for this part of the burning

$$\Delta P = (2\rho/d)\, E\,(E-1)\, rS_u^2$$

i.e. the pressure fluctuation depends linearly on the instantaneous flame radius and on the square of the burning velocity. This gives the amplitude of the sound or noise level; the intensity varies as the square of this. After the flame reaches the surface of the expanded soap bubble, burning ceases and dr falls to zero; because of the momentum of the expanding gas it takes a short but finite time for dr/dt to fall to zero, and during this time d^2r/dt^2 has a large negative value so that the term involving this quantity in equation (7.2) is important and ΔP becomes negative, corresponding to a rarefaction.

Thus for these centrally-ignited spherical bubbles, the pressure difference should rise linearly during the steady burning period and then drop sharply negative at the completion of combustion. Fig. 7.1 illustrates this diagrammatically; (a) shows the radius of the burnt volume, as deduced from the streak schlieren photograph; (*b*) and (*c*) show the first and second derivatives; (*e*) is the calculated pressure pulse using equation (7.1) and (*f*) is the pressure pulse as observed by the microphone. The observed pulse shows some final fluctuations which are presumably due to pulsations of the bubble of hot burnt gas.

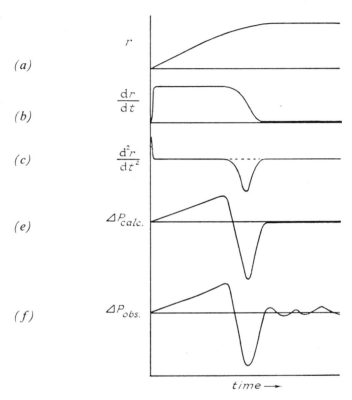

Fig. 7.1. Derivation of shape of pressure wave generated by a spherically expanding flame. [*Redrawn from Thomas & Williams*

It was shown by Thomas & Williams that the sound pulse was much weaker when the bubble was ignited non-centrally and fell to only about 1/20 of that for central ignition when the igniting spark was set near the bubble surface. This may be explained by the longer time taken for the flame to reach the extremity of the bubble and by the different flow pattern of the expanding burnt gas. They conclude that the pressure wave depends on the rate of change of the rate of increase in the volume of the gas. They also confirmed the strong dependence on burning velocity, the peak pressure difference varying nearly as the square of S_u, as predicted by (7.2); the noise intensity should thus vary as S_u^4.

For non-spherical flames it is more difficult to estimate the combustion rate from streak-schlieren photographs and Hurle, Price, Sugden & Thomas (1968) and Hurle, Price & Pye (1968) have extended the work by using the light emission from the flame front as a measure of the rate of combustion. The radiation of the C_2 Swan and of the CH bands comes only from the reaction zone and is due to chemiluminescent processes;

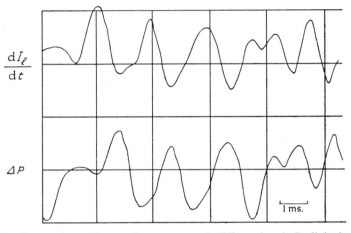

Fig. 7.2. Comparison of sound-pressure and differentiated C_2 light-intensity signals from a turbulent premixed ethylene/air flame.
[*Redrawn from Hurle* et al. (1968)]

thus for a mixture of fixed composition the total light intensity for C_2 or CH should be proportional to the actual surface area of the flame front. Hurle *et al.* confirmed this by repeating the spherical bubble experiments but using light intensity instead of streak-schlieren photography to measure the burning rate; they also showed that for premixed flames a plot of C_2 light intensity against flow rate was linear through the laminar flow region, through the transition region and into fairly strong turbulent flow, only falling off perhaps very slightly at the fastest flows studied. They then showed that for turbulent flames the pressure fluctuations, observed with a microphone, closely followed the differentiated light intensity signal dI_l/dt. Fig. 7.2 shows tracings of their oscillograph record: the lag in the pressure signal is due to the time taken for the sound to reach the microphone.

It will be seen that the correlation between the sound pressure and the variation in light intensity, which is equivalent to variation in overall combustion rate, is extremely good. It is concluded that the turbulent flame may be considered as a collection of burning elements of the combustible gas (i.e. the turbulence balls) and acoustically is equivalent to a collection of monopole sound sources. Since the pulses from the monopole sources tend to interfere with each other, the total noise intensity will fluctuate in a random manner but will be lower than that from a similar quantity of combustible mixture burnt in a spherical explosion flame. In later work Price, Hurle & Sugden (1969) have shown that a similar technique, using C_2 light emission, can be applied to the study of turbulent diffusion flames and flames of burning sprays.

Some estimates have been made of the actual energy in the noise of turbulent flames. We may define η as the ratio of the noise energy to the energy released by the combustion process. Powell (1963) studied a diffusion flame on a small Primus burner and found $\eta = 3 \times 10^{-9}$ and estimated a value of between 1×10^{-8} and 3×10^{-8} for premixed flames. Smith & Kilham (1963) found $\eta = 8.2 \times 10^{-8}$ for a turbulent flame of 6% ethylene with air. Bragg (1963) made some calculations which gave η to be of the order 10^{-6} for a typical hydrocarbon burning with preheated air. Thomas & Williams (1966) used their measurements of pressure pulses from centrally ignited bubble explosions to derive the energy flux (equation (7.1)); they found that for a hydrocarbon/air flame with a burning velocity of 60 cm/sec the ratio of noise to thermal energy was 2×10^{-6}. Because of the fourth-power law dependence on burning velocity the noise of fast-burning mixtures with oxygen instead of air will be very much greater. It is not clear whether this fourth-power law should be applied generally to explosions; in the extreme case of a detonation the propagation velocity depends on the speed of sound and the heat release rather than on the burning velocity. Detonations are, of course, extremely noisy because of the strong associated shock wave.

Some of the technical aspects of combustion-generated noise have been discussed by Putnam & Brown (1974), while Gupta, Syred & Beer (1975) have shown that the noise from turbulent flames on swirl burners may be reduced, by reflection effects at the hot-gas boundary, if the main flame is surrounded by a second outer flame.

Jet noise

Some theoretical and practical studies have recently been made on the noise from highly turbulent gas jets, especially from gas-turbine engines, and it seems of some interest briefly to record these. The basic theoretical treatment by Lighthill (1952; 1954; 1962) shows that the fluctuating shear stresses due to turbulence behave as quadrupole sources of sound. We have previously described a monopole source as like a pulsating balloon. Two monopole sources side by side and out of phase behave as a dipole, the sound radiation from which is strongly directional, falling to zero in a plane perpendicular to the axis of the dipole; a vibrating solid normally behaves as a dipole source. Two equal and opposite dipoles behave as the more complex quadrupole source, which gives a clover-leaf distribution of sound.

Lighthill's treatment shows that to a first approximation the noise intensity from a subsonic turbulent jet stream should be proportional to the eighth power of the jet velocity. Experimental studies appear generally to confirm Lighthill's eighth-power law. The quadrupole sources move with the gas flow and the intensity has a maximum in the down-stream direction. Observations show that the actual intensity distribution is roughly heart-shaped; Ribner (1959; 1960; 1967) has discussed this work and emphasised that sound refraction in the hot gas plays a part in modifying the distribution of sound intensity.

Bragg (1963) has made calculations, based on Lighthill's theoretical work, for a typical gas turbine engine which indicate that the ratio of noise energy to combustion energy η is around 4×10^{-9}, which is two orders lower than his calculation of the combustion noise. However, the combustion noise will usually be reduced because the flame is enclosed. In actual aircraft it seems that the jet noise is particularly important at take-off, while the noise from the compressor blades is heard most on landing. The eighth-power law suggests that jet noise will become of increasing importance in supersonic aircraft using higher jet thrust. It is probable that the eighth-power law applies up to a Mach number of 1, but for supersonic jets the noise may vary as M^3. It is to be hoped that further research on combustion and jet noise will lead to their better control.

Singing flames

So far we have dealt with the random noise, of unspecified frequency, from flames. We now deal with the production of sound and vibrations of selected frequencies. The so-called singing flames have been known for over 150 years, and the literature of the last century contains numerous papers on the subject. The effects occur for both diffusion-type and premixed flames, and appear to be associated with resonance between some vibration of the flame on its burner with that of a surrounding shield or enclosure.

For a diffusion flame, such as a jet of hydrogen or other combustible, burning on a narrow tube enclosed within a wider tube (which may be either open or closed at the bottom), it is found that a more or less pure note may be emitted for selected mass flows and positions of the jet within the outer tube. Toepler (1866) investigated these flames with a stroboscope and with his 'schlieren' apparatus, and his pictures can hardly be improved upon. Fig. 7.3 shows stroboscopic pictures of various phases in his singing diffusion flame of hydrogen. At a certain phase the flame begins to grow bigger and bigger and then to separate into two, a small flame remaining on the burner orifice, the outer part lifting and burning away. The next cycle starts with a new flame growing from that remaining at the orifice. Under some conditions vibrations may become so intense that the flame is blown off at once.

Lord Rayleigh (1896) gave the first satisfactory theory, and this has been confirmed experimentally by Richardson (1923). The vibrations may be started by a periodic heat supply of the flame, and the vibrations will be self-exciting if the heat supply occurs at a favourable moment in the cycle. The gas in the outer tube vibrates with its characteristic organ-pipe frequency and induces vibrations in the gas in the supply tube, which thus causes the issue of gas from the tube to be periodic. If the issue of gas is at its maximum during the moment of pressure-maximum in the outer tube, then the combustion of this extra supply of issuing gas will further raise the pressure and make the vibration self-exciting. The important role of the inner supply tube can be demonstrated by partly plugging it with cotton-wool to damp the vibration, causing the flame to stop singing. The occur-

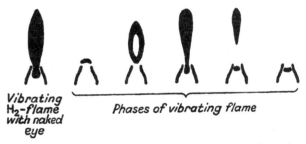

Fig. 7.3. Visual impression of vibrating hydrogen diffusion flame (on left), and five phases as seen with stroboscope.

rence of vibrations depends on the length of the supply tube and the coupling between it and the outer tube.

The outer tube carrying the main vibration acts like an organ pipe, or if the outer vessel is shaped more like a bottle with a neck then it behaves like a Helmholtz resonator. This main vibration produces forced vibrations in the supply tube; singing only occurs if the maximum issue of gas from the supply tube falls into a pressure maximum of the outer tube. The flame will therefore sing only when the outer tube is not in resonance with the supply tube. For an outer tube which is open at both ends, the flame will sing only when the orifice of the supply tube is not too far from the centre of the outer tube. If the orifice is near one of the open ends the flame will not sing, because at the open ends there are only velocity changes, not any pressure changes.

For premixed flames the conditions for the excitation of vibration are slightly different (Sander & Wolfhard, unpublished work) as the main heat release will not occur when the gas is issuing from the tube at its maximum rate, but will occur when the flame front has its greatest area. Plate 14 shows the vibrating flame without a stroboscope, and also gives a fivefold picture, obtained by running the stroboscope five times too fast. We see that the flame front rises first from a small cone which does not fill the orifice. Both the centre and the sides of the cone rise, so that the flame front takes on a rather hemispherical form. The flame then becomes compressed from the sides, so that these become nearly tube-shaped. This flame is then bound to collapse, as we have a vertical flame front propagating inwards from all sides. We are finally left with the small flame on the orifice, and the cycle is then repeated.

The effects can be explained in terms of two overlapping flow patterns: (i) a parabolic flow distribution of the gas out of the tube (this is only fully developed for a long tube) and (ii) a vertical vibration, which is of constant amplitude across the burner mouth. The result of these movements is that at the centre we always have a flow in the outward direction, though of periodically changing velocity, but at the rim of the burner the flow will be sometimes outwards and sometimes inwards. It is at the moment when the portion of the flame front near the rim is being sucked in, while the centre is still held by the gas flow, that the flame front collapses. It can be shown that the area of the flame front is not at its greatest when the outward mass flow is at its maximum. This occurs at a later phase in the cycle, at 240°. The flame sings when there is resonance between inner and outer tubes. Fig. 7.4 shows the relation between the vibrations at different phases.

Considering the vibration in the supply tube, the flame will be at its lowest point when the gas vibration has zero velocity, so that the pressure in the supply tube is at its maximum. The pressure in the supply tube is therefore 90° in advance of the velocity cycle. Let us now consider the pressure cycle in the outer tube which causes the forced vibrations in the supply tube. There are three possibilities: (i) the length l of the supply tube is less than one-quarter the wavelength, λ, for the outer tube; in this case, if damping is small, there will be no difference in phase between forced and exciting vibrations; from the study of the phase diagrams it can then be shown that the flame front has its maximum area some 240° after the point of maximum downward flow; the biggest heat release will then fall into a pressure minimum and the vibrations will be self-damping. (ii) If l exceeds $\lambda/4$, the pressure in the resonance tube will have a phase difference of 180° with the pressure in the supply tube. The biggest heat release will then fall just into a region of positive pressure in the outer tube, but the excitation of vibrations will not be very vigorous. (iii) If $l = \lambda/4$ we have resonance between outer tube and supply tube; the pressure in the former will then be 90° in advance of that in the latter, and the biggest heat release falls very nearly into the pressure maximum. There will thus be vigorous excitation of vibration if there is resonance, $l = \lambda/4$. If there is some damping by friction the phase relationships may be modified very slightly. These conclusions for a premixed flame may be contrasted with those for a diffusion flame in which resonance between outer and supply tube is not required. For premixed flames the exact calculation of the sound frequency and resonance condition is not easy as the temperature of the gas will of course affect the velocity of sound.

Vibrating flames may take some curious shapes. Sometimes the collapse of the flame front leaves a ball of gas separated from the rest of the unburnt gases. Bunsen flames show this effect quite well, Plate 14 showing such a ball in the act of being separated. Rich benzene flames often give soot deposition from the tip of the inner cone, and if such a flame is made to sing it may eject puffs of luminous carbon, and the flame may give the amusing impression of a machine-gun firing tracer bullets. Plate 14 shows such a flame by direct photography and through a stroboscope.

These rhythmic vibrations of the type encountered in singing flames are

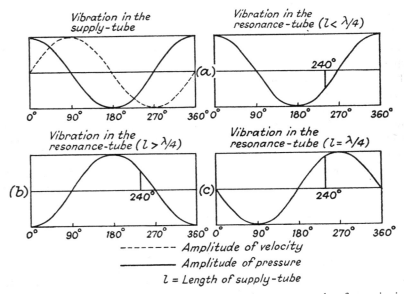

Fig. 7.4. Phase relationships between supply and resonator tubes for a singing premixed flame.

troublesome because they reduce flame stability and produce unpleasant noise. They can usually be avoided by plugging the supply tube; in some cases glass beads or lead shot may be used to partly fill the supply tube to detune it. In a few cases the vibrations have been made use of in flame studies; Eltenton (1947), in his early mass spectrometric studies, made use of the vibration of his low-pressure flames to scan the reaction zone; Graiff (1964) has made similar use of a singing flame as ionisation detector in gas chromatography, by employing A.C. amplification on the fluctuating signal.

Combustion chamber oscillations

Periodic vibrations in large combustion systems (rocket motors, gas-turbine reheat systems, etc.) are undesirable and even dangerous because of the intense sound, loss of flame stability and destruction of metal parts, including fatigue failure of metals. The subject is too technical for full discussion here, but a brief introduction to the principal causes of these vibrations is given. It is largely based on articles by Guderley (1938), Putnam & Dennis (1956), Bragg (1963), Tsuji & Takeno (1965) and Porter (1967).

Usually an enclosed combustion chamber will have characteristic acoustic frequencies which are determined by its dimensions and the velocity of sound through the burning gases. Acoustic vibrations may couple with and be driven by a varying heat-release rate. This will happen if the combustion processes are affected by the periodic changes in pressure or gas velocity and the changes in the combustion have an associated time factor which can resonate with the acoustic frequency. Generally the Rayleigh criterion that the varying rate of heat release must be in phase with the varying pressure will have to be satisfied.

Among the factors which may cause variation in heat-release rate, we may note the following,

(i) Direct effect of pressure on heat release; for premixed gases the burning velocity usually does not vary rapidly with pressure so that the heat release rate tends to be proportional to overall pressure; however for most practical systems the heat release is limited not by the rate of combustion reactions but by diffusion and mixing processes; in droplet and spray combustion Hall & Diederichsen (1953) have noted that there is a slight dependence of fuel consumption rate on pressure.

(ii) Change in flame area; this may be caused by varying supply rate of the combustion mixture under the influence of the pressure fluctuations or by flame movement relative to the tube walls.

(iii) Variation of fuel/air ratio; pressure fluctuations may affect the supply rate of fuel and air differently; fluctuation in air flow rate past a spray injector may alter the droplet concentration in the mixture.

(iv) Variation in mixing rate; increased turbulence due to gas motion may improve mixing; in some cases variation of angle of attack of a gas flow may alter vortex shedding from the rim of a flame holder and affect the mixing processes.

The time factors associated with these variations in combustion processes are more difficult to specify. They may include (i) the time for chemical reaction to occur, (ii) the time for the flame area to increase by flame propagation into new regions, (iii) time for droplet evaporation and ignition, (iv) time for an increased flow of fuel to build up following an increased pressure drop at the injector point, (v) time for a change in mixing conditions to become effective and (vi) time for establishment of vortices or turbulence.

Various types of acoustic oscillation may occur. The simplest is the lowest mode of a resonant cavity or the fundamental organ-pipe mode of a long combustion tube. Tsuji & Takeno found that the second and third harmonic modes as well as the fundamental could be excited. The alarming 'screaming' of rocket motors at around 1000 cycles per second often involves the first tangential or 'swashing mode'; in this the associated cross velocities increase the rate of mixing of propellant and oxidant and thus the rate of combustion, and pressure amplitudes may then exceed 25% of the chamber pressure, with disastrous results; pressure fluctuations of this magnitude propagate faster than normal sound and tend to steepen up to sharp-fronted shock waves.

Pulsations in motors are not always associated with acoustic resonance. Thus, according to Bragg, 'chugging' in liquid-propellant rocket engines occurs if the rate of supply of propellant depends on the pressure drop across the injector; a momentary reduction in chamber pressure results in a transient increase in propellant flow, which after a little delay due to mixing, evaporation and ignition processes results in a pressure increase at a time when the chamber pressure has already regained its mean value; this in turn causes a pressure overshoot followed by a flow decrease and thus a further transient pressure decrease in the combustion chamber. Pressure fluctuations of around 10% may result. Any resonance with an acoustic frequency tends, however, to amplify the effect. Chugging, which usually has a frequency of between 40 and 200 Hz, does not increase combustion efficiency, whereas screaming, although in other ways highly undesirable, does materially increase combustion rates. Many combustion oscillations can be cured by modifying the chamber dimensions so as to avoid resonance effects or by using high injection pressure across small inlets so that pressure fluctuations in the chamber have a relatively small effect on fuel flow. There is, of course, at least one infamous example of use being made of pulsating combustion in the V.1 Flying Bomb which was driven by a long (Schmidt) tube, open at the rear end and closed by flap valves at the front; the tube resonated in an organ-pipe mode with a pressure node at the front, so that air and fuel entered through the valves when the pressure was low, was ignited by residual hot gases and burnt at the pressure maximum, thus satisfying the Rayleigh criterion.

Flame flicker and other instabilities

The flickering of moderate-sized diffusion flames and candles is a familiar phenomenon. This flicker may under favourable conditions lead to quite strong low-frequency flame oscillations and some investigations have been made by Kimura (1965) and Toong *et al.* (1965). The oscillations start when the fuel flow exceeds some critical value and do not appear to depend on the length or diameter of the fuel supply pipe. They can be suppressed by putting the flame into an air stream flowing in the same direction as the flame gases; this at once suggests they are associated with shear forces between the fuel jet and the surrounding air. The oscillations are thus attributed to instability in the laminar fuel jet which may develop a sinusoidal oscillation and form periodic vortices. These cause the flame to flicker and when strong may result in tongues of flame periodically breaking off from the main flame or may result in the flame developing a neck and then perhaps breaking off above the neck. Toong *et al.* have discussed the jet instability at the boundary layer and interpreted the oscillations as due to the so-called travelling Tollmien-Schlichting disturbance waves. The frequency of the oscillations seems to be almost fixed, and Kimura quotes a frequency of 10 to 15 Hz for propane jets burning in air and around 18 Hz for city gas (containing 40% of primary air) in air. Interaction with any acoustic frequency of the burner or surrounds or with an externally applied sound wave may modify the flame oscillations; thus there is a connection with both sensitive flames and singing flames. Both the papers cited show some good photographs of flickering flames.

Further detailed studies of flicker in laminar diffusion flames using laser-Doppler anemometry to follow gas movements and special fast-response thermocouples to record temperature fluctuations (Durão & Whitelaw, 1974) have confirmed that for methane, hydrogen and towns' gas flames there is a basic frequency close to 11 Hz irrespective of burner dimensions or gas flow. The effects are again attributed to aerodynamic instabilities of the Tollmien–Schlichting type. The propagation velocity of the flicker up the flame corresponds to the bulk flow velocity. For these slow oscillations, around 11 Hz, the wavelength of sound is very long and resonance with natural frequencies of small burner tubes or combustion chambers is unlikely. Other recent papers on flame vibrations are by Deckker & Sampath (1971), Leyer & Manson (1971), Grant & Jones (1975) and H. Jones (1977).

The flicker of turbulent flames is, of course, different from that of the laminar flames discussed above, and depends on the character of the turbulent flow. However, as discussed in Chapter V, page 134, coherent vortices of large scale also exist. These flicker effects in large turbulent industrial flames, such as power-station boilers using oil or pulverised fuel, have been used to monitor whether a given flame in a multi-flame system is alight or not. Two telescopes view a point at the edge of a selected flame from a small angle, and the near infra-red light being received through the telescopes is detected by lead sulphide photocells, and electronic cross-correlation techniques are used to separate flicker at the point at the edge of the selected flame from the general background flicker (B. G. Gaydon, 1976; Noltingk, Robinson & Gaydon, 1975).

In Chapter II we discussed the instabilities leading to the formation of cellular and polyhedral flames. Behrens (1951b) and Markstein & Somers (1953) found experimentally that for flames in vessels closed at both ends or open at the upper end, vibrations tended to occur for those mixtures which also tended to show abnormal flame structure. The cellular or polyhedral structure of flames obviously affects the flame area and thus the rate of heat release, and also has associated with it a time factor – probably a time for diffusion – and flames frequently change rapidly from one cellular pattern to another. Thus there may be a mechanism for flame vibration which is connected with instabilities in the flame front. Kaskan (1953) has shown some good photographs of cellular structure developing in the front of flames which are starting to blow-back down a tube. Flames travelling in tubes do often oscillate and violent vibrations of the flame front frequently precede the initiation of detonation.

We have noted (page 24) that under some conditions the higher hydrocarbons, aldehydes and ethers show cool flame phenomena and sometimes successive cool flames cause oscillations in the temperature and pressure of the reacting mixture. The periodicity of these oscillations has been accounted for by specific chemical reactions in the case of acetaldehyde (Halstead *et al.*, 1971). Later these thermokinetic oscillations have been explained in more general terms by a mathematical model (Halstead *et al.*, 1975) based on degenerate branching chains involving two termination processes, one linear and the other quadratic in free radical concentration. Two routes are required for formation of the branching agent, one involving intermediate products of oxidation; the branched chain reactions must have different activation energies.

The phenomenon of 'knock' in internal-combustion engines is due to pre-ignition of the mixture ahead of the normal flame front, which should propagate out smoothly from the igniting spark. Knock is basically a thermal ignition due to the heating of the gas by the compression in the cylinder as the piston moves in, this heating being increased by the pressure rise due to the combustion, but it occurs most readily with those fuels which show cool flames and two-stage ignition (*see* page 25); chemical aspects have been discussed by Lewis & von Elbe (1961). Knock is important because it limits the maximum compression ratio which can usefully be employed in an engine. It makes the engine noisy, reduces its efficiency and greatly increases cylinder and piston wear. The resistance of a fuel to knock is expressed by its octane number; iso-octane is particularly good, whereas *n*-heptane is particularly bad and knocks readily; the fuel is compared with a mixture of octane and heptane and its rating expressed as the octane percentage in the matching mixture. Lead tetra-ethyl is found to suppress the ignition ahead of the flame front, which constitutes the knock, and is commercially added to automobile petroleum. The instability now generally known as knock was sometimes referred to in the older literature as detonation, but is not usually connected with the true detonations mentioned above.

Sensitive flames

Diffusion flames emanating from small burners may be very sensitive to sound. Normally, if shielded from all draughts, these flames will be very tall, but slight sounds, such as the tinkling of a bunch of keys, may cause them to 'duck'. These effects have been known for a very long time, and many burners which produce sensitive flames have been described (e.g. Rayleigh, 1882). Brown (1932; 1935) has been able to explain the main principles causing the phenomenon.

If the supply of coal-gas through a smooth burner of between 0·5 and 5 mm diameter is slowly increased we first have a laminar diffusion flame which increases in height. For a certain flow the flame will become turbulent, and it is just before the flame reaches this turbulent state that the flame is most sensitive. If sound is applied the flame seems to become prematurely turbulent and the rate of combustion increases and the flame becomes shorter. Even if the flame is already turbulent, however, some sensitivity remains; this may be due to a decrease in the height to turbulence (*see* page 160).

A flame is not sensitive to all sound frequencies; if we use pure notes and vary the frequency we find that the flame is sensitive to a few frequencies within a limited range, but not to other frequencies. If we use a rectangular burner to produce a sheet of flame, we also find that the sensitivity varies with the direction from which the sound comes, the flame height showing maxima and minima as the source is moved round. There seem to be a number of universal frequencies to which a flame tends to be sensitive, these frequencies being about 5850, 4600, 3300 and 2400 Hz (*see* Brown).

The frequencies to which a flame is sensitive are independent of mass flow, but the size and shape of the burner and the type of fuel determine in which of the possible ranges the greatest sensitivity will occur. The phenomenon of sensitive flames appears to be closely associated with flame flicker, discussed in the last section, and it would seem natural to expect a flame to be sensitive to frequencies which affect the rate of vortex shedding. If the supply tube of a sensitive flame is partly plugged it does not alter the sensitivity, in contrast to singing flames which are so altered.

It seems that the sensitivity has really nothing to do with the flame, but is just a property of the jet of gas, the flame merely serving to show visually the change in the state of the jet. This has been shown by studying jets of air containing smoke. These simple jets of air are sensitive to much lower frequencies than the flames, but where overlapping occurs it seems that air jets and coal-gas flames have the same sensitive frequencies.

Brown used a stroboscope coupled with the sound frequency to view the smoke jets. He observed waves travelling up the gas stream and found that at a certain point the jet forked. Near the frequencies of maximum sensitivity the amplitude of the waves increased and the forking point moved down towards the burner orifice. The observations also showed that jets which were already turbulent did not develop the wavy structure, whereas sensitive laminar jets, although near the point of turbulence, did develop distinct periodic sinuosities.

It must be borne in mind that jets of air, made visible with smoke, have different characteristics according to whether the orifice is circular or rectangular. The rectangular jet will tend to develop cyclindrical vortices as shown in Fig. 7.5, whereas the circular jet gives ring-shaped vortices. The smoke jet, unlike a flame, has no definite upper boundary. The sensitivity of the smoke jet is only seen to the naked eye as an opening-up of the jet in a fan-like manner. The angle of the fan varies from a few degrees to a maximum of 90°, and the structure as seen through a stroboscope is seen in Fig. 7.5. The flow remains straight-sided up to a certain height and then a disturbance appears first on one side and then on the other. These disturbances travel upwards to form incipient vortices, curling round to entrain air. The sensitivity begins at Reynolds numbers (referred to the diameter of the orifice) which are very much smaller than the Reynolds numbers required to give turbulence in the main gas stream.

Brown measured the angular velocity, ω, of the vortices caused by sound of frequency n, and found that

$$\omega = \pi n.$$

This means that the frequency of rotation of the vortices is the same as the sound frequency, indicating that the vortices are triggered by the sound

Fig. 7.5. Stroboscopic impression of sensitive air jet, made visible with smoke to show vortices in alternating positions.   [*Redrawn from Brown* (1935)]

impinging on the base of the jet. The fact that only certain sound frequencies can produce the vortices means that only vortices with certain angular velocities are physically possible. This statement is only true for vortices which are placed alternately, and not for symmetrically placed vortices. These vortex phenomena have to be clearly distinguished from turbulence, which has been treated mathematically by Tollmien (1926). There appear to be three stages in diffusion flames: (i) a laminar or non-turbulent flame, (ii) a semi-turbulent flame in which the turbulence is only induced when sound is applied; the entrainment of air then shortens the flame, and (iii) the fully turbulent flame, where the Reynolds number is high enough to produce turbulence in the absence of sound.

Zickendraht (1941) found that flames which are sensitive to sound are also sensitive to alternating electric fields. For this, the ionisation in the flame should be increased by the addition of Na or K. It was found by Dubois (1949; 1950) that hydrogen jets and flames are sensitive to very high sound frequencies, up to 70 000 and 400 000 Hz respectively.

The influence of sound on premixed flames

It is of interest to know whether sound waves change the normal flame velocity in flames of premixed gases. The subject has been discussed in two papers. In the first Hahnemann & Ehret (1943) used sound of about 5000 Hz frequency with an amplitude of up to 0·3 mm, corresponding to a sound intensity in the reaction zone of 12·5 watts cm^{-2}; it is interesting to note that the amplitude may in this case exceed the thickness of the reaction zone. In the second paper, by Loshaek, Fein & Olsen (1949), work with higher frequency (12 500 Hz), but very much lower amplitude, is described. Both investigations agree in concluding that if a stable flame can be maintained then the normal flame velocity is unchanged, but that the flame stability is affected, the region of conditions for a stable flame (i.e. stability region) being narrowed.

It is important to realise that the energy imparted to the flame by the sound is very small compared with the energy released in the combustion, and unless the sound can trigger some instability, such as the vortices in the sensitive flames, we should not expect the sound to produce any marked change. However, although the energy contribution is small, the amplitude of the movements due to sound is not so small, and there are some interesting changes in flame shape.

A propane/air flame burning on a nozzle gives a straight-sided cone, as previously discussed (page 65). If, however, the gas flow to the nozzle comes from a large reservoir containing an ultrasonic oscillator which applies a sound wave from below the reaction zone, we may study the effect of sound of varying amplitude on the flame. Plate 14 shows the flame without and with applied sound. Below an amplitude of 0·01 mm (for 5000 Hz) there is no effect on the flame. At higher intensity the centre of the reaction zone becomes blurred and the base of the cone becomes wider than the nozzle diameter. For still higher intensity the reaction zone becomes brush-like, similar in appearance to that of a turbulent flame. A further increase in intensity leads to a quiet flame of strange shape. The intensities and frequencies for which this form of flame develops depend somewhat on fuel concentration and mass flow, but the dependence is otherwise of a general character.

In this work of Hahnemann & Ehret the flame was shielded from draughts with a casing, and we think that the unstable intermediary shapes could be due in part to resonance vibrations between the casing and gas reservoir, as discussed previously. The flat flame in the fourth photo of Plate 14, is, however, obviously free from vibration of this type and is due solely to the effect of the ultrasonic vibration. Such flat flames are produced by sound of intensities between 0·14 and 1·5 watts cm^{-2}. Still more intense sound lifts and extinguishes the flame.

To understand these changes in flame shape, we must consider the interaction between three factors: (i) the initial flow pattern due to gas movement up the burner, (ii) the potential flow pattern due to the sound, and (iii) the ponderomotive effects of the sound (radiation pressure and sound wind). For a frequency of 5000 Hz and an amplitude of either 0·03 or 0·1 mm the intensity will be 0·14 or 1·5 watts cm^{-2} respectively, and the average velocity of the vibrating gas, $u = 12\pi v f \sqrt{5}$ (where v is frequency and f is amplitude), will be 253 or 844 cm/sec respectively. In the range covered by Hahnemann & Ehret's experiments, the gas flow is only in the range 40 to 200 cm/sec, so we may regard the gas flow as only a disturbance of the flow pattern due to the sound. The sound wind will only be important for still greater intensity.

We may consider an example of a flow pattern which illustrates this assumption. If a gas or liquid is suddenly set in motion through a nozzle, it does not at first emerge as a coherent jet, but tends to creep initially round the edge of the burner, a potential flow pattern being established to begin with. Later the momentum of the issuing gas will cause it to form a coherent jet, but a so-called starting vortex will be left behind (*see* Plate 13*e* and *f*). With applied sound the flow will be interrupted, and in its outward direction will never get beyond the potential flow state. The assumption that the general flow may be considered as a disturbance to a stronger sound vibration therefore leads to the conclusion that the gas flow is no longer in the form of a jet, but that the gas will flow outwards in all directions according to a potential flow pattern and the direction of flame propagation will always be directly opposed to the flow of the gas. Hahnemann & Ehret were able to support their interpretation by an ingenious electrical analogue which enabled them to compare the potential flow pattern of the nozzle and the flame position and to show that the flame front was indeed parallel to the equipotential lines. We think that this may not be the only example of a potential flow pattern; for very slow flows near limiting-mixture conditions when the Reynolds number is very small we think a similar effect may occur, with the direction of flame propagation opposite to the flow direction. The burning velocities measured by Hahnemann & Ehret for flames subjected to sound are the same as for the normal flame. There is appreciable scatter in the measurements of velocity, probably because the ponderomotive forces cause entrainment of air and thus alteration in the mixture strength.

Flame oscillations may also respond to applied alternating electric fields of suitable frequency. Some investigations (Babcock, Baker & Cattaneo, 1967; Wenaas & McChesney, 1970) have indicated that electric fields may be used either to enhance or suppress oscillations in flames and combustion chambers; a servo-mechanism involving a microphone to pick up the sound and an electrical oscillator was used.

Chapter VIII

Solid Carbon in Flames

Introduction

The formation of solid carbon particles in flames is of considerable industrial importance. Up till about 1952, very little fundamental work had, however, been done, such experimental facts as were available having been gathered in a very casual manner, while in industry the regulation of carbon and soot formation has been done empiricially. In recent years much systematic work has been carried out and a large number of valuable papers have appeared and although the lay-out of the sections is similar to that of previous editions, this fourth edition discusses fully the new information contained in these papers.

In automobile engines 'coking' due to deposition of hard carbon deposits on the piston head and in the cylinder is a familiar nuisance, now fairly well under control. Air pollution from motor and diesel exhaust gases has become more serious (Agnew, 1968; Burt & Thomas, 1968) and has led to recent legislation for the control of exhaust soot. In the combustion chambers of gas turbines the formation of solid carbon is undesirable because of the greater time taken subsequently to burn the particles and because of the danger of soot deposition on the turbine blades. In most open-hearth furnaces it is important to have a luminous flame, to give high heat transfer to the metal, but the deposition of soot will certainly not be tolerated. Knowledge of the mechanism of carbon formation in flames also appears to be important for the carbon-black industry. In this case the size as well as the quantity of carbon particles formed is important because it affects the physical properties of rubber when carbon black is incorporated.

In addition to the problem of the formation of carbon particles, we must also consider that of the combustion of the particles. In flames the relation between these two processes will determine whether or not soot is actually set free. The combustion of the particles is also of interest for the burning of coal dust, extensively used as pulverised fuel in power stations.

The difficulty in reading the older literature on carbon formation is that often no clear distinction was made between diffusion flames and premixed flames. The main features of the two types of combustion have been discussed in previous chapters. In diffusion flames, processes are allowed a fairly long time for their completion, but there is little or no contact between fuel and oxygen. In premixed flames fuel and oxygen are in direct contact, but the processes may be limited by the very short time of passage through the reaction zone. It may therefore be that the processes of formation and combustion of carbon in diffusion and premixed flames may be different, and we must be careful at all times to distinguish between effects in the two general types of flame.

It seems that Davy (*see* Bone & Townend, 1927) was one of the first to realise that carbon particles were the cause of the luminosity of flames. Clear flames show, of course, mainly banded emission spectra due to simple molecules, and some weak continuous radiation, due to chemical processes or recombination of ions, may also be emitted, but the typical emission from the ordinary yellowish luminous flames is due to carbon. The intensity distribution in the continuous spectrum of these flames is near that of a Planckian radiator. The evidence that the emission is due to solid particles of carbon seems quite conclusive. In some flames the particles actually appear at the top of the flame as soot. Even from flames which do not actually give soot, it is usually possible to obtain a deposit of carbon by passing a cold probe through the luminous zone, and the particles can then be examined either chemically or with the electron microscope. Also the particles scatter light in a way which obeys the Rayleigh law for small solid particles. Plate 15*a* shows the track of a strong focused beam of light from the anode of a carbon arc passing through the flame of a Hefner candle. The scattered light is quite strong and could only be given by solid particles. Sir George Stokes was apparently the first to show that the scattered radiation was polarised and therefore due to particles. Although carbonaceous particles are undoubtedly responsible for the main continuous radiation, there may be some contribution from banded emission by large carbon and hydrocarbon molecules for the region close to the reaction zone and especially for the 'mantle' region of oxy-acetylene flames; Echigo, Nishiwaki & Hirata (1967) have supported this view, although their infra-red observations are far from conclusive because in a flame of non-uniform temperature the CO_2 and H_2O emission and absorption distort the spectrum in a complex way. It has also recently been found by Müller-Dethlefs (unpublished) that this region of the flame shows fluorescence in a strong laser beam; the spectrum appears continuous, but must presumably be assigned to complex molecular species.

A number of possible mechanisms for the formation of soot have been suggested. There is probably no single all-embracing route by which organic compounds dehydrogenate and polymerise to form large carbonaceous particles, but considerable progress in understanding the processes has been made in recent years. Here we shall endeavour to sort out the experimental material first, then to state the various theories and to give our own views on the subject. The final section is a brief survey of the carbon-black process, which does not fit easily into the main development of the subject.

Diffusion flames

We are all familiar with a simple diffusion flame with its pale bluish base and luminous tip (e.g. Plate 1). Flames of this type, with luminous tips, are given by almost all organic gases and vapours; exceptions are carbon monoxide, methyl alcohol, formaldehyde, formic acid, carbon disulphide, and, rather surprisingly, cyanogen.* In some cases black soot is formed at the tip of the flame.

Very small diffusion flames on circular burners do not show carbon formation, but if the mass flow is increased, then at a fairly definite height the flame develops a luminous tip. The luminous region actually first appears at a point within the otherwise blue flame area and not at the tip. As the flow is increased the luminous part gets longer and longer until it extends beyond the boundary of the blue flame. With further increase in flow, sooting may start at the top. We are badly in need of some measure of the tendency of a fuel to form carbon, and although the height at which luminosity or soot formation commences may depend on some other factors, this appears the most convenient measure to take. Usually the height at which the flame first becomes luminous is inconveniently small, but the height of flame for which soot formation just begins is easy to measure. The height depends on the diameter of burner or type of wick used, but for a burner of fixed dimensions it is possible to compare the relative tendency to smoke for various fuels. Measurements of this type have been reported by Minchin (1935) and Clarke, Hunter & Garner (1946). The tendency to smoke is inversely related to the height h; the value of h may be recorded directly or the tendency to smoke may be expressed as a/h where a is a constant.

Measurements of the tendency to smoke for various homologous series, from Minchin, are given in Fig. 8.1. These are in general agreement with results by Clarke *et al.* For paraffins, the tendency to smoke increases with molecular weight, but the reverse is true for the olefine, di-olefine, benzene and naphthalene series. Methyl alcohol does not soot at all, and the tendency increases with molecular weight for primary alcohols. Secondary alcohols soot more easily than primary alcohols.

* Only very large flames of C_2N_2 give carbon.

The C/H ratio is obviously an important parameter, but not the only one. The structure of the molecule appears to be important; thus branched chain paraffins smoke more readily than the corresponding normal isomers, although the C/H ratio is of course the same. The more highly branched the paraffin, the greater is the tendency to smoke. The position of the branching does not seem to matter.

Unsaturation increases the tendency to smoke, but higher members of unsaturated series tend to get more paraffinic. Generally it seems that the tendency to smoke increases with compactness.

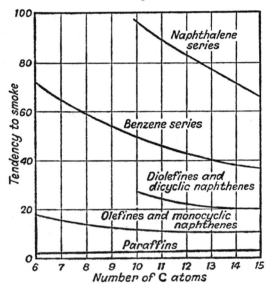

Fig. 8.1. Tendency of various fuels to smoke.
[*After Clarke, Hunter & Garner* (1946)

The luminosity of a flame is also related to the amount of carbon formation. It depends on the amount of carbon and on the temperature. For many hydrocarbons the maximum flame temperatures are about the same, and we may then find the luminosity a better measure of carbon formation than the tendency to smoke, because the formation of smoke requires the passage of the carbon out of the flame and thus also involves the competing oxidation processes. Coward & Woodhead (1949) measured the luminosity of liquid organic fuels in a miner's spirit lamp and found, in agreement with the previous results, that the structure of the molecule is important. Thus iso-octane gave a much brighter flame than *n*-octane. Benzene and toluene gave flames of low luminosity, probably because so much soot was liberated that the flame temperature was reduced. Alcohols gave a close relation between luminosity and tendency to smoke; for these the flame temperatures were all about the same. The following table shows the tendency to smoke, expressed as the height in centimetres and the candle-power for the alcohols studied by both Coward & Woodhead and by Clarke, Hunter & Garner.

Alcohol	Height to smoke point, cm	Candle power
Methyl	—	<0·01
Ethyl	37·7	<0·01
n-propyl	27·7	0·08
n-butyl	23·6	0·18
iso-butyl	18·4	0·30
sec-butyl	19·1	0·33
tert-butyl	10·6	0·62
n-hexyl	16·9	0·28
2-ethyl-butyl	12·8	0·42
1, 3-dimethyl-butyl	11·2	0·42
Allyl	8·3	0·83

We have already discussed in Chapter VI the structure of a diffusion flame in some detail, Fig. 6.2 on page 151 shows the region of the flame in which the carbon luminosity develops. The blue reaction zone forms at the very base of the flame, being separated from the burner only by a very thin, dead space. The carbon zone does not, however, develop until further up the flame, usually a distance of several millimetres. We have seen also that for flames with oxygen the fuel is decomposed in the preheating zone and there is no direct contact between undecomposed fuel and oxygen, the two being separated by a layer of combustion products. It is not clear at this stage whether the reason for the carbon only being formed after the gases have risen several millimetres is because the processes require time, or because the normal structure of the flame, with the wedge of combustion products separating fuel and oxygen, is only attained in the upper part of the flame; at the base there are no combustion products to separate fuel from oxygen, so some premixing occurs, the presence of the oxygen perhaps preventing carbon formation.

The luminosity of the carbon formed and its after history (i.e. liberation as smoke or consumption by oxygen) may be expected to depend to some extent on the position in which it is formed. For fuels such as methane which require a high temperature for their break-up, we may expect the carbon to be formed in a region of relatively high temperature and relatively near the oxygen side; in such cases the particles may be hot and luminous but fairly quickly consumed so that soot is not formed. For fuels like acetylene which decompose at lower temperature, we may expect the carbon particles to be formed in a cooler region more into the fuel side, where they will be cooler and more likely to form soot because of the lower concentration of oxygen-containing substances.

Although the tendency to carbon and soot formation increases with the gas flow for diffusion flames, if the flame becomes turbulent, then air is entrained at the base and less soot is formed. Powerful, lifted turbulent flames, even of acetylene, do not form any free soot. Many industrial flames are turbulent, and this turbulence no doubt helps to reduce carbon formation.

It has been shown, page 170, that large oil droplets usually burn with a surrounding diffusion flame, and burning oil sprays are usually brightly luminous because of the carbon formation in the diffusion flames surrounding the larger droplets. The extent of carbon luminosity and soot formation in oil sprays depends on droplet size, better atomisation reducing carbon, and on the relative velocity of the oil droplets and the air. Sjögren (1973) has found that there is a critical 'extinction' velocity above which the diffusion flames surrounding each droplet are blown out; in pure air this extinction velocity is of the order of 100 cm/sec, but it depends on the oxygen content of the surrounding atmosphere and dilution with carbon dioxide is quite effective in lowering the extinction velocity and so reducing carbon formation; this agrees with observations that recirculation of flue gases reduces soot and luminosity. Similarly in stirred reactors of the Longwell type (*see* page 128) recirculation of product gases reduces soot by up to 90%, although the critical mixture ratio for onset of soot formation is less affected, by about 10 to 20% (Wright, 1970).

Premixed flames

For any given fuel and mixture strength there is a rather limited range of burner size for which a stable non-turbulent flame can be obtained, so that comparison of a range of fuels or mixture strengths on a fixed burner is difficult. The only true results are obviously those in which premixed gases burn in a surrounding atmosphere of their own burnt products. In the ordinary small laboratory flames of the Bunsen type, diffusion of surrounding air into the flame produces a very much greater effect than is commonly realised. Premixed oxygen/fuel flames of very rich mixtures burning in open air often show a fairly luminous 'mantle' just above the inner cone; this may extend for a millimetre or so and has a clearly defined outer edge. If the flame is then enclosed to exclude air, the mantle becomes very much thicker, extending quite high into the flame and fading out gradually.

Generally in premixed flames we should not expect carbon formation for mixtures in which there is more than sufficient oxygen to burn the carbon to carbon monoxide, while if the total oxygen is insufficient to convert all the carbon to CO then carbon should be thrown out. These equilibria will be discussed later in more detail. In actual flames, the time factor and the freezing out of the equilibrium may be important. For flames in open air, the interdiffusion of air may alter the mixture strength and prevent carbon formation, or consume particles already formed, so that the carbon will only appear if it is formed quickly, near the reaction zone.

Measurements of the onset of carbon formation in premixed flames, in which air entrainment was prevented by an outer jacket, have been made by Street & Thomas (1955), and in most cases carbon luminosity occurs at mixtures which are much less rich than those required to liberate free carbon under equilibrium conditions. Results for a selection of flames with air are given in Table 8.1. Only for acetylene is the observed carbon point anywhere near that expected for equilibrium.

TABLE 8.1

Onset of carbon luminosity in selected premixed fuel/air flames (from Street and Thomas)

Fuel	Mean critical air/fuel ratio by weight	λ, mixture strength. Actual air as fraction of stoichiometric	λ for O/C=1, i.e. theoretical carbon point	O/C at actual carbon point
Ethane	9.7	0.60	0.285	2.1
Propane	10.1	0.64	0.300	2.1
n-pentane	10.4	0.68	0.313	2.2
n-octane	10.8	0.72	0.320	2.2
Acetylene	6.4	0.48	0.40	1.20
Ethylene	8.1	0.55	0.33	1.67
Benzene	9.3	0.70	0.40	1.75
Ethyl alcohol	6.0	0.66	0.167	2.5
Acetaldehyde	4.3	0.55	0.20	1.9
Diethyl ether	6.4	0.58	0.25	2.0

These measurements are a considerable improvement on previous studies of premixed flames, as effects of air entrainment are avoided, but there is probably still some dependence on burner dimensions and flow rates, and selective diffusion effects of the type producing polyhedral flames could still occur. It is, however, very clear that carbon formation occurs more readily than might have been expected, and varies from fuel to fuel. On an air/fuel ratio basis, the tendency to form carbon increases in the order: aldehydes; ketones; ethers; alcohols; acetylene; light aromatic compounds; olefins; isoparaffins; paraffins, heavier monocyclic aromatic compounds and then naphthalene derivatives. For the heavier compounds, the air/fuel ratio approaches a value of about 11. If the critical carbon formation point is expressed by its mixture strength, λ, the order is modified to: acetylene; aldehydes, ketones and ethers; olefins; isoparaffins; alcohols; monocyclic aromatic hydrocarbons and anilines; naphthalene derivatives.

Further work on the onset of carbon formation has been done by Fenimore, Jones & Moore (1957), using an inverted porous plate as burner. This gave a flat flame, and carbon formation was estimated by the onset of actual soot formation. Results again indicate that carbon often occurs with O/C ratios in the range 1·5 to 2·0 instead of at the expected 1·0. It may be noted that for many fuels (e.g. methane, ethane, propane, butane) carbon formation is observed well before the rich limit, although in equilibrium carbon should not be formed even at the rich limit. Using an Egerton-Powling type flat flame, Dr. G. N. Spokes found that some fuels (dimethyl ether, diethyl ether, acetaldehyde) have a range of mixture strength in which the carbon luminosity appears, but that carbon is not formed for very rich mixtures near the limit.

In flat fuel-rich premixed flames there is usually a blue reaction zone close to the burner surface, with a relatively dark zone, about as thick as the blue zone, above the latter, and then the thick yellow carbon zone. Sampling of gases from the dark zone shows that some acetylene, methane and traces of other hydrocarbons are present, as well as water, hydrogen and carbon monoxide. Fenimore, Jones & Moore (1957) have attempted to correlate the subsequent formation of the soot with the gas composition in this dark zone. From sampling they find empirically that carbon begins to appear when

$$\frac{(2P_{C_2H_2} + P_{CH_4})(P_{H_2})^{\frac{1}{2}}}{P_{H_2O}} \equiv R = 0{\cdot}6, \tag{8.1}$$

where pressures are expressed in cmHg; this expression holds fairly well for many fuels, and even for the addition of various additives, but for flames containing benzene it is necessary to add an extra term of $100 P_{C_6H_6}$ to the first bracket in (8.1). This empirical treatment is likely to be superseded by the more fundamental and quantitative approach used by Jensen (1974) which is discussed later.

In small premixed flames on burners, development of cellular structure or polyhedral flames greatly affects the position at which carbon appears (*see* pages 38 and 151; also the excellent review by Homann, 1967) and effects of selective diffusion will modify the local gas composition and stoichiometry so that the condition for soot formation at a point in the flame is not simply related to the initial mixture composition. Homann has followed Behren's grouping of flames into two types, an acetylene type in which carbon forms throughout the whole outer cone, and a benzene type in which it forms at the tip of the inner cone (*see also* page 34).

The effect of pressure

Generally speaking, high pressure seems to favour soot formation and low pressure to reduce it. However, the apparent decrease in soot at reduced pressure may really be due to change in the flame caused by the greater wall quenching at low pressure, and it is necessary to examine the observations rather critically.

For premixed flames, Smith (1940a) made some early studies of carbon formation and of flame spectra over a fairly wide range of pressure. For ethylene, he found that above atmospheric pressure there was a marked increase in soot and decrease in the C_2 and CH radiation. At reduced pressure, there was less sooting and at some mixture strengths the C_2 bands became very dominant giving the so-called 'green flames'. For the high-pressure work Smith used either explosions in a 'bomb' or impinging jets and at low pressure either a travelling flame or a stationary flame held in a quartz tube; thus although a fair pressure range was examined conditions over this range were not strictly comparable and wall quenching no doubt became increasingly important at low pressure. Many years ago we ourselves studied low-pressure flames of acetylene down to about 1/100 atm, taking care to keep the burner diameter inversely proportional to the pressure; we found very little variation with pressure of the critical mixture strength at which carbon luminosity just became visible.

Fenimore et al. (1957) studied a number of hydrocarbons at reduced pressure with a cooled inverted porous-plate burner.* Although their results (Fig. 8.2) appear to indicate that the threshold for carbon formation moves to lower oxygen/carbon ratios at lower pressure, comparison of the results with the 1·6 and 3·3 cm burners shows that there is a strong effect of burner quenching; for a meaningful comparison both burner diameter and flow rate should be increased at low pressure.

Bonne & Wagner (1965) used a large burner, of 19 cm diameter, keeping the linear gas velocity constant and studying the onset of carbon formation on the burner axis; they found that it was almost independent of pressure. Macfarlane, Holderness & Whitcher (1964) studied various hydrocarbons (pentane, hexane, benzene, etc.) up to 20 atm using turbulent and laminar flames. Their graphs again indicate that there was practically no change in the threshold mixture strength at which soot appeared. Their measurements also showed that heat loss to the burner depended on pressure and that changes in flame temperature could affect the carbon formation.

It seems that for premixed flames the threshold for the first appearance of soot does not in fact vary much with pressure. However, once carbon is set free the proportion of fuel so converted does seem to increase with increasing pressure.

Carbon formation in diffusion flames at reduced pressure was studied by Parker & Wolfhard (1950); the burner consisted of two concentric tubes,

* i.e. a downward flow of the gas mixture.

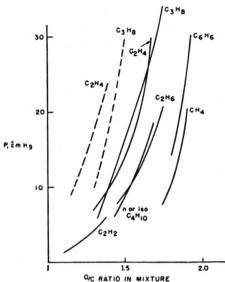

Fig. 8.2. Pressure at which soot is just formed in various hydrocarbon/oxygen flames. Solid curves for 3·3 cm burner; dashed curves for 1·6 cm burner. Flow (reduced to S.T.P.) 25 cc/sec for both burners.

[*Fenimore, Jones & Moore* (1957)

and the mass flows of fuel and air or oxygen were adjusted so that the velocities of fuel and air or oxygen were the same. At very low pressure no carbon formation was observed at all (*see* Fig. 6.2, page 151), but C_2 and CH radiation was strong. As the pressure was raised a luminous spot first appeared near the centre of the flame, this increasing in size as the pressure was further raised. In these experiments the mass flow was kept constant, and it should be realised that under these conditions the flame height should be independent of pressure because the change in gas velocity with pressure is exactly balanced by the change in diffusion coefficient. The height of the luminous carbon zone, was, however, very sensitive to pressure. This may be partially explained by the fact that as we go to lower pressure the decreased rate of reaction allows more time for diffusion and causes the flame to approach premixed conditions, and oxidation takes place throughout the body of the flame instead of only in a thin region on the surface of the flame. This is not, however, the complete explanation. If it was only necessary to establish the diffusion-flame type reaction to shield the carbon zone from oxygen we should expect that the low-pressure limit for carbon formation would be inversely proportional to burner diameter. Actually, increasing the burner diameter does lower the pressure limit, but only slightly. Also the pressure for which luminosity begins for a certain burner is almost independent of mass flow. We must therefore conclude that for diffusion flames there is a real decrease in carbon formation with decreasing pressure. The soot also becomes noticeably denser with increasing pressure, as in premixed flames.

The effect of temperature

Change of flame temperature has a complex effect on soot formation. Near the threshold, higher temperature appears to suppress soot formation, whereas for very rich flames it may increase the amount of soot deposited.

When the gases of a premixed flame are preheated to around 400°C the final flame temperature is increased by around 160°C and Street & Thomas (1955) found that this shifted the threshold for carbon formation to slightly richer mixtures. In large flat premixed ethylene/air flames on a porous burner Millikan (1962) found that increasing the flow rate raised the flame temperature by up to 150° and again shifted the threshold value to a higher carbon/oxygen ratio. In experiments on premixed flames at 15 atm pressure Macfarlane *et al.* (1964) found that replacing the nitrogen in the air by argon raised the temperature by up to 400°; near the soot threshold this reduced carbon formation, but at richer mixtures it produced a very marked increase in soot deposition. However, their observation that raising the flame temperature by installing a radiation shield increased the soot formation at all mixture strengths appears inconsistent with the argon experiment and emphasises the complexity of the problem. The subject has been discussed by Homann (1967) and Homann & Wagner (1968).

Near the threshold, higher temperature probably decreases soot formation because it has a greater effect on the competing oxidation process than on the soot formation process; Millikan deduced a difference of 34 ± 10 kcal/mole in the activation energies, the oxidation being favoured at the higher temperature. In very rich mixtures, however, the competing oxidation will be less important and the high temperature will accelerate the now dominant pyrolysis reactions.

Explosion flames

Many of the older observations on carbon formation were made with explosion flames, analysis of the products being made after the completion of all reaction. In diffusion flames we saw that the reactions tend to have ample time, but are influenced by subsequent combustion of the carbon. In stationary premixed flames, the processes may be dominated by the short time available in the reaction zone. In explosions the conditions may be nearer to those for premixed flames, but since the gases are shielded from surrounding air and remain hot for an appreciable time, we might expect a better chance for the products to reach true equilibrium. In closed vessel explosions the much higher final pressure and the heating of the mixture by adiabatic compression before the initiation of the reactions in the flame front may influence the composition of the product gases.

Bone & Townend (1927) made some analyses of the products from a number of mixtures exploded at high pressure. The experiments fall into series in which certain parameters like the C/H or C/O ratios were kept constant. They were mainly concerned with disproving the theory of preferential combustion of hydrogen or carbon but the experiments are very relevant to the problem of carbon formation.

With methane, the mixture $2CH_4 + O_2$ at around 10 atm, which we might expect to burn mainly to $2CO + 4H_2$, actually gave one-fifth of the carbon as free soot, and incidentally also showed a fair amount of water in the products, with some CH_4 surviving. With a slight increase in the oxygen content of the mixture to $60\% \ CH_4 + 40\% \ O_2$, no carbon was formed and the amount of water was, surprisingly, found to decrease, the formation of more CO apparently reducing the amount of oxygen available to form water. Townend (1927) found that for $2CH_4 + O_2$ and $5CH_4 + 2O_2$ at very high pressure, up to 100 atm, the amount of soot formation became very small, although it was considerable at lower pressure.

For $C_2H_6 + O_2$, which would be expected to yield $2CO + 3H_2$, it was found, at atmospheric pressure, that soot was again formed in appreciable quantity, with some water and 10% of CH_4 in the products. To test the influence of surface effects, the experiments were repeated in a large spherical vessel and in a long tube. The spherical vessel gave plenty of soot, while the tube gave very little. This apparently showed that the carbon formation is not due to surface suppression of oxidation but rather the reverse, the surface checking the pyrolysis. For these ethane explosions the carbon formation was reduced from 27% at 1 atm to only $3\cdot4\%$ at 25 atm initial pressure. A detonation through the mixture yielded about 3% of the carbon as soot.

On the other hand for acetylene, ethylene and higher olefines mixed with oxygen in the proportion just to burn to CO and H_2, explosions at initial pressures of a little below 1 atm did not give any solid carbon. Experiments do not appear to have been carried out at higher pressure.

The results of these closed-vessel explosions are thus quite surprising, the saturated hydrocarbons methane and ethane apparently showing a greater tendency to form carbon than do acetylene and olefines. It is probably desirable to repeat these important experiments. In explosions the time factor may be such that once free carbon is formed it cannot be removed by reaction with water sufficiently quickly, so that its formation may be regarded as a competing process with oxidation. If carbon is formed it leaves some oxygen over to react with hydrogen, forming water which cannot then react back with the carbon before the fall in temperature freezes the equilibrium. Surface effects, e.g. condensation of water, may also be important in these closed-vessel explosions. With methane it may be that in the competition between carbon formation and oxidation, the high-ignition temperature of methane checks the oxidation and therefore favours carbon formation.

Although cyanogen will not give any carbon formation in either diffusion or premixed flames of normal size, we have obtained soot quite readily from explosions of cyanogen/oxygen mixtures in open-ended tubes; carbon is also set free in detonations of rich cyanogen/oxygen mixtures (Munday, Ubbelohde & Wood, 1968).

The effect of non-metallic additives

In considering the effect of additives, it is necessary again to distinguish between diffusion and premixed flames. This was not always done in the early literature and there was some confusion and apparently contradictory statements were made.

The most outstanding case of a small amount of a non-metal having a marked effect on soot formation is that of sulphur trioxide. It is known (Gaydon & Whittingham, 1947) that the addition of as little as 0·1% of SO_3 to the air supply of a fully aerated Bunsen, burning town's gas, will cause the flame to show strong carbon luminosity. This is a very striking effect and clearly shows that carbon liberation in premixed flames is not solely determined by equilibrium considerations; SO_3 is a strong oxidiser and its addition will increase the O/C ratio. Fenimore, Jones & Moore (1957) found that 0·2 mole % of SO_3 added to isobutane/air increased carbon by 40% but that in the dark zone below the carbon luminosity the gas composition was changed so that the relationship of equation (8.1) (page 202) was still satisfied at the carbon threshold. For diffusion flames, on the other hand (Wolfhard & Parker, 1950), SO_3 has little effect, tending if anything to reduce carbon formation.

Both SO_2 and H_2S, unlike SO_3, tend to decrease carbon formation in premixed flames, although a fair amount of these additives, of the order 4% is required. These additives also tend to decrease soot in diffusion flames.

If either CO or H_2 is added to a premixed flame which is just at the point of carbon formation, then we have found that carbon formation increases; the addition, of course, makes the mixture richer, but if we assume carbon formation should set in when there is just too much carbon for reaction to CO and H_2 only, then further addition of CO or H_2 would not affect the equilibrium. If at the threshold point, some of the fuel (C_2H_2) is replaced by H_2 or CO in the correct amount to keep the mixture strength the same, then the carbon formation is decreased, the decrease being more marked for H_2 replacement than for CO replacement. Nitrogen dilution tends slightly to increase carbon formation in flames of benzene or kerosene with air (Street & Thomas, 1955); this is probably a temperature effect.

Our own observations and those of Street & Thomas show that the majority of substances, when added to premixed flames in small quantities, have little effect, although there may be minor changes in the intensity of the carbon luminosity or of its size or position. Diethyl peroxide and methyl ethyl ketone produce hardly any effect. NO and NO_2 tend to reduce carbon formation if anything, but amyl nitrate increases it a little. Halogens or halogenated compounds (e.g. chlorine and methyl bromide) make the inner cone of the flame much greener, due to stronger C_2 emission, but there is only a slight increase in carbon luminosity.

In diffusion flames there seems a general tendency for inactive diluents to reduce carbon formation if added in sufficient quantity. Thus Arthur (1950) reports that the addition of 45% CO_2 will stop carbon formation in a diffusion flame of methane, this amount being independent of flame size. The effect is not due to lowering of temperature, as the amount of CO_2 required was compensated by preheating the gases. With CO_2 addition the equilibrium between solid carbon, CO and CO_2 will, however, be altered. Nitrogen also reduces carbon formation in diffusion flames, in contrast to its effect in premixed flames. Dilution with an inert gas may be equivalent to a reduction in pressure and we have seen that this does reduce carbon in diffusion flames. Carbon formation may also be reduced on an industrial scale by recirculating part of the flame gases through the fuel bed. Similar effects on oil sprays and in stirred reactors have already been noted (page 200).

Halogens and halogenated compounds (CH_3Br, CCl_4, $POCl_3$, etc.) tend to increase carbon formation in diffusion flames. These substances are all flame inhibitors and will in some circumstances lead to complete extinction of diffusion flames. Ibiricu & Gaydon (1964) have studied the effect of these compounds on counter-flow diffusion flames of methane and ethylene, and have also studied the change in intensity of the band spectra of OH, C_2, CH and HCO; OH emission is reduced while the C_2 radiation and carbon luminosity is increased on adding the halogen. These observations can be explained by assuming that the inhibitor removes excess hydroxyl radicals, by reactions of the type

$$OH + HBr = H_2O + Br,$$

and this slows the competing oxidation processes and enables pyrolysis, polymerisation and carbon formation to proceed better.

Durie (1952b), working with one of us, has studied flames of organic compounds burning with fluorine, in the absence of air or oxygen. These flames are of the diffusion type, but are so small that they probably correspond more nearly to the premixed type. Carbon formation and emission of the C_2 Swan bands occur in much the same way as for combustion with oxygen. Methane again gives carbon formation, but methyl alcohol does not. All the organic halides also give carbon formation with F_2 except CCl_4, which is the only one of those studied which does not contain any hydrogen. This gives a pale non-luminous flame, but if hydrogen in fair amount is added, then the flame of CCl_4 also becomes luminous due to carbon. Further addition of H_2 causes the flame to turn bright green with strong C_2 emission. The interpretation of this result is not clear, because the addition of hydrogen will affect both the equilibrium composition and the flame temperature. It may, however, be of some significance that CCl_4, which like C_2N_2 does not contain hydrogen, does not readily give carbon, whereas $CHCl_3$ does give carbon.

The effect of metallic additives

Many easily ionised metals when introduced into flames, both diffusion and premixed, have an important effect on soot formation. Alkaline–earth metals, especially barium, and alkali metals are most effective but some others such as molybdenum and tungsten also influence the amount of soot formed. This subject was briefly referred to in the third edition of this book, but has now become of increased importance both for the understanding of soot formation processes and in the commercial control of smoke emission from Diesel engines.

The effects are quite complicated and addition of salts of most easily ionised metals may either decrease or increase soot formation depending on the circumstances. Bartholomé & Sachse (1949) noted that addition of alkaline–earth salts (and also nickel) reduced or stopped carbon formation in premixed methane/air flames, and Arthur (1950) found that sodium chloride reduced the luminosity of methane diffusion flames. Among more recent papers we note those by Addecott & Nutt (1969) who studied counter-flow diffusion flames and reversed rich premixed flames, Cotton, Friswell & Jenkins (1971) who used a propane diffusion flame, and Bulewicz, Evans & Padley (1975) who used oxy-propane and oxy-acetylene flames. Usually salts of alkaline–earth and alkali metals reduce soot formation but Cotton et al. and Bulewicz et al. found that in very low concentrations these additives had a pro-soot action, the anti-soot effect only occurring with higher concentration of additive. Also in very rich mixtures these additives tend to have a pro-soot action at all concentrations. Barium salts are the most potent of this type of additive but Sr, Ca, Cs, Mo and W have a strong effect.

Salooja (1972; 1973) used a wire probe coated with various additives to study both premixed and diffusion flames and found that the influence on soot formation depended on where in the flame the probe was inserted. When inserted into the base of the flames it tended to have an anti-soot action but if inserted higher up the flame it could increase soot formation. Salooja also noted that addition of two metals in some cases produced a greater effect than the sum of the separate additions of these metals.

Explanations for these complex effects have been linked with various theories of the process of soot formation. Early views were that these additives reduced the concentration of OH radicals in the flame, and this role of OH was supported by the work of Cotton *et al.* Addecott & Nutt (1969), Mayo & Weinberg (1970) and Bulewicz *et al.* (1975) produced evidence for an ionic mechanism. At first it was thought that small positive ions served as nuclei for soot formation. Howard (1969) discussed the electrical forces between charged and neutral soot particles and their effect on the rate of agglomeration of the particles. Bowser & Weinberg (1974) have now explained most of the phenomena associated with the introduction of salts of these easily ionised metal additives as due to changes in the rate of particle agglomeration. When only a few particles are positively charged these attract neutral particles by a dipole effect and agglomerate with the neutral particles, but when most of the particles become positively charged then mutual repulsion between particles reduces the rate of agglomeration. Thus there tends to be a maximum agglomeration rate when about half of the particles are charged. When metal additives are present the concentration of free electrons is increased and since these have a high diffusion rate they tend to neutralise the predominantly positive charge on carbon particles. By applying an electric potential to the probe Bowser & Weinberg were able to confirm that the influence of additives on soot formation was associated with the emission of electrons from the coated probe. Detailed interpretation of the effects is complicated by the abnormally high ionisation naturally present in organic flames (*see* Chapter XIII).

Work on the effect of barium salts and similar additives on combustion in Diesel engines has been reported by Pegg & Ramsden (1966) and Golothan (1967). It seems that a useful reduction in smoke emission can be achieved and also barium addition may alternatively have a beneficial effect on engine power by improving the completeness of combustion.

Electrical effects

Soot formation and carbon luminosity in flames may be changed very markedly by applied electrical fields, and Prof. Weinberg and colleagues have made a most valuable study of these effects (Payne & Weinberg, 1959; Place & Weinberg, 1966; 1967; Weinberg, 1968; Lawton &

PLATE 3

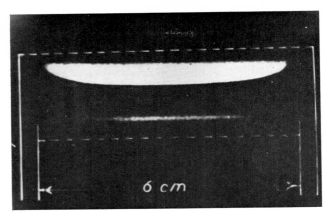

(a) Rich ether/air flame on rectified-flow burner, showing thin cool flame well below main flame front. The broken white lines indicate position of burner rim and stabilising grid. (*Thabet* 1951)

(b) The entrainment of air into a town's gas diffusion flame, as revealed by MgO particles. (*Simmonds and Wilson* 1951)

PLATE 4

(a) Polyhedral flame of rich benzene/air. Note carbon formation at tip. (*From Behrens*)

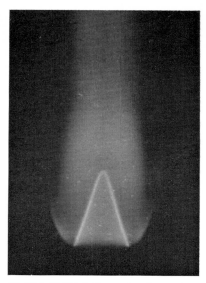

(b) Rich C_2H_2/air flame showing carbon formation beginning at outer edge. (*From Behrens*)

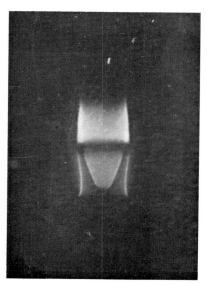

(c) Partially inverted flame of benzene/air showing carbon no longer at tip. (*From Behrens*)

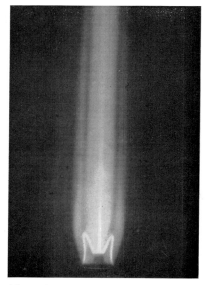

(d) Partially inverted flame of C_2H_2/air showing carbon now at tip. (*From Behrens*)

PLATE 5

(a) Flame threads in lean premixed $H_2/O_2/CO_2$ on burner. (*By Behrens*)

(b) Rich propane/air on flat-flame burner, showing cellular structure and carbon luminosity. (*Thabet* 1951)

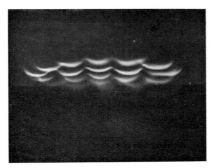

(c) As (b) but with air deficient in O_2. (*Thabet* 1951)

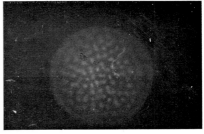

(d) $H_2/O_2/CO_2$ mixture burning above sintered glass disc; lean mixture showing cell structure. (*By Behrens*)

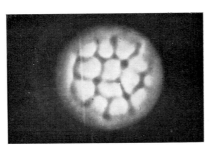

(e) Similar to (d) but lean $CH_4/H_2/$air. (*By Behrens*)

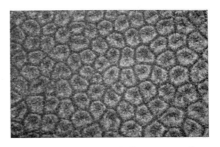

(f) Cellular structure in heat convection of liquid containing Al flakes. (*From Siedentopf after Prandtl*)

PLATE 6

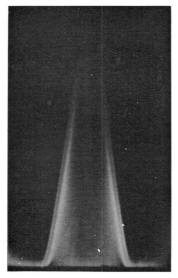

(a) Superposed direct and schlieren pictures, showing schlieren inside luminous cone.

(b) Particle tracks in inverted-cone flame, showing effect of back pressure. (*From Lewis and von Elbe*)

(c) Particle tracks through the inner cone of a flame. (*From Lewis and von Elbe*)

(d) Enlargement of right-hand side of (c).

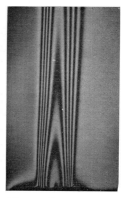

(e) Interferometer picture of a candle flame. (*From Hübner*)

PLATE 7

(b) Direct photograph of turbulent flat flame. (*From Snyder* 1962)

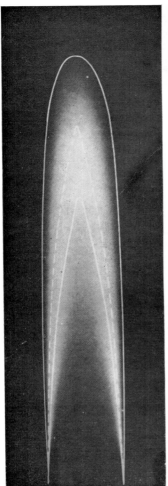

(a) Direct photograph of turbulent premixed flame. White lines indicate the inner and outer limits of the flame brush. (*From Bollinger and Williams* 1949)

(c) Instantaneous schlieren photograph of the top of an axially symmetrical turbulent flame of stoichiometric propane-air, showing 'cusp' formation in the wrinkled flame surface. (*From Fox and Weinberg* 1962)

PLATE 8

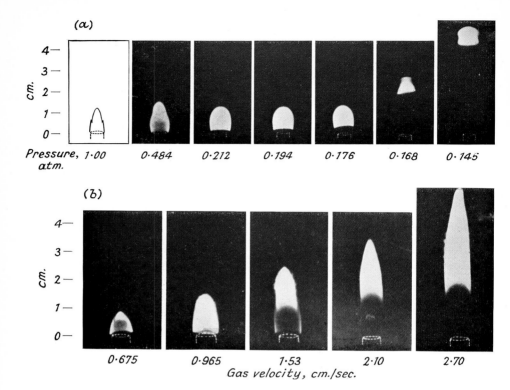

(a) Photographs of diffusion flame (*Barr and Mullins*) showing effect of pressure on flame height. Air velocity × pressure constant at 9·4 cm. atm./sec. Fuel velocity × pressure = 1·18 cm. atm./sec. For 1 atm., a diagram from another series is substituted.

(b) Diffusion flames (*from Barr and Mullins*) burning in air with reduced oxygen content (18%), showing effect of fuel velocity.

PLATE 9

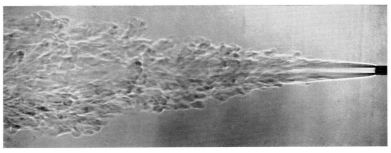

(d) Schlieren of (c).

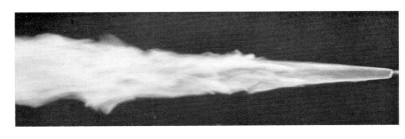

(c) Turbulent diffusion flame of city gas.

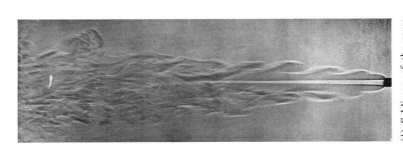

(b) Schlieren of the same flame as (a) showing long laminar fuel jet in centre.

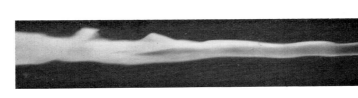

(a) Direct photograph of laminar city-gas diffusion flame. (*From Wohl*)

PLATE 10

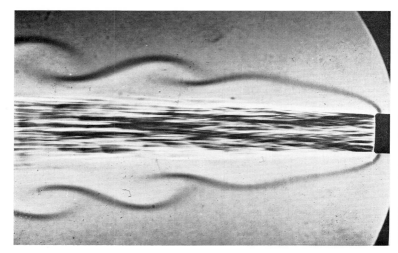

(b) Schlieren photograph of same flame taken with longer exposure (arc source); the streaks show that the turbulence balls retain their identity for an appreciable time.

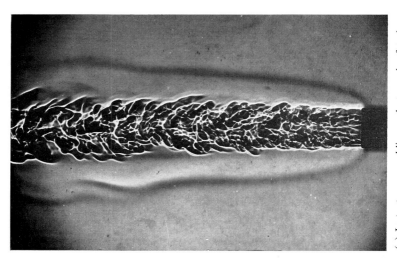

(a) Instantaneous schlieren photograph of turbulent butane diffusion flame showing turbulence balls. (*From Wohl*)

PLATE 11

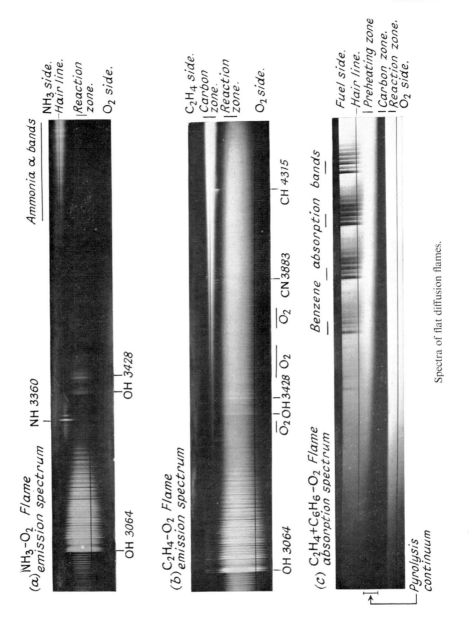

Spectra of flat diffusion flames.

PLATE 12

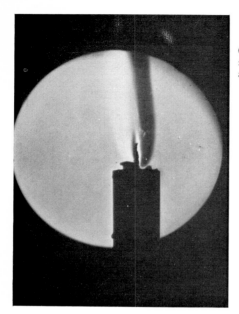

(*a*) Schlieren picture of candle flame, showing ascending hydrocarbon vapour around wick.

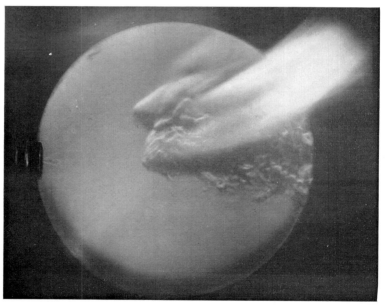

(*b*) Schlieren picture of burning kerosene spray. The spray is from left to right, with the luminous tail of the flame showing, in its own light, on the right. The schlieren show a continuous flame front.

PLATE 13

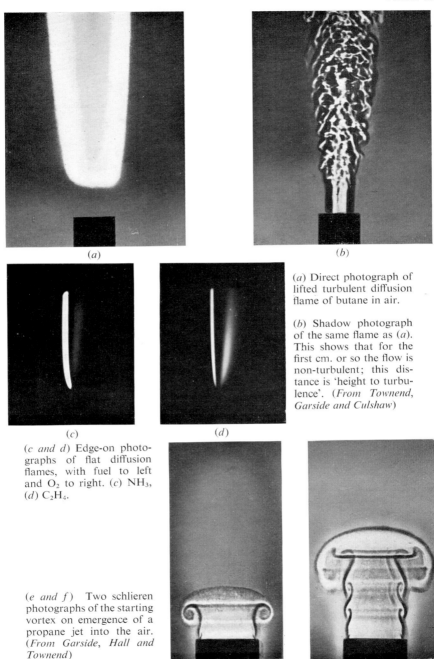

(a) Direct photograph of lifted turbulent diffusion flame of butane in air.

(b) Shadow photograph of the same flame as (a). This shows that for the first cm. or so the flow is non-turbulent; this distance is 'height to turbulence'. (*From Townend, Garside and Culshaw*)

(c and d) Edge-on photographs of flat diffusion flames, with fuel to left and O_2 to right. (c) NH_3, (d) C_2H_4.

(e and f) Two schlieren photographs of the starting vortex on emergence of a propane jet into the air. (*From Garside, Hall and Townend*)

PLATE 14

(*a to d*) Effect of increasing sound intensity on shape of a premixed flame. See p. 193.

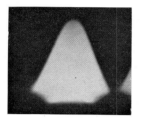 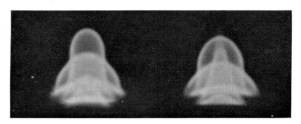

(*e*) Direct photograph of singing premixed flame.

(*f*) Fivefold stroboscopic photos of singing flame. See p. 184.

(*g*) Stroboscope photo showing ball separating from vibrating Bunsen flame. See p. 185

(*h and i*) Vibrating polyhydral benzene flame, sooting at tip of cone. (*h*) Direct, (*i*) with stroboscope. See p. 185.

PLATE 15

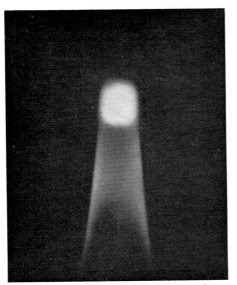

(a) Luminous Hefner candle, burning a mixture of amyl acetate and benzene, with transverse beam of light from carbon arc through top, showing scattered light from solid carbon particles.

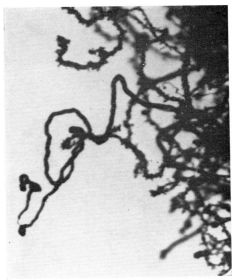

(b) Electron micrograph of carbon filaments collected on a probe inserted into the fuel side of a flat hydrocarbon flame. (*From Parker and Broatch*)

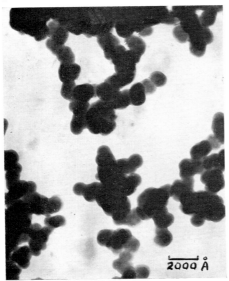

(c) Electron micrograph of particles from an ethylene counter-flow diffusion flame. (*By courtesy of F. J. Weinberg and R. Place*)

(d) Electron micrograph of particles from an ethylene flame (same as (c)) with an applied potential of 10 kV. (*F. J. Weinberg and R. Place*)

PLATE 16

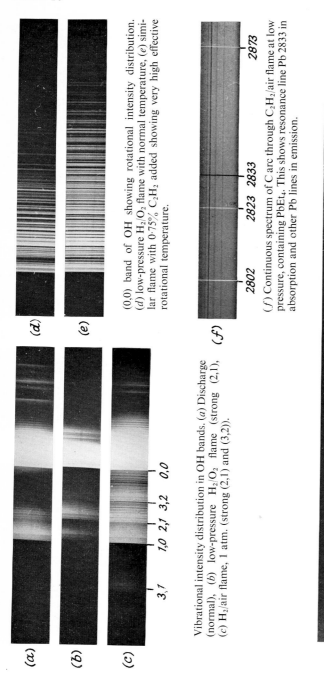

Vibrational intensity distribution in OH bands. (a) Discharge (normal), (b) low-pressure H_2/O_2 flame (strong (2,1), (c) H_2/air flame, 1 atm. (strong (2,1) and (3,2)).

(0,0) band of OH showing rotational intensity distribution. (d) low-pressure H_2/O_2 flame with normal temperature, (e) similar flame with 0.75% C_2H_2 added showing very high effective rotational temperature.

(f) Continuous spectrum of C arc through C_2H_2/air flame at low pressure, containing PbEt$_4$. This shows resonance line Pb 2833 in absorption and other Pb lines in emission.

(g) Spectrum of low-pressure flame of methanol/O_2 containing Fe(CO)$_5$. This shows full Fe I spectrum in reaction zone, at bottom of strip and strong OH, but interconal gases, at top, show only resonance lines of Fe and weak OH. (h) Spectrum of iron arc, showing similar intensity distribution of Fe I lines to reaction zone of flame.

PLATE 17

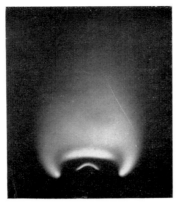

(a) Flat premixed hydrocarbon/NO₂ flame.

(b) Flame of boron dust (230 mgm./l. size 0·7 μm) in 39% O₂, 61% N₂.

(c) Flame of graphite dust (size 0 to 7 μm) with O₂, on burner.

(d) Al dust (size 0 to 7 μm 225 mgm./l.) with air. Note dark zone below r.z., possibly due to vaporization of particles.

(e) Ethane/chlorine on 9 mm. diameter burner. Note copious carbon formation.

PLATE 18

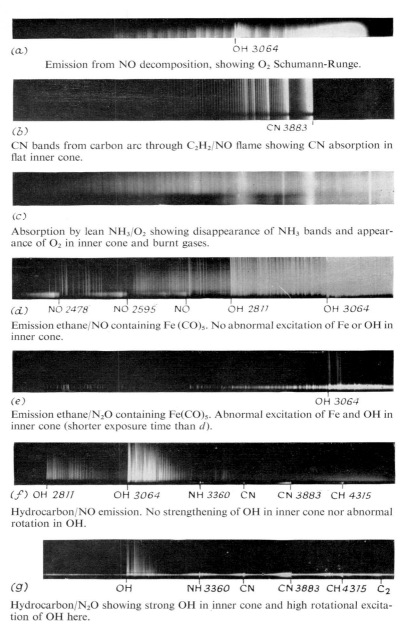

(a) OH 3064
Emission from NO decomposition, showing O_2 Schumann-Runge.

(b) CN 3883
CN bands from carbon arc through C_2H_2/NO flame showing CN absorption in flat inner cone.

(c) Absorption by lean NH_3/O_2 showing disappearance of NH_3 bands and appearance of O_2 in inner cone and burnt gases.

(d) NO 2478   NO 2595   NO   OH 2811   OH 3064
Emission ethane/NO containing $Fe(CO)_5$. No abnormal excitation of Fe or OH in inner cone.

(e) OH 3064
Emission ethane/N_2O containing $Fe(CO)_5$. Abnormal excitation of Fe and OH in inner cone (shorter exposure time than d).

(f) OH 2811   OH 3064   NH 3360   CN   CN 3883   CH 4315
Hydrocarbon/NO emission. No strengthening of OH in inner cone nor abnormal rotation in OH.

(g) OH   NH 3360   CN   CN 3883   CH 4315 C_2
Hydrocarbon/N_2O showing strong OH in inner cone and high rotational excitation of OH here.

Spectra of flames involving nitrogen compounds. The spectra were all taken with a focused image of the flame on the slit and are mounted longer wavelength to right, with top of flame upwards.

PLATE 19

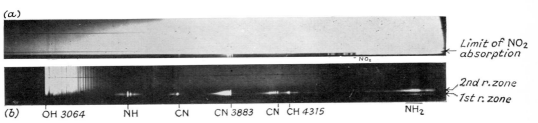

CH$_4$/NO$_2$ flame. (a) In absorption, with carbon arc as background. (b) In emission. The hairlines across the slit enable the NO$_2$ absorption limit in (a) and the reaction zones in (b) to be related.

CH$_4$/NO$_2$ flame in absorption in far ultra-violet. NO$_2$ is visible, and the appearance and disappearance of NO.

CO/NO$_2$ in emission on long-range spectrum plate. Only the first reaction zone is visible, showing a continuum.

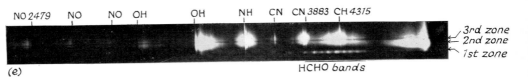

Methyl nitrite/O$_2$ emission, showing three reaction zones.

Methyl nitrate decomposition flame.

PLATE 20

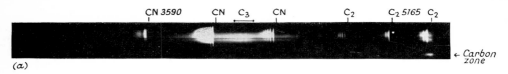

(a)

Emission spectrum of diffusion flame of CH₄ (1·8 c.c./s.) and H₂ (18 c.c./s.) mixture in ClF₃ (containing a trace of N₂).

(b)

As (a), but CH₄ flow increased to 29 c.c./s.

(c)

As (a), but CH₄ flow 29 c.c./s. and no H₂.

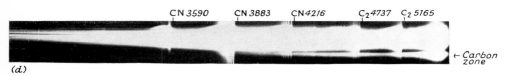

(d)

Diffusion flame of CH₄ + H₂ mixture in ClF₃. Spread of C₂ and CN bands is due to scattering of over-exposed spectrum.

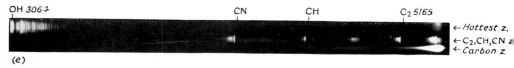

(e)

As (d), but with trace of O₂ added to ClF₃. Exposure time unchanged. Note appearance of CH and OH and weakening of C₂ and CN.

PLATE 21

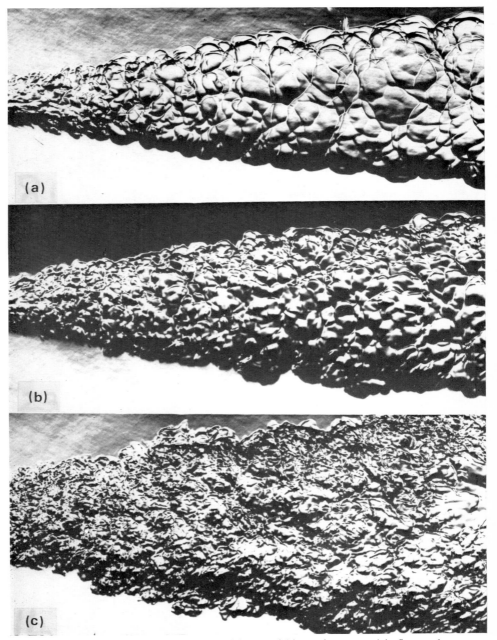

Schlieren photographs of flame spread in a stoichiometric propane/air flow at low Reynolds number approach-stream turbulence. $U = 43$ ft/sec. (*a*) Turbulence intensity 2%, (*b*) 5%, (*c*) 10%. The flame spread enables the turbulent burning velocity to be determined. (*From Lefebvre and Reid*, 1966)

PLATE 22

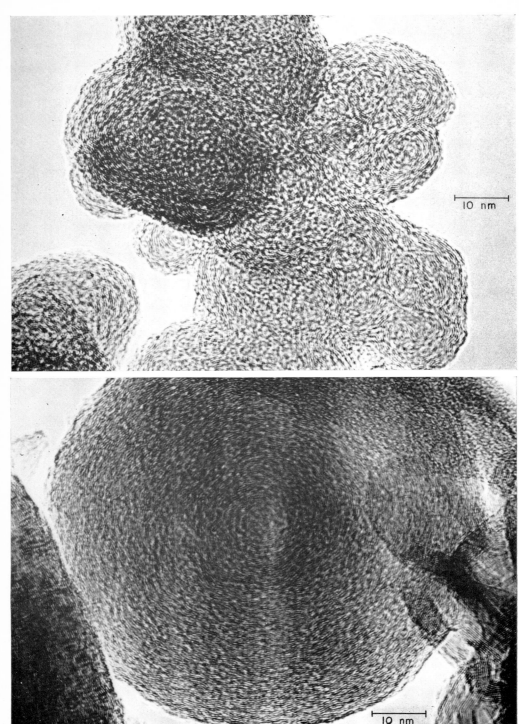

(Upper) Phase contrast electron micrograph of an N-347 furnace black. (Lower) Phase contrast electron micrograph of MT thermal black. (*From Marsh, Voet, Mullens and Price,* 1971)

Weinberg, 1969). Early studies were made in rich premixed ethylene flames and showed that the positively charged particles were repelled from a positive electrode plate; with the plate negative the soot deposition on it was much more voluminous, but detailed study of the bulkier deposit showed that it was much fluffier and less dense so that the total weight of carbon collected was rather less than with the uncharged plate.

Later work has been done with a counter-flow diffusion flame (*see* page 162) of ethylene and air or of (ethylene + nitrogen) and (oxygen + nitrogen). The main effects of a fairly strong electric field of either sign are to reduce the carbon luminosity, to reduce the size of the soot particles and to reduce the weight of soot collected. Fig. 8.3 shows the variation of OH, C_2, CH and solid carbon (C_s) radiation with applied voltage. The weight of soot collected may be reduced to 2% of that with no field. Colour photographs (Weinberg, 1968) show strikingly the change in flame appearance from yellow to blue. Plate 15c and d shows the reduction in particle size which in this work appears to be from about 1100 Å (i.e. 110 nm) to rather less than 100 Å. In later work, using slightly different flame conditions, Mayo & Weinberg (1970) found that the particles in the field-free flame had a diameter of about 270 Å and that the particle size decreased

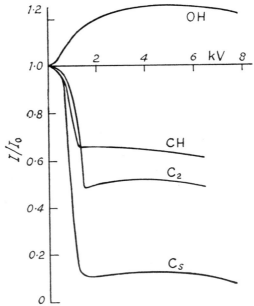

Fig. 8.3. Variation of relative intensity, I/I_o, for OH, C_2, CH and solid carbon emission with applied voltage (fuel electrode negative) in an ethylene counter-flow diffusion flame. [*After Place & Weinberg* (1966)]

rapidly with applied potential but reached a constant value of 92 Å for applied potentials above 3 kV.

With a suitable small field in the correct direction it is possible to hold the positively charged particles practically still in the reaction zone against the gas flow so that their time of passage through the carbon forming region is increased. Under these conditions there is rapid growth to large macroscopic particles which occasionally collect to form a network of soot across the flame.

In the later work by Mayo & Weinberg the particle mobilities were measured, these varying from 10^{-3} to 3×10^{-2} cm^2 V^{-1} according to the applied potential. It was shown that nearly all particles have unit positive charge. They also determined the number of particles produced, from measurement of size and weight collected. Fig. 8.4 shows the variation of the rate of particle formation with electric current; with low currents the rate of formation increases only slightly, but there is a sharp increase when the current reaches the saturation value at which removal of flame ions by the field equals their rate of formation in the flame.

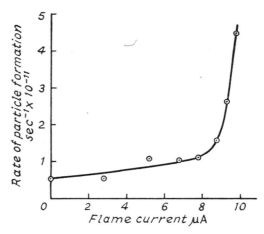

Fig. 8.4. Effect of electric current on rate of carbon particle formation in a counter-flow ethylene diffusion flame. [*After Place & Weinberg* (1967)]

These results show that flame ions can serve as nuclei for formation of soot particles but that some other mechanism is also required for most of the carbon. This is presumably growth on uncharged nuclei. Neutral particles later acquire charge by an attachment process, or, probably less often, by thermal electron emission.

Most of the above work was done with ethylene/air flames. We note that Homann (1967*b*) found that premixed acetylene flames did not show a similar marked reduction of carbon when a weak field was applied unless an actual electrical discharge occurred.

The nature of carbon particles and deposits

Particles may be collected from both sooting flames and from luminous flames which are not actually smoking by passing a cold probe or grid through the flame gases. They are too small to be examined with an ordinary microscope, but can be studied with an electron microscope. Particle sizes are usually in the range 50 to 2500 Å; each particle is roughly spherical, and the particles are usually clustered together in chains. Plate 15c shows an electron micrograph, by Place & Weinberg, of soot from an ethylene/air counter-flow diffusion flame; these particles are about 1100 Å in diameter. Generally, diffusion flames and the slower-burning premixed flames give particles in the larger size range, while fast-burning premixed flames give smaller particles. Early observations did not indicate much growth of particles up the flame, but later work close to the reaction zone shows that there is growth. Homann (1967) used a rich oxy-acetylene flame at low pressure, to increase the thickness of the reaction zone, and made electron microscope measurements of particles collected at various heights in the flame; his results are shown in Fig. 8.5. Howard, Wersborg & Williams (1973) also used oxy-acetylene at 20 torr, employing molecular-beam sampling with associated techniques and found that in the first 4 cm of flame the particles grew to about 100 Å diameter, and after this coagulation into chain-like clusters took place; each 100 Å particle consisted of about 1000 crystallites. D'Alessio et al. (1975) used a light scattering method to study particles in the flame, thus avoiding sampling problems; for an oxy-methane flame with $CH_4:O_2$ ratio of 0·95 they found a particle diameter of 50 Å in the reaction zone and 150 Å some 18 mm higher up the flame; for $CH_4:O_2 = 1·27$ growth was more marked, the size increasing from 50 Å to 1530 Å. In earlier work using light scattering, Dalzell (1969) found that care was needed in interpretation of results as the observations could otherwise lead to giving the size of clusters formed by coagulation. This coagulation into chains has also been studied by Jones & Wong (1975) who found that long chains become oriented in the gas flow and cause a polarisation effect known as streaming birefringence.

Using the light scattering method Dalzell, Williams & Hottel (1970) studied the burn-out of particles in a turbulent acetylene diffusion flame, and were unable to detect any decrease in particle size, concluding that burn-out occurred so quickly in the turbulent eddies that it was impossible to detect partly consumed particles. The burn-out time thus depended on the properties of the turbulent flow rather than on the chemical reaction rate, and in their experiments the half-life for burn-out was $7·4 \times 10^{-3}$ sec.

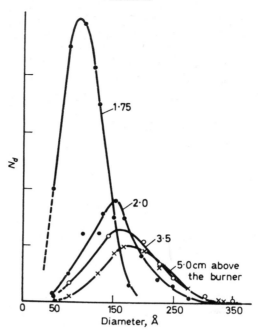

Fig. 8.5. Distribution of diameters of soot particles at different heights in a flat flame, $C_2H_2/O_2 = 1\cdot 6$, at 20 torr pressure.
[*From Bonne, Homann & Wagner, after Homann* (1967a)]

Chemical analysis shows that the soot particles usually contain from 1 to 3% by weight of hydrogen, i.e. usually over 12% of hydrogen in terms of atom numbers. Electron-spin-resonance studies show that the particles have many unpaired electrons, presumably because they contain free radicals (A. Thomas, 1962).

X-ray diffraction studies show that the particles have a basically graphitic structure but that each particle consists of a large number ($\sim 10^3$) of small crystallites of around 20 to 30 Å in size; Hofmann & Wilm (1936) gave a value of 13 Å for the height and 21 Å for the average length of the crystallites; D. B. Thomas (1962) quotes sizes up to 47·5 Å for crystallites in acetylene black and shows that the crystallites are comparable in size and hydrogen content with a polybenzenoid hydrocarbon such as circumanthracene, $C_{40}H_{16}$.

The lattice constants for the soot crystals are slightly different from those of pure graphite. For soot $a = 4\cdot 21$ and $c = 7\cdot 1$ Å, against $a = 4\cdot 25$ and $c = 6\cdot 69$ for Ceylon graphite. This slight distortion may be due to the presence of hydrogen, which is probably situated between the graphite planes.

These planes still appear to be parallel to each other in soot, but placed one on top of the other oriented at random.

Electron diffraction studies of carbon formed by pyrolysis of pure methane (Grisdale, 1953) show that the graphite platelets are oriented preferentially; when the carbon is deposited on a flat surface, the platelets are parallel to the surface; in carbon formed in the gas phase, they are oriented as though formed from spherical drops with the platelets tangential to the drop surface.

High-resolution phase-contrast electron microscopy of particles of furnace black (*see* page 236) by Marsh *et al.* (1971) has shown that each particle has a number of randomly oriented nuclei and that the graphite layers are curved or bent, indicating growth round the nuclei. A reproduction of one of their photographs is shown in Plate 22*a*. The electron micrographs show a significantly greater spacing between graphite layers than do X-ray studies because of this curvature of the layers. For thermal black (Plate 22*b*), however, growth appears to occur round a single nucleus.

It is possible to extract a number of compounds from soot particles with suitable solvents. In early studies of diffusion flames, Thorp, Long & Garner (1951) observed long-chain olefins, substances with a fulvenic structure (i.e. a five-membered carbon ring), and compounds containing a carbonyl group; only benzene flames were at first found to give polycyclic products such as diphenyl and anthracene. Later, however, soot extracts from diffusion flames of propagas were found (Ray & Long, 1964) to contain 49% pyrene, 25% phenanthrene, 18% fluoranthene and 17% anthracene. In the soot from reversed diffusion flames of methane, Arthur *et al.* (1958) found 35% pyrene, 15% phenanthrene, 12% fluoranthene and 12% anthracene, while flames of ethylene gave 59% acenaphtylene, 8% fluoranthene, 8% anthracene and 7% phenanthene and small quantities of many other products. Mass spectrometric study (Homann & Wagner, 1967) of the volatile components of soot from premixed acetylene flames showed substances with mass numbers to at least 500, pyrene, acenaphthalene, phenanthrene, naphthalene and coronene being prominent. The composition of the volatile components changed with the height in the flame from which they were collected, larger mass numbers occurring high in the flame.

If a probe is held for some while in a flame, long filaments of carbon grow from the probe. These may have a very unusual appearance (Plate 15*b*). It seems that these long filaments must grow from the tip rather than in all directions. Electrical effects may be important.

Normal surface deposits of carbon are of quite a different character from soot. In diffusion flames of hydrocarbons with oxygen, when the burner rim gets very hot, the fuel may be cracked and build up a cylindrical ring of carbon on the burner rim. The growth may be quite rapid, perhaps

1 to 2 mm in 10 sec. Surface-deposited carbon of this type usually has a hard, shiny surface and is called vitreous carbon. Its mode of formation appears to be different from that of soot.

The equilibrium between carbon and the burnt gases

As long as there are no solid particles present we may consider only the equilibrium between gases such as CO, CO_2, H_2, H_2O, N_2, NO and their gaseous dissociation products such as O, H and OH. This has been done in Chapter XII. As long as the ratio of carbon atoms to oxygen atoms is less than unity ($n_C/n_O < 1$) we may assume that all the carbon goes to CO and consider only the various gaseous equilibria. As soon, however, as $n_C/n_O > 1$ we shall expect free carbon. We propose to examine the conditions under which free carbon will be present in equilibrium in some detail.

The relationship between CO, CO_2 and solid carbon will be governed by the so-called Boudouard reaction

$$CO \rightleftharpoons \tfrac{1}{2}CO_2 + \tfrac{1}{2}C_{solid} + 20 \text{ kcal/mole}$$

This reaction is, of course, very important for the formation of producer gas. In actual flames further equilibria involving H_2O, H_2, etc., will also have to be satisfied as well, but this will not affect the existence of the above equilibrium.

The equilibrium constant for the Boudouard reaction is

$$K_p^B = [p_{CO}]^2/p_{CO_2}.$$

Values given in the following table have been calculated by us from other equilibrium constants listed by Gurvich et al. (1962),

T(K)	K_p^B	T(K)	K_p^B	T(K)	K_p^B
400	4.96×10^{-14}	900	0.184	1500	1.55×10^3
600	1.78×10^{-6}	1000	1.81	2000	4.01×10^4
700	2.55×10^{-4}	1100	11.6	2500	2.61×10^5
800	0.0104	1200	54.2	3000	8.60×10^5

CO is therefore unstable below about 1000 K and should dissociate to give CO_2 and solid carbon. Actually we know that this does not happen, the reaction rate becoming so small, especially in the absence of surfaces, that the equilibrium freezes out at a high temperature at which CO alone is for all practical purposes the only constituent. Prof. Lennard-Jones informed us that he had, on occasions, observed decomposition of CO to carbon on heated surfaces contaminated with iron. If the mixture at high temperature is free from soot particles then freezing of the equilibrium may occur more readily than when solid nuclei are already present. The freezing in the gas phase usually occurs at about 1500 K (Behrens, 1950a; 1951a).

If, in a flame, solid carbon is formed in more than equilibrium amounts, then it will burn at its surface by reaction with CO_2, H_2O and OH. These surface reactions may, however, be rather slow. The effect of pressure in reducing the excessive formation of carbon in methane explosions, reported by Bone & Townend, may be due to slower cooling and faster surface reactions at high pressure.

In calculating the composition and temperature of a burnt gas mixture (Chapter XII) we use the various equilibrium constants and the ratios of numbers of the various atoms initially present. As we go to richer mixtures the ratio $[p_{CO}]^2/p_{CO_2}$ becomes larger. Now it cannot exceed the value corresponding to K_p^B at the gas temperature. If it does then carbon will separate out and reduce n_C for the gas phase. We thus lose one of the equations, the ratio n_C/n_O for our equilibria, and have to substitute the additional equation $[p_{CO}]^2/p_{CO_2} = K_p^B$.

Thus carbon formation should, for equilibrium, set in for mixtures which are slightly richer than for $n_C/n_O = 1$. We find experimentally that an acetylene/air flame begins to show luminosity below $n_C/n_O = 0.9$ and many other fuels show luminosity with n_C/n_O around 0·5 (page 201). The actual equilibrium point at which carbon should, in equilibrium, be liberated will depend slightly on the gas temperature as well as on n_C/n_O.

Table 8.2 shows the equilibrium composition of rich oxy-acetylene flame gases. It will be seen that free carbon should start at $n_C/n_O = 1.08$ in this very hot flame. For ethylene it should start at $n_C/n_O = 1.01$ and for all other ordinary hydrocarbons and for all flames with air it should start very close to 1·00. Gay et al. (1961) considered the presence of higher carbon molecules, C_5, C_7, etc.; however, other work (Drowart et al., 1959) on the mass spectroscopic analysis of carbon vapour indicates that the concentration of these larger carbon molecules is very small. It may be noted that equation (8.1) (page 202) is not satisfied at the equilibrium condition for solid carbon just to appear; R is then about 2×10^5.

In a hydrocarbon/air diffusion flame the fuel will be slowly heated as it diffuses towards the reaction zone, and as the oxygen-bearing constituents will be in very low concentration in this region, n_C/n_O will be very high, and we should expect carbon formation as soon as the temperature is high enough to decompose the fuel. As the carbon particles rise vertically, they will penetrate deeper into the wedge of burnt gases containing H_2O, CO_2 and CO and in this region will be in a gas of lower n_C/n_O; there will therefore be a tendency, provided the temperature is above about 1000 K, for the particles to react with water and CO_2 to give CO. Whether the particles are entirely consumed in this way or whether they cool to a temperature

TABLE 8.2

Equilibrium temperatures and partial pressures, in atm, of some carbon species and major constituents of acetylene/oxygen flames at 1 atm

C/O	0·67	0·91	1·00	1·03	1·08	1·11	1·25	2·0
T (K)	3435	3383	3325	3304	3275	3266	3233	3122
C(gas)	4.9×10^{-9}	2.2×10^{-8}	1.9×10^{-5}	3.8×10^{-4}	4.8×10^{-4}	4.5×10^{-4}	3.5×10^{-4}	1.3×10^{-4}
C_2	1.3×10^{-14}	3.8×10^{-13}	3.8×10^{-7}	1.8×10^{-4}	3.6×10^{-4}	3.2×10^{-4}	2.4×10^{-4}	8.2×10^{-5}
C_3	0	2.5×10^{-17}	3.1×10^{-8}	4.5×10^{-4}	9.5×10^{-4}	1.1×10^{-4}	7.4×10^{-4}	2.7×10^{-4}
C(solid)*	0	0	0	0	0·0009	0·0197	0·0945	0·414
CH	3.1×10^{-10}	2.1×10^{-9}	2.1×10^{-6}	4.5×10^{-5}	6.1×10^{-5}	5.8×10^{-5}	4.9×10^{-5}	2.8×10^{-5}
C_2H_2	1.2×10^{-13}	9×10^{-12}	1.6×10^{-5}	9×10^{-3}	0·0215	0·0226	0·0234	0·0257
CO	0·50	0·58	0·61	0·60	0·60	0·59	0·57	0·47
CO_2	0·049	0·011	1.0×10^{-5}	4.3×10^{-7}	2.7×10^{-7}	2.7×10^{-7}	2.6×10^{-7}	2.1×10^{-7}
H_2	0·092	0·17	0·21	0·22	0·22	0·23	0·25	0·37
H	0·15	0·19	0·18	0·17	0·16	0·16	0·15	0·14
H_2O	0·075	0·027	2.8×10^{-5}	1.2×10^{-6}	8.0×10^{-7}	8.3×10^{-7}	9.3×10^{-7}	1.3×10^{-6}
OH	0·059	0·013	1.1×10^{-5}	4.3×10^{-7}	2.5×10^{-7}	2.5×10^{-7}	2.4×10^{-7}	1.9×10^{-7}

* The solid carbon is expressed as moles of carbon per mole of gaseous products. Partial pressures of C_2 and C_3 are from Jessen & Gaydon (1969) using $D(C_2) = 143$ kcal/mole and $D(C_3) = 321$ kcal/mole (from Gaydon, 1968). Other data are from Gay et al. (1961), but rounded to two significant figures as the accuracy of the equilibrium constants does not justify more. For C/O = 0·67 the partial pressures of O_2 and O are 0·017 and 0·057. For very rich flames, C/O > 1·03, methane has a partial pressure around 2×10^{-6} and ethylene 5×10^{-7}

below that at which the equilibrium again freezes, so giving sooting or smoke emission, depends on the geometry of the flame.

Detection of intermediate products

Chemical sampling of unstable intermediaries in hot flame gases is obviously difficult. Stable species, such as methane, acetylene, ethane and ethylene may readily be detected by probe sampling and subsequent gas chromatographic analysis, although interference by reactions in the probe can be troublesome. In recent years good progress has been made in mass-spectrometric analysis of samples and some free radicals as well as stable species have been determined; probe disturbances have been reduced by sampling through a small (100 μm) pinhole at the top of a quartz probe under critical flow conditions so that further reaction is rapidly quenched by the expansion of the gas to very low pressure. Some species (C_2, CH, C_3, CHO) are detectable by their emission spectra, and some progress has been made in estimation of species by absorption spectroscopy.

Before discussing results, we must emphasise that the stationary concentration of an intermediate species in a reacting system depends both on its rate of formation and on its rate of removal. Presence of a species even in relatively high concentration does not necessarily indicate that it is an active precursor of carbon. Indeed it may be that the active precursors are quickly removed and never attain high concentrations, while species which are readily formed but which are then unreactive are those which are easiest to detect. For example the 5-carbon fulvenes detected in soot extracts may well be of this type, being substances which cannot readily break down to graphite.

Spectroscopic study of premixed flames (Gaydon, 1974) shows that the C_2 and CH emission, especially the former, is strong for rich mixtures, although their intensity begins to fall before the threshold for carbon luminosity is reached. In low-pressure flames the C_2 is found slightly below CH, probably because CH is formed by reaction of C_2 with OH. The 4050-Å (comet-head) band of C_3 has been observed in the mantle of rich oxy-acetylene flames. In diffusion flames C_2 and CH emission occurs on the oxygen side of the carbon zone, not in the pyrolysis zone.

Studies of the absorption spectra of premixed flames are difficult because of the thinness of the reaction zone where the active species occur, because of light deflections due to the high refractive-index gradient in this zone and because of the strong chemiluminescent emission of some of the interesting species. Advances in techniques for studying absorption spectra of flames have been fully described by Gaydon (1974). Using a multiple-reflection system to increase the optical path, Spokes & Gaydon (1958) reported benzene absorption bands in the preheating zone of rich flames of hexane and ether; a flat flame, of the Egerton-Powling type, was used. Spokes & Gaydon also detected CH_3 and C_2H_5 radicals and C_2 absorption in low-pressure flames. Both C_2 and CH are notoriously difficult to obtain in absorption, but Bleekrode & Nieuwpoort (1965) have observed them in a low-pressure oxy-acetylene flame using a light-chopper technique. Subsequently, Jessen & Gaydon (1969) combined this with an improved multiple-reflection system (which uses a focused image and thus avoids light-deflection troubles) to study the mantle of oxy-acetylene flames at 1 atm near the carbon threshold; they find C_2, C_3 and CH absorption; the C_2 concentration reaches about 2×10^{14} molecules/cc, about the expected value for equilibrium with soot particles, but in the reaction zone it is about fifty times higher than equilibrium; C_3 has a concentration of about 1×10^{15} in the mantle, again consistent with the equilibrium value, but it was probably no higher, if present at all, in the reaction zone. The C_3 absorption band is diffuse in these high temperature sources and the ultra-violet part of the band appears stronger than in the band observed by Brewer in carbon vapour. It is likely that some other species contribute to the absorption; Merer (1967) has reported C_3H_2 (probably linear HCCCH band structure in this region in flash photolysis studies. Unfortunately in flame sources bands of many of these poly-atomic species tend to become complex and apparently diffuse and therefore difficult to identify because of the large number of rotational and vibrational energy levels which are populated at these high temperatures. We have already noted (*see* page 196) that this mantle region of flames shows fluorescence, presumably of molecular origin, but with a diffuse, apparently continuous, spectrum.

In flat diffusion flames Parker & Wolfhard (1950) and Wolfhard & Parker (1952) found an apparently continuous absorption spectrum just on the fuel side of the luminous carbon zone (*see* Plate 11c). This continuum appeared (probably because of falling plate sensitivity in the ultra-violet) to be strongest at short wavelengths, spreading towards the visible spectrum with increasing strength closer to the carbon zone. However, quantitative measurements of absorption (Laud & Gaydon, 1971) using a light-chopper technique and a counter-flow diffusion flame showed clearly that the absorption, although without any fine structure or marked band heads, was not a pure continuum of the type which could be caused by absorption or scattering losses from small particles or droplets. Curves of absorption against wavelength for ethylene diffusion flames on counter-flow and Wolfhard–Parker burners are shown in Fig. 8.6. This absorption is therefore due to molecular intermediates preceding the soot formation, but because of its diffuse, featureless character it is not possible to suggest the nature of the absorbing species. This absorption is, however, different from

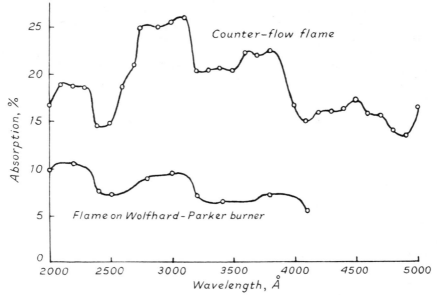

Fig. 8.6. Absorption, at the position of maximum strength, for ethylene diffusion flames on counter-flow and Wolfhard–Parker burners. For the counter-flow flame the fuel flow was 10 ml/sec C_2H_4 + 35 ml/sec N_2, and the oxidant flow was 20 ml/sec O_2 + 25 ml/sec N_2; the effective path length was 55 mm. For the Wolfhard–Parker flame pure ethylene and oxygen were used with a path length of 25 mm.

that found near 4050 Å in the mantle of rich oxy-acetylene flames, described in the previous paragraph.

When a little benzene is added to a flat ethylene/oxygen diffusion flame, the C_6H_6 absorption bands are detectable on the fuel side (Plate 11c) and it has been noted (*see* page 162) that these bands disappear, due to removal of the benzene, before soot is formed.

Mass spectrometric studies of low-pressure premixed flames (Bonne, Homann & Wagner, 1965) give exciting information about the intermediate species. Fig. 8.7 shows the variation with height above the burner rim of the concentration of some of the most important species in an acetylene flame. Other species which were detected include HCO, H_2CO, CH_3OH, C_2H_4 (superimposed on CO), C_3H_6, C_4H_6, C_5H_6, C_6H_6, C_6H_4, C_8H_6, $C_{12}H_8$, CH_3, C_3H_3, C_4H_3 and of course stable species like CO, CO_2, H_2 and excess C_2H_2.

The most abundant species is diacetylene C_4H_2, which persists high into the flame. The radical C_2H, methane and C_3H_4 appear first, then vinyl acetylene (C_4H_4) followed by the polyacetylenes C_4H_2, C_6H_2 and C_8H_2. Are

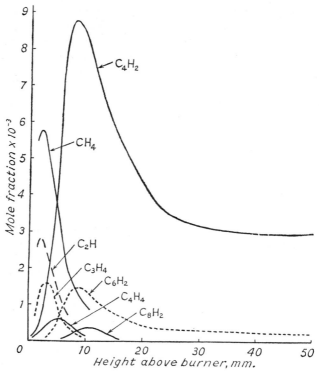

Fig. 8.7. Concentrations of carbon–hydrogen species in oxy-acetylene flame ($C_2H_2/O_2 = 0.95$) at 20 torr.

[*Reconstructed from figures by Bonne, Homann & Wagner*

these polyacetylenes important precursors of carbon or just abortive side products? Benzene was just detectable but other aromatics were not observed in appreciable concentration.

Similarly for premixed methane/oxygen d'Alessio *et al.* (1975) detected by mass spectrometry the following main species in order of attaining maximum concentration: C_2H_6, C_2H_4, C_2H_2, C_4H_2, C_6H_6 and C_4H_4. They also found traces of polycyclic aromatics, and classified these into two groups. The first group, of reactive species, showed a clear maximum of concentration with height in the flame and included coronene, 3,4 benzopyrene, 1,2 benzopyrene and benzperylene. The second group, of unreactive species, increased in concentration with height and did not show a clear maximum, and included pyrene, fluoranthene, anthracene and phenanthrene. The reactive species, however, were thought to account for only a small part of the soot formation in this premixed flame, while the unreactive species may be regarded just as by-products.

Direct mass-spectrometric sampling of the positive ions drawn from flames (Knewstubb & Sugden, 1958) shows a variety of ionised carbon–hydrogen molecules with mass numbers up to 150. Molecules with from 3 to 10 carbon atoms occur, those with an odd mass-number and an odd number of carbon atoms being most abundant. In acetylene flames the first species to appear are CHO^+, H_3O^+ and $C_3H_3^+$. In the mantle of a rich oxy-acetylene flame the commonest ions are $C_3H_3^+$ and $C_5H_3^+$; $C_3H_4^+$ and $C_7H_3^+$ are also strong. In other fuels the occurrence and abundances of the species are different; ethane gives mainly $C_3H_3^+$ and $C_3H_5^+$ while methane gives relatively few hydrocarbon ions although $C_3H_3^+$ and some others are observed.

Isotope tracer studies

The first work of this type on carbon was by Ferguson (1955) who used C^{13}-enriched acetylene mixed with normal acetylene and measured the relative strength of the Swan emission bands of C^{12}_2, $C^{12}C^{13}$ and C^{13}_2. The results indicated almost complete randomisation of C^{12} and C^{13} in the excited C_2 molecules, showing that they were not formed directly from C_2H_2 by dehydrogenation reactions of the type

$$C_2H_2 + H = C_2H + H_2 \quad (\text{or } C_2H_2 + OH = C_2H + H_2O)$$

followed by $C_2H + H = C_2^* + H_2$.

The nearly complete randomisation could not be explained either, by any simple polymerisation mechanism followed by decomposition of the polymer, and Ferguson expressed the opinion that the main route to formation of electronically excited C_2 must involve radicals containing a single carbon atom.

Later Ferguson (1957) studied explosions of rich propane-2-C^{13} mixed with oxygen and measured mass spectroscopically the C^{13} content of the soot particles. Again almost complete randomisation of C^{12} and C^{13} was found; if the soot had been formed via acetylene by breaking off an end C^{12} atom, leaving $C^{13}C^{12}$ to form acetylene and the particles, then there would have been an enrichment of C^{13} in the soot. In these experiments some acetylene was also formed in the gas phase and sampling of this again showed randomisation of the isotopes. Addition of ordinary carbon monoxide to the flame did, however, show that in this case randomisation was not obtained, the CO not being involved in the soot formation process or its growth.

Radioactive carbon (C^{14}) has been used as tracer to study soot formation from an ethanol/air diffusion flame (Lieb & Roblee, 1970) and the results show that there is some preferential formation from the non-hydroxylated carbon atom, in the ratio 2:1. This has been explained as due to breaking of the CH_3–CH_2OH bond; the CH_2OH radicals react to form inactive formaldehyde while the CH_3 radicals (and also H atoms) react with more ethanol to form C_2H_4 and then soot.

The pyrolysis of hydrocarbons

Davy, and later Marchand & Bertholet (*see* Bone & Townend, 1927) suggested that the formation of carbon particles in flames was due to the thermal decomposition of the hydrocarbons. They observed that at high temperature these decomposed largely into carbon and methane. Acetylene, which is very endothermic, may be decomposed explosively to carbon and hydrogen. Pyrolysis processes may occur in the preheating zones of flames and are thus of interest for elucidating the mechanism of carbon formation.

Experiments by Tropsch & Egloff (1935) have shown that at high temperature the pyrolysis may occur in a very short time. They passed pure hydrocarbons through a heated tube of 0·3 mm bore; temperatures ranged from 1100°C to 1400°C with contact times of only around 10^{-3} sec, comparable to those in the preheating zones of flames. Their results for ethane are shown in Fig. 8.8. Ethylene was formed first, but acetylene was the main product. For methane results are similar; using a porcelain tube at 1000°C, Gordon (1948) found the main products to be ethane, ethylene and acetylene, with a little benzene and traces of naphthalene, anthracene, phenanthrene, pyrene and methyl acetylene. In this pyrolysis of methane it was also found that acetylene had a marked autocatalytic action.

The pyrolysis of acetylene is particularly interesting because of its ready formation from other hydrocarbons and its suggested role in soot formation. The pyrolysis has been studied, among others, by Cullis & Franklin (1964) who found that vinyl acetylene was the sole initial product; diacety-

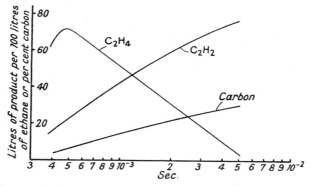

Fig. 8.8. Time variation of products of pyrolysis of ethane in a heated tube at 1400°c. [*After Tropsch & Egloff*

lene, benzene, ethane, hydrogen, ethylene and methane were formed later. They showed that the kinetics were second order in acetylene and probably involved the formation of acetylene in an excited triplet electronic state, perhaps at the tube surface. Cullis, Read & Trimm (1967) also studied the pyrolysis of vinyl acetylene and acetylene using the carbon isotope C^{14} as tracer; they concluded that vinyl acetylene mainly reacted to give polymers and was not an important intermediate in the formation of surface carbon from acetylene.

The absorption spectra of hydrocarbons undergoing pyrolysis in a tube heated to 1000°C have been studied by Parker & Wolfhard (1950; 1953); saturated hydrocarbons, initially transparent, developed continuous absorption, first at the ultra-violet end of the spectrum, then spreading to longer wavelengths with increasing temperature, finally causing the gases to become yellow. At the highest temperature a mist formed in the tube. This continuous absorption was very similar to that obtained just on the fuel side of the carbon zone of flat diffusion flames (*see* page 162). This pyrolysis continuum was observed with all hydrocarbons and all alcohols except methanol, which happens to be one of the few fuels which does not give carbon in flames.

These flow experiments are all open to the criticism that surface reactions may be involved, and in some cases (e.g. Tropsch & Egloff) the carbon formed is the hard vitreous form. Contact times were usually longer and temperatures usually lower than in flames, although the results of Tropsch & Egloff do show that carbon may be formed at temperatures and contact times similar to those in flames. Shock-tube methods (for a review of these see Gaydon & Hurle, 1963) enable pyrolysis to be studied in the gas phase at high temperature. The products may be quenched out by sudden cooling in a chemical shock-tube and then analysed by gas chromatography or infra-red spectroscopy. Alternatively a time-resolved study of the products may be made with a scanning time-of-flight mass spectrometer sampling the hot products direct from the shock-tube.

Acetylene has been studied most (e.g. Aten & Greene, 1956; Gay *et al.*, 1965); the main product, prior to carbon and hydrogen, is diacetylene, but Aten & Greene also found vinyl acetylene. Time-resolved studies showed C_4H_3 as first product, followed successively by C_4H_2, C_6H_2, C_8H_2 and on to soot and hydrogen. The activation energy for decomposition of acetylene is about 46 kcal/mole and the process is second order; it may involve free radicals or an excited triplet state of acetylene.

Similar studies on methane (e.g. Greene, Taylor & Patterson, 1958; Glick, 1959) show mainly acetylene, but also some diacetylene and vinyl acetylene preceding the soot. The activation energy for methane is higher, values from 93 to 103 kcal being quoted. For ethylene (Gay *et al.*, 1966) the main product is again acetylene, with some diacetylene; the activation energy is 50·5 kcal and again it is not certain whether an excited triplet state or a radical mechanism is involved.

The shock-tube pyrolysis of aromatics is especially interesting and has been examined by Graham, Homer & Rosenfeld (1975). They studied benzene, toluene, ethyl benzene and indene, all diluted with argon and heated in incident shock waves. Soot formation was measured by light absorption at various wavelengths to give particle size and concentration, and observations were made as a function of time at various shock temperatures. They found that at a time of 2 ms soot yield rose to almost 100% at 1750 to 1800 K, but then surprisingly fell again at higher temperatures to below 5% at 2300 K. They concluded that there must be two routes to soot formation for these aromatics, (a) a condensation to higher aromatics and graphitisation of these, and (b) a decomposition of the initial aromatic to smaller non-aromatic fragments and then polymerisation of these, possibly via polyacetylenes. The first process was dominant at low temperature, but at higher temperatures the break-up of the initial aromatics occurred quickly and then the subsequent soot formation from the simpler fragments was relatively slow, and so, although the initial pyrolysis of the aromatics was quicker, the soot formation rate was much slower. They also found, during the low-temperature pyrolysis, some spectroscopic evidence for light absorption by large gas-phase molecules.

The problem of the formation and growth of soot particles

We have seen that it should be possible to predict from equilibrium considerations the amount of carbon present in the solid form in a mixture of any composition at high temperature. However, in practice, freezing of the equilibria becomes dominant, so that the rates of the various reactions leading to formation and removal of carbon, rather than the true equilibrium conditions, determine whether or not a flame is luminous and whether sooting occurs.

In forming soot we start with a small hydrocarbon or organic molecule and end with a relatively large particle containing many thousand atoms and a much higher carbon/hydrogen ratio. Thus we must obviously have both dehydrogenation and growth or condensation. The exact route from small hydrocarbon to large soot particle is, however, uncertain. In one extreme we might have dehydrogenation to atomic carbon or C_2 radicals followed by condensation to solid carbon; in the other extreme, polymerisation to a very large hydrocarbon molecule which then loses hydrogen and graphitises from within. Many intermediate routes appear possible. A 'diagonal' route would be simultaneous polymerisation and dehydrogenation all the way. Other more step-like routes might occur; breakdown to atomic carbon or C_2 and condensation to a small nucleus of carbon might be followed by surface decomposition of hydrocarbons on this nucleus, with consequent growth to a large particle; polymerisation to a medium-size hydrocarbon, ring-closure to an aromatic structure and then further growth might occur. A number of reviews on the mechanism of carbon formation have been published in recent years, and some of these have already been referred to. In discussing some of the detailed mechanisms we must emphasise that the occurrence of one route does not necessarily exclude the simultaneous occurrence of others, and that the actual routes followed no doubt depend a lot on flame conditions – temperature, pressure, oxygen content, nature of initial fuel, time factor.

It seems that both polymerisation and oxidation proceed by chain reactions, usually involving free radicals. At some stages in these chain reactions there may be several possible products; of these products one or more may then lead readily to soot formation, others may continue an oxidation chain and some may be abortive, leading to less reactive intermediates which persist for some while before either oxidising or breaking down to carbon.

Dehydrogenation and polymerisation via acetylenes. Acetylene is the commonest stable intermediate observed in rich premixed flames of other hydrocarbons; it has also been obtained in pyrolysis of hydrocarbons, both at low temperatures and in shock-tubes. Acetylene, being so unsaturated, appears to give soot very readily, and is used in the carbon-black industry. Thus it is fairly obviously a candidate for consideration as an intermediate in the break-down of other hydrocarbons (Porter, 1953). We note, however, that actual measurements (*see* page 201) on premixed flames show that on a mixture-strength or n_C/n_O ratio basis acetylene is a fuel with one of the lowest tendencies to form soot.

We have seen that in rich flames of acetylene itself, and also in pyrolysis, diacetylene, C_4H_2, and other polyacetylenes C_6H_2 and C_8H_2 occur. It is possible that these really are intermediates in the formation of soot from acetylene itself, although the initial steps may involve either excited triplet acetylene or radicals such as C_2H; vinyl acetylene and C_4H_3 also occur before diacetylene.

For other hydrocarbons, some acetylene is certainly formed during their break-down; this acetylene may then form carbon either through the polyacetylene route or by surface decomposition on existing carbon nuclei. Although some fraction of the soot is formed this way, we see no real evidence that this is the quickest or main route, since most hydrocarbons give carbon more readily than acetylene itself does. Recent studies (Cullis, 1975) of the kinetics of the reactions of ethynyl radicals (CH_2) with low molecular weight acetylenes also lead to the conclusion that although polyacetylenes can be formed, other considerably more facile reactions must be responsible for the rapid formation of soot in flames.

Polymerisation via aromatics. Rummel & Veh (1941) suggested that the formation of aromatics and polycyclic hydrocarbons preceded the formation of carbon. This is a very attractive hypothesis as the benzene ring has nearly the same dimensions as that of graphite, and for a group of adjoining benzene rings we have essentially a graphite nucleus. Thomas (1962) apparently was inclined to favour this mechanism and compared the properties of soot and circumanthracene. Benzene flames smoke very readily and naphthalenes give soot most readily of all hydrocarbons in both premixed and diffusion flames.

There is now some direct evidence for the presence of aromatics in rich-limit flames, although amounts always appear to be very small. Spokes & Gaydon (1958) reported benzene absorption bands in rich hexane and ether flames. As already noted, Bonne et al. (1965), by mass spectrometry, were just able to detect benzene but not other aromatics, but d'Alessio et al. (1975) did detect some polycyclic compounds and classified these as reactive and unreactive ones. Most aromatics show banded absorption spectra at low temperatures, but at flame temperature these banded features disappear and the spectrum becomes nearly continuous; this nearly continuous type of spectrum could contribute to the pyrolysis continuum reported by Parker & Wolfhard and the slightly banded structure found by Laud & Gaydon. Extracts from soot particles contain such aromatics as pyrene, phenanthrene, fluoranthene and anthracene; these are just the group of aromatics which d'Alessio et al. classified as unreactive and they probably occur because their relative stability enables them to persist and be absorbed on the surface of the growing soot particles.

If graphite is to be formed from straight-chain hydrocarbons then ring closure must occur at some stage. Branched-chain hydrocarbons, such as iso-paraffins, form soot more readily in flames than do normal paraffins; this is also true for iso- and normal alcohols. It is certainly rather difficult to visualise formation of an aromatic ring from a branched chain. It is known, however, that normal paraffins oxidise more readily than the iso-paraffins, so the difference in sooting properties is probably attributable to the stronger competition from oxidation.

Polycyclic aromatics have also been detected in ovens burning domestic oil (Herlan, 1978) and in addition to pure hydrocarbons some oxygenated aromatics are also formed. These compounds are of importance because they could possibly include carcinogens.

Condensation and graphitisation. There is good evidence that with burning oil sprays the larger droplets may crack to carbon, sometimes forming xenospheres which pass through the flame front. In the pyrolysis of methane and other hydrocarbons Parker & Wolfhard observed the formation of a mist. If the products of the pyrolysis were burnt as a diffusion flame these large mist droplets did not vaporise, but cracked to carbon as they passed through the flame. Parker & Wolfhard thus suggest that a route for carbon formation may be via polymers or other large molecules with a low vapour pressure; when the saturation vapour pressure is exceeded these condense to the liquid droplets which then graphitise.

Hofmann & Wilm (1936) used X-rays to follow the graphitisation of sugar at 1000°C. They concluded that nuclei of graphite formed within the sugar and crystallites grew from these nuclei. The size of the crystallites was determined by the initial number of nuclei. Below 1400°C the crystalites did not grow at the expense of each other; at higher temperature there was a very slow growth. If liquid droplets are formed and graphitise in a similar way to sugar then it would account for the soot particles from diffusion flames being all spherical, of similar size, and each composed of around 10^3 crystallites. Grisdale's observation (page 215) the carbon particles formed by pyrolysis of methane had graphite crystallites orientated with their planes tangential to the particle surface, fits in well with this mechanism.

If this condensation to droplets occurs at all it is most likely in diffusion flames where the residence time is relatively long and the temperature may be fairly low. It can hardly contribute to soot formation in premixed flames where the time of passage through the reaction zone is so short and we have also seen that the particle size increases rapidly with height. Even for diffusion flames the effects of applied electric fields indicate that the particle size depends on residence time in the pyrolysis zone, and the simplest explanation is that the soot particles themselves are growing; it could be, however, that it is the droplets which are growing when held by the field for long times in the pyrolysis zone, if the droplets are charged, perhaps as a result of initial nucleation on a flame ion.

Nucleation, particle growth and coagulation. It is known that carbon, often in the hard vitreous form, is formed on the surface when a hydrocarbon vapour is passed through a hot tube. Thus once a suitable surface at high temperature is available in the form of a carbon particle, its growth will continue.

It is well known that nucleation in aerosols usually occurs on dust particles already present, on positive or negative ions, or even on other large molecules when the vapour reaches a certain supersaturation. In pure vapour a much higher supersaturation is necessary for condensation to start. A summary of work on this type of process is given by Dunning (1973), and there are other interesting papers on nucleation processes in the same Faraday Symposium. In flames we know from observations of the spectrum that carbonaceous molecules such as C_2 and C_3 are present and these have been suggested as likely nuclei (*see* later section on C_2). Jensen (1974) has considered whether growth on nuclei such as C_2, C_3 and C_2H could occur and has concluded that it is indeed possible; this conclusion depends on a calculation of the critical particle radius and makes assumptions, probably reasonable ones, about the value of the surface free energy and degree of supersaturation.

Tesner (1959) and colleagues studied the growth of carbon particles held in a furnace through which hydrocarbons are flowing. He shows a very convincing series of electron micrographs demonstrating the growth of individual particles under these conditions. He says that growth occurs by actual decomposition of the original hydrocarbon on the surface, not via some intermediate polymer, and supports this by showing that an equivalent amount of hydrogen is always set free. If this is generally true, then it would be an argument against this process in flames as we know that soot always contains over 12% of hydrogen atoms by number.

Tesner observed very different rates of growth for methane, benzene and cyclohexane; benzene requires a much higher temperature. The temperature dependence of the rate of growth also has a characteristic value for each hydrocarbon. For isolated particles in a flame decomposition of an endothermic compound like acetylene might raise the temperature of the particle and assist the process, whereas an exothermic molecule like methane would cool the particle when it was decomposed. Watson (1948) in a paper on electron micrography techniques took as example a statistical study of the size of carbon-black particles formed from Shawinigan acetylene; Tesner (1962) discussed this and other work on acetylene decomposition and concluded that nucleation occurred in the temperature range 500°C to 800°C, that the activation energy was 60 to 70 kcal/mole and that the activation energy for surface decomposition of the acetylene on the nucleus was probably zero and definitely less than 8 kcal; this seems rather a lot of information to squeeze out of a study of a mixed sample of acetylene black, but does appear to support a nucleation and growth mechanism for acetylene decomposition, and thus perhaps for acetylene flames as well.

We have noted (page 216) that carbon monoxide is unstable and breaks down to carbon and CO_2 below about 1000°C, but that in the gas-phase the Boudouard equilibrium is frozen. It would appear possible that carbon particles might grow by surface decomposition of carbon from CO in the low-temperature regions of flames. However Tesner has informed us that carbon particles in contact with CO show no growth whatever at any temperature, and the C^{14} tracer work confirms this.

If satisfactory nucleation occurs, the first nuclei formed will grow rapidly and the degree of supersaturation of the vapour must then tend to fall. Further pyrolysis may, of course, maintain the supersaturation, but eventually the rate of nucleation will fall to zero as the supersaturation goes below the critical level, and then, as it continues to fall, small incipient particles will evaporate while large ones continue to grow. The number of nuclei forming will depend on the rate of temperature rise in the system, a high rate causing rapid supersaturation of the vapour and generation of a large number of nuclei. As the supersaturation of the vapour falls even the larger particles will cease to grow, and then coagulation to chains will become important. In flames, competing oxidation is also important and may tend to favour removal of small active nuclei.

Tesner (1973) and colleagues (Tesner, Snegiriova & Knorre, 1971; Tesner et al., 1971) have measured the maximum rate of nucleation in acetylene, variously given as 10^{14} or 10^{15} particles ml^{-1} sec^{-1}. They also determined the activation energy for nucleation of 140 to 180 kcal/mole, this value differing from the carbon-black value already referred to. Similarly, the activation energy for growth is given as 50 kcal/mole (Tesner, 1973). There does not yet seem to be a consensus of opinion about the values of these activation energies; there are probably several routes involving different species, and the type of surface on which growth occurs may also affect the value.

The role of C_2. We may rule out formation of solid carbon by condensation from gaseous atomic carbon because all processes leading to initial formation of the atomic carbon are so forbiddingly endothermic. However, we know spectroscopically that the diatomic species, C_2, is present in the reaction zone of rich flames. Smith (1940a) suggested that condensation of C_2 to solid carbon might occur. We shall see below that this is unlikely. However, they are probably important as nuclei.

Fairbairn & Gaydon (1955) found that the afterglow of carbon monoxide in a discharge tube contained C_2 and probably atomic carbon, and that this afterglowing gas when mixed with acetylene was able to produce solid carbonaceous deposits on the walls. This did not occur with several other hydrocarbons tested, such as methane, ethane or hexane; ethylene and benzene gave a little carbon. The decomposition of hydrocarbons on hot surfaces had been discussed, but this was probably the first experimental evidence that a nucleus as small as C_2 or C could liberate carbon from a gas at relatively low temperature.

Since earlier editions of this book were written, absorption spectroscopy has yielded actual concentrations of C_2 radicals in some flames, and these are higher than we had at first thought. For rich oxy-acetylene flames, the luminous mantle region shows a concentration of C_2 which is close to the equilibrium value of about 5×10^{14} ml^{-1} (Jessen & Gaydon, 1969), but in the reaction zone the C_2 concentration was around 50 times higher. This degree of supersaturation could presumably lead to nucleation and some growth directly from C_2. However, the rate of growth will depend not only on supersaturation, which is the ratio of actual vapour pressure to equilibrium vapour pressure, but on the actual vapour pressure, because the rate depends on the number of molecules diffusing to the particle surface. In acetylene flames the acetylene concentration will obviously be three or four orders greater than that of C_2, and C_2H_2 should be the main growth species. Other flames are much less hot and usually less rich in carbon, and the concentration of C_2 is expected to be very much lower, so again C_2 is probably not an important growth species. Acetylene, formed by pyrolysis of other fuels, and perhaps the polyacetylenes, are likely to be more important than C_2 for growth. Also growth from C_2 alone would give a soot consisting of pure carbon, whereas we have seen that actual soot contains a considerable amount of hydrogen. We also note that the flame of rich cyanogen/oxygen which is very hot and shows very strong C_2 emission bands does not normally form soot at all.

Although C_2 is favoured for nucleating hot premixed flames, this process is less likely in cooler flames and in diffusion flames. We note especially that in flat diffusion flames the C_2 emission occurs only on the oxygen side of the carbon zone, not in the pre-carbon zone where the continuous absorption is found.

Ions as nuclei. The strong effect of an electric field in controlling soot deposition and in modifying the size of the particles has been described (page 211). Most of these effects may be explained by movement of the charged particles under the influence of the field and by variation of residence time in the carbon-depositing zone. However, it is well known that ions may serve as nuclei for droplets growing from supersaturated vapours, as in the Wilson cloud chamber, and we need to consider whether ions could serve as nuclei for carbon particles. Mayo & Weinberg (*see* page 210) studied the effect of applied fields and of seeding with CsCl, using mostly an ethylene counter-flow diffusion flame, and they concluded that ions could serve as nuclei, but that most of the soot was produced by some other mechanism.

Howard (1969) considered nucleation by larger flame ions of the general type $C_nH_m^+$, which are known from mass spectrometric studies to be present. He considered the possibility of these nuclei growing to form crystallites of 20 to 30 Å size, the mutual repulsion due to the positive charge preventing coagulation at this stage. For larger particles agglomeration could occur to form the larger spherical granules.

The present view seems to be that although ions can serve as nuclei, other species are actually responsible for most of the nucleation. Electrical effects are more important at the coagulation stage, where, as we have seen, easily ionisable metal additives may either reduce or increase soot deposition. In diffusion flames there is probably little ionisation in the pre-carbon zone. Carbon formation also occurs in the pyrolysis of hydrocarbons in the absence of oxygen; there is some evidence for chemi-ionisation during pyrolysis (Bowser & Weinberg, 1976), but the amount is probably small because the main chemi-ionisation processes are known (*see* page 360) to involve reactions of oxygen such as $CH + O = CHO^+ + e^-$. Apart from serving directly as nuclei, ions could also serve to assist condensation of large polymers to mist droplets, as previously discussed.

The role of oxygen; particle combustion. Oxygen in sufficient quantity suppresses soot formation and carbon luminosity. It does, however, play a more positive role in causing the heat release which promotes the pyrolysis; it is possible to run some flames (e.g. that of ether) so rich that the flame is not hot enough to pyrolyse the fuel to carbon. Also the polymerisation reactions which may precede carbon formation usually occur by chain reactions and these may be catalysed by oxygen. Gaydon & Whittingham (1947) discussed the effect of SO_3 in promoting carbon formation and discussed radical chain mechanisms and quoted work showing the strong catalytic effect of oxygen in the polymerisation of ethylene.

Oxygen may, in excess, reduce carbon formation in two ways. It may interfere with the polymerisation process by breaking the reaction chains, providing an alternative path to oxygenated products, such as aldehydes instead of large unsaturated polymers. Alternatively it may burn away the carbon particles as they are formed. The combustion of coal particles has been discussed previously (page 173); if some of the hydrogenated part of the soot is volatile, then soot might burn in the same way as very small coal particles. We may also note some old work by Gumz (1939) on the rates of burning of solid carbon particles, Fig. 8.9 shows the time to burn particles. It will be noted that the particles are consumed more rapidly in a water-gas atmosphere than in air and that the effect of temperature is not very great; this is because the rate of combustion depends mainly on the rate of mass transport to the surface rather than on the chemical reaction rate. However, below a certain temperature, the chemical reaction rate will be the limiting factor, and work by Matsui, Koyama & Uehara (1975) shows that the surface combustion of carbon is very sensitive to temperature in the range 1050–1250 K; the effect of gas flow rate has also been studied.

Summary. Since the first edition of this book some twenty-five years ago, considerable progress has been made in understanding soot-formation processes, and even since the third edition some valuable attempts have been made to put the subject on a really quantitative basis. However, of the routes suggested it is still not possible to pick out one and say this is

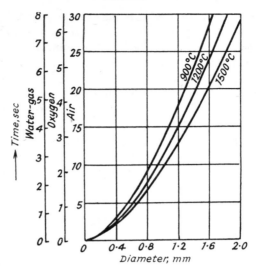

Fig. 8.9. Time for complete combustion of carbon particles as a function of initial diameter, for various temperatures and atmospheres.

[*After Gumz*

the way soot is formed. We must stress again that the route depends on the type of flame and there is probably no universal one. However, progress has been made in obtaining reliable rate data for many of the reactions and in computer handling of this data.

Of the six routes considered here – polymerisation via acetylene and polyacetylenes – ring closure to form aromatics and polycyclic compounds – condensation and graphitisation – nucleation by a small neutral fragment and surface growth on this – nucleation by ions – and polymerisation of C_2 – it seems that there is little support for the last two, even though the reactions themselves are possible.

In hot premixed flames nucleation by small neutral fragments, probably C_2 or C_2H, has received strong support, especially from the quantitative studies of Jensen (1974). He has made a computer study of the radical production, nucleation, coagulation, growth and oxidation processes using actual values of rate constants; thus he finds it necessary, for methane, to use rates for a minimum of ten reactions to produce the necessary radicals. He has also considered growth and coagulation of solid particles by considering collisions between particles of fifteen size groups. His calculations of the variation with time of species concentrations and size distribution apparently agree with experimental data for the pyrolysis of methane, and he has also compared further calculations with observations on smoke production in the practical case of a rocket motor using liquid isopropyl nitrate as monopropellant and found satisfactory agreement. Jensen concluded that initial nucleation was by C_2, C_3 or C_2H, with the growth species mainly C_2H_2 with, perhaps, some growth from polyacetylenes as well.

For diffusion flames and normal pyrolysis, available reaction times are longer than for premixed flames and temperatures in the soot-forming zone are usually lower. Although Jensen's work on methane pyrolysis seems to favour a nucleation mechanism also, there is a lot of evidence for formation of large molecules including aromatics in the pre-carbon zone. Also the likely nucleating species, such as C_2, are not detected spectroscopically in the pre-carbon zone of flat diffusion flames as they only occur on the oxygen side of the carbon zone at the base of this type of flame. There is obviously need for more quantitative work on diffusion flames. The work of Graham, Homer & Rosenfeld (1975) has shown that in the pyrolysis of aromatics there are two routes, via polycyclic compounds at low temperature and via C_2-type fragments at high temperature. There could even be a difference between diffusion flames with oxygen and those with air because of the different temperatures of these flames. Again high-temperature pyrolysis in shock waves appears to involve polyacetylenes and may differ from normal pyrolysis which occurs more slowly at lower temperature.

There are also other unsolved problems. In the nucleation process Jensen suggested C_2, C_3 and C_2H as possible species, and some of his calculations seem to favour C_2H rather than C_2 because the former gives a higher growth rate, using the reaction

$$C_2H + C_2H_2 = C_4H + H_2,$$

which might also account for the presence of some hydrogen in the soot. However, Cullis (1975) studied the kinetics of various reactions of C_2H, which he produced photochemically, and found that hydrogen abstraction reactions of the type

$$C_2H + C_2H_2 = C_2H_2 + C_2H$$

were able to compete with the chain-lengthening hydrogen-transfer

$$C_2H + C_2H_2 = C_4H_2 + H$$

and he concluded that the present kinetic results suggest that the growth of polyacetylenic species does not occur sufficiently rapidly to account for the almost instantaneous formation of carbon in sooting flames. Although C_3 is detected spectroscopically in the mantle of rich oxy-acetylene flames it is not a strong feature of the spectrum and the measurements by Jessen

& Gaydon showed that C_3 did not attain a high supersaturation in the reaction zone as was found for C_2. C_3 appears to us to be unlikely as a main nucleating species, but it is difficult to decide between C_2 and C_2H.

Another problem is to explain the structure of individual soot particles. We have seen that these appear to be spherical but at the same time to be composed of a number of crystallites. It is not clear whether this structure can be explained by coagulation of separate growing nuclei, or whether it requires crystal growth within an already formed particle.

The carbon-black process

The most important positive application of the formation of carbon in flames is the carbon-black industry. The soot, or carbon black, is mainly used as a filler in rubber-making, the carbon increasing its strength and resistance to abrasion. It is also being used as a filler in plastics to give protection against photochemical degradation and sometimes to impart electrical conductivity. Other applications are in printer's inks, black paper, paints and microphones. The value depends to some extent on the particle size, forms with small particle size being the more highly prized because they are blacker.

The following are the principle types of carbon black, classified mainly by their method of manufacture. For details see Ellis (1937) and Kirk & Othmer (1964).

(a) Channel black. A hydrocarbon, usually natural gas, is burnt as a diffusion flame in a deficient air supply. Thin flat flames, of the type given by a batswing burner, are employed, and the flames impinge on a cooled surface from which the carbon is periodically collected. The yield of carbon depends to some extent on the air supply and size of flame, but with natural gas it is always low: from 1% to 5% of the possible carbon content of the fuel. The finest particles with an average size of only 90 Å are obtained with the low yield, whereas the higher yield gives particles around 300 Å. The maximum flame temperature is around 1400 K and the rate of heating of the fuel is rather low. One can easily perceive the extreme inefficiency of this process when flying over Texas, seeing the black clouds obscuring the landscape.

(b) Lampblack. This is the oldest form, and is made by burning petroleum residues or tar oils in open pans. The smoke is ducted off and allowed to settle under gravity. Particle sizes range from 650 to 1000 Å.

(c) Furnace black. This is made by burning a hydrocarbon, either natural gas or oil, with a deficient air supply in a furnace or retort; in some forms a portion of the fuel is completely burnt and then more fuel is fed into the hot products. The hot gases are partly cooled with a water spray and the soot is collected either by electrostatic separation or by centrifuging in a cyclone. Average particle sizes are usually around 600 to 800 Å and the yield is higher than for the channel process, ranging from 10% to 30%.

(d) Thermal black. This is obtained by pure pyrolysis. In some early forms natural gas was blown through a heated tube, surface deposition, which gives an undesirable hard carbon, being prevented by forcing an inert gas to diffuse through the porous walls of the tube. The more common practice now is to pass natural gas through a large furnace vessel packed with refractory bricks which are initially heated to about 1300°C; normally a pair of furnaces are used alternately, the bricks in one being heated by a flame of roughly stoichiometric mixture while the other furnace is used for the pyrolysis. The soot is separated in cyclones. The particle size is rather large, mean values ranging from 1200 Å to 5000 Å, but the efficiency may reach 40 to 50%. The structure of particles of thermal black has been discussed by Marsh *et al.* (1971), and Plate 22*b* shows a phase-contrast electron micrograph.

(e) Acetylene black. This is usually made by continuous pyrolysis at around 800°C, the exothermic decomposition serving to maintain the temperature without auxiliary heating. Alternatively it may be made by explosive decomposition of acetylene or by the channel process. The average particle size is fairly small, 400 Å to 500 Å. The thermal and explosive acetylene blacks have a very low hydrogen content (no volatile or extractable components), have low surface activity and cannot be wetted by water.

The size of the particles, and the way they adhere together in chains, affects the quality of the product when they are compounded with rubber. The colour of the carbon black also varies with particle size and agglomeration. The structure of the particles and its effect on electrical conductivity, especially in relation to its use in microphones, has been discussed by Grisdale (1953). The ignition temperature of carbon black in air is quite low, about 500°C.

Chapter IX

Radiation Processes in Flames

The radiation from any system in complete equilibrium will be continuous and the same as for a black body at the same temperature as the system. Fully aerated flames usually show an emission spectrum of discrete bands, while even luminous flames of fuel-rich mixtures normally have a much lower emissivity than 1, i.e. radiate less than a black body. Thus flame gases are never in complete equilibrium. This simple fact is easily overlooked.

In small flames the emission of radiation by molecules and particles is not balanced by the absorption of radiation, so that there is a steady de-activation of excited molecules and radiation cooling of particles which has to be made up by collision processes within the flame gases. If these collision processes are not sufficiently efficient then the distribution of energy among excited molecules or the temperature of solid particles may differ from that for equilibrium at the temperature of the flame gases. It is thus necessary to examine in detail the processes of radiation and of excitation to the high-energy states from which radiation occurs. The radiation from flames is frequently used, by methods to be discussed in the following chapters, for measuring the flame temperature, especially for very hot flames. The radiation itself is also important in some cases, either because of its role in heat transfer as in furnaces, or for illumination.

The nature of radiation

The distribution of radiation with wavelength is given by Planck's radiation law. In a constant temperature enclosure at temperature T, the radiation density between wavelengths λ and $\lambda + d\lambda$ is equal to

$$8\pi hc\lambda^{-5} d\lambda/(e^{c_2/\lambda T} - 1),$$

where c is the velocity of light, h is Planck's constant and c_2 is the second radiation constant. This expression may be rewritten for convenience as

$$8\pi hc \cdot P, \text{ where } P \text{ is } \lambda^{-5} d\lambda/(e^{c_2/\lambda T} - 1).$$

This expression covers radiation in all directions, and the radiation density in a small solid angle $d\omega$ is

$$8\pi hcP\, d\omega/4\pi = 2hcP d\omega.$$

Thus the radiation emitted per unit time in a small solid angle normal to the surface of a black body of area A will be

$$c.A.2hcP d\omega = 2hc^2 PA d\omega.$$

For radiation in a direction making an angle α with the normal the effective area will be foreshortened and the radiation will be

$$2hc^2 PA d\omega \cdot \cos\alpha.$$

For the total hemispherical radiation we have

$$2hc^2 PA \int_0^{\pi/2} 2\pi \sin\alpha \cdot \cos\alpha \cdot d\alpha = 2\pi hc^2 PA.$$

It is not always realised that the radiation per unit solid angle normal to the surface is $1/\pi$, *not* $1/2\pi$, times the total hemispherical radiation, and care must be taken in using the appropriate value of the first radiation constant not to make an error by a factor of 2. The correct form of expression is given by Harrison, Lord & Loofbourow (1949) and Unsöld (1955). We are most frequently interested in the radiation normal to the surface, and for this we may express the Planck radiation law in the form

$$I_\lambda = \frac{2E_\lambda hc^2 \lambda^{-5}}{e^{c_2/\lambda T} - 1} \cdot A d\lambda = \frac{2E_\lambda c_1 \lambda^{-5}}{e^{c_2/\lambda T} - 1} A d\lambda. \qquad (9.1)$$

E_λ is the emissivity at wavelength λ and c_1 is the first radiation constant.*
c_1 has the value 0.588×10^{-5} erg cm^2 sec^{-1} or 0.588×10^{-12} watts cm^{-2}, c_2 is equal to 1.438 cm deg, and λ is in cm and A in cm^2. I_λ is the intensity, in ergs/sec.or watts, per unit solid angle, normal to the surface. Curves for black-body radiation at 2000, 2500 and 3000 K are plotted in Fig. 9.1. These show the very rapid variation of radiation with temperature in the visible region.

For practical use a close approximation to this is given by Wien's law

$$I_\lambda = 2E_\lambda A c_1 \lambda^{-5} e^{-c_2/\lambda T} d\lambda. \qquad (9.2)$$

This holds with sufficient accuracy as long as λT is less than 0.2 cm deg. It begins to fail for long wavelengths or very high temperatures.

For a black body the emissivity E_λ is equal to 1 for all wavelengths; if the value of E_λ is less than 1 but constant for all wavelengths we have a grey body. In practice, flames do not usually approximate to either of these

* Some books give $c_1 = 2\pi c^2 h$, which is 2π times the c_1 used here.

conditions. For fully aerated flames E_λ is close to zero for most values of λ but may reach fairly high values, near 1, for a limited number of emission bands, the strongest of which are in the infra-red. Even for luminous flames, in which the radiation is due to emission from solid carbon particles, the emissivity varies to some extent with wavelength.

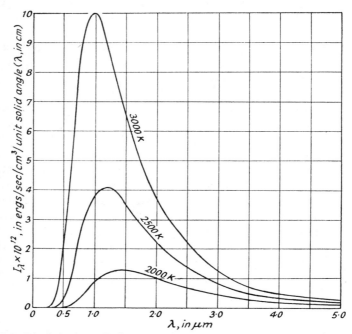

Fig. 9.1. Black body radiation at various temperatures, as a function of wavelength.

For a black body the wavelength λ_m for maximum intensity varies with temperature, and is given by Wien's displacement law, $\lambda_m \cdot T = 0\cdot 289$ cm K, or with λ_m in angstroms,

$$\lambda_m T = 2\cdot 89 \times 10^7 \tag{9.3}$$

The total radiation (over a hemisphere) from unit area of a black body at temperature T is given by the Stefan-Boltzmann law,

$$W = \sigma T^4 = 5\cdot 67 \times 10^{-12} \times T^4 \text{ watts cm}^{-2}. \tag{9.4}$$

These four fundamental laws are most useful for dealing with bodies emitting a continuous spectrum, but their application to small flames, which emit discrete banded spectra, is limited. However, it should be remembered that if the flame gases are in thermodynamic equilibrium, in which the energy is equally partitioned among its various possible forms, then the emissivity at any wavelength E_λ cannot be greater than 1 and the Planck radiation law and Wien's law may be used to predict the maximum possible brightness at any wavelength.

Continuous spectra are in general emitted by hot solid bodies, but in some cases regions of continuous emission may be due to processes such as recombination of ions, or associations of atoms or radicals. For molecules in the gaseous state, however, the energy of each molecule is restricted to a limited number of possible values, i.e. is quantised, and the wavelengths of radiation which the molecule can emit or absorb are limited to a relatively small number of lines in the spectrum.

Spectra in the visible and ultra-violet regions are generally due to changes of electronic energy, i.e. to a transition from one configuration of electrons about the molecule to another configuration. This change determines the position of the band system as a whole. Accompanying changes in the vibrational energy of the atoms of the molecule determine the position of individual bands within the band system. Accompanying changes of rotational energy of the molecule as a whole determine the fine structure of individual bands, i.e. the fine line structure of the band. For a more detailed discussion of the theory of molecular spectra *see* Herzberg (1945; 1950; 1966; 1971) or Gaydon (1968). Band spectra in the near infra-red are due to changes of vibrational and rotational energy of the molecules only, while spectra in the far infra-red are due to changes of rotational energy only.

None of the ordinary molecules which are the stable products of combustion, such as H_2O, CO_2, CO, O_2 or N_2 possess electronic energy levels situated so as to enable these molecules to give spectra of appreciable strength in the visible or ultra-violet regions. The only product of combustion which has an appreciable equilibrium concentration which does give a strong spectrum is the hydroxyl radical, OH, which gives a band system in the ultra-violet, with the strongest head at 3064 Å. There is some weak chemiluminescent emission of the Fourth Positive system of CO below 2300 Å. In some very hot flames there may be weak emission of the Schumann-Runge bands of O_2 giving band structure from the blue into the ultra-violet, and in the combustion of carbon monoxide the mainly continuous spectrum is accompanied by a weak banded structure due to CO_2, but generally speaking we may say that burnt flame gases do not give appreciable emission or absorption of radiation in the visible region and the only strong feature in the ultra-violet is the OH band system. Thus the emission and absorption by the interconal gases in the visible and ultra-violet is, quantitatively, very small indeed as long as no carbon is present. In the reaction zones of flames, corresponding to the inner cone of the Bunsen flame, many unstable radicals such as CH, C_2, HCO, NH, NH_2, etc., may be formed and these do give appreciable emission in the visible and near ultra-violet. The spectra of flames is fully discussed by Gaydon (1974) in *The Spectroscopy of Flames*.

H_2O has strong vibration bands at 1·8, 2·7 and 6·3 μm and a rotational band that covers the region from 10 to about 100 μm. CO_2 emits at 2·8, 4·3 and 15 μm. Other vibrational bands of interest are CO at 2·3 and 4·5 μm, NO at 2·6 and 5·2 μm and the OH vibrational bands that cover the near infra-red area to about 4 μm. It follows from Wien's displacement law (9.3), that at flame temperatures the maximum emission will tend to be in the near infra-red between about 2 μm and 1 μm according to the temperature of the flame.

For small flames of the Bunsen type the radiation is mainly in the near infra-red. For high aeration about 10% of the heat of combustion is lost by radiation, rising to about 18% for normal full aeration, then falling slightly with richer mixtures, until the flame becomes luminous owing to carbon formation, when the radiation rises to about 19% (Hartley, 1932). These figures serve to indicate the magnitude of the heat loss by radiation from a Bunsen flame, but have little precise meaning; the radiation from the reaction zone may depend mainly on the chemical processes and amount of gas burnt, but the bulk of the radiation comes from the hot burnt gases, and its amount will depend on whether these gases are given time to radiate; if the burnt gases are protected, either by the nature of the gas flow or by deliberate screening, then the amount of radiation may be large, whereas if there is rapid turbulent mixing of the burnt gases with surrounding cool air, then the radiation may be reduced considerably.

As already stated the main emission is in the infra-red. For clear flames the radiation in the visible and ultra-violet accounts usually for less than 0·4% of the heat of combustion. This visible radiation comes mainly from the inner cone, while the infra-red radiation comes from the main body of the gases, both the interconal gases and the burnt products. This is well illustrated by some photographs by Bonhoeffer & Eggert (1939); using blue, green or yellow light the photograph of the flame shows only the inner cone clearly, but using infra-red light of wavelength 8500 or 10050 Å, the inner cone is invisible but the interconal gases show up. For luminous flames, radiation from hot carbon particles will increase the radiation in the visible. For the much hotter flames supported by oxygen, we should expect an increase in radiation; it is possible that to some extent the shift of the wavelength for maximum radiation, given by Wien's law, towards the visible where the emissivity is very low, may limit the amount of radiation; also with the faster burning mixtures the volume of flame gases tends to be smaller; we are not aware of any quantitative work on the total amount of radiation from flames supported by oxygen.

The equipartition of energy

In any system the available energy may exist in many forms, such as internal energy of the molecules, kinetic energy of motion, chemical energy and as radiation. In any system in complete thermodynamic equilibrium the amount of energy in any one form will be statistically constant, and it can be shown that the energy will be equally partitioned among the various possible forms. For any form, or 'degree of freedom', for which a continuous range of energy is possible, the amount of energy will be proportional to the absolute temperature T, and equal to $\frac{1}{2}kT$ per molecule, where k is the Boltzmann constant $= 1\cdot38054 \times 10^{-23}$ J deg^{-1} or $\frac{1}{2}RT$ per gram molecule, where R the gas constant $= 8\cdot3143$ J deg^{-1} mole^{-1} or $1\cdot9872$ cal deg^{-1} mole^{-1}. Thus there are three degrees of freedom for translational energy. For the various forms of internal energy of the molecules, such as energy of rotation, of internal vibration and of electronic excitation, the energy of a molecule may only take certain discrete values because of quantisation restrictions. For rotation of the molecule the possible energy levels are usually sufficiently close for these restrictions to be unimportant and then each rotational degree of freedom (two for a diatomic molecule, three for a polyatomic one) will take up its full $\frac{1}{2}kT$. For the vibrational and electronic energies, however, the quantum restrictions reduce the amount of energy which can go into these degrees of freedom. The distribution of molecules among the possible energy states will then be given by the Maxwell-Boltzmann distribution law. If for a particular molecular species there are possible energy states with energies E_0, E_1, E_2, E_3, etc., then the number of molecules with these energies will be proportional to

$$e^{-E_0/kT},\ e^{-E_1/kT},\ e^{-E_2/kT},\ e^{-E_3/kT},\ \text{etc.}$$

If for any one energy value there are a number g of physically distinguishable states of the molecule, all having the same energy E, then the proportion of molecules with this energy E will be increased by a factor g to $g \cdot e^{-E/kT}$. g is known as the statistical weight, and its value can be assigned if the type of energy term is known from the spectroscopic data (*see* Gaydon, 1968, page 153).

In equilibrium the rate of conversion of energy of one form into that of another must be the same as for the reverse process. Thus, for example, deactivation of an electronically excited molecule by collision must be balanced by excitation by collision; deactivation by emission of radiation must be balanced by absorption of radiation; release of translational energy following chemical reaction must be balanced by loss of translational energy in activating the reverse chemical reaction. This is sometimes generalised as the principle of micro-reversibility, and is often useful for the detailed study of departures from equilibrium (e.g. Perrin, 1930; Gilmore, Bauer & McGowan, 1969).

The attainment of vibrational equilibrium within a closed system requires time. This time is usually measured in microseconds at a reference pressure of one atmosphere (μsec-atm). This is equivalent to a statement that the relaxation process requires a given number of collisions. The probability of vibrational excitation or de-excitation in one collision, for example for $CO_2 (v_2) + M \rightleftharpoons CO_2 + M$, is about 10^{-4} at room temperature and about unity at 8000 K (Taylor & Bitterman, 1969). Rotational relaxation is even faster and requires only a few collisions. In general the fastest energy transfer is translational $\rightleftharpoons$ rotational, followed by vibrational $\rightleftharpoons$ vibrational, and the slowest is translational $\rightleftharpoons$ vibrational. Electronic excitation will be considered separately below (*see also* Bauer, 1969).

It can be seen that for laboratory flames at one atmosphere the vibrational and rotational relaxation processes are so fast as to be unobservable. However, for flames at low pressure or at hypersonic speed they are very important.

The speed of chemical reactions depends greatly on temperature. At room temperature a hydrogen–oxygen mixture can be stored indefinitely. Above the ignition temperature equilibrium will be attained very rapidly. There are some reactions that require large activation energies, for example formation of NO. As nitrogen does not participate in the reactions of many flames (for example, hydrogen–air), and the dissociation of N_2 must precede the formation of NO, the latter will be formed very slowly even at 1 atm (*see* page 130). Similarly NO once formed in the flame will not decay and nitric oxide will be present in the burnt gases after they have cooled to room temperature. Also, as most atoms can only recombine by three-body collisions which are rare at low pressures, one can expect large excess populations of atoms in the cooled burnt gases of low pressure flames. For conditions in hydrogen flames see Wolfhard & Hinck (1967).

Although electronic excitation will be discussed in the following section, we will consider here some aspects of electronic relaxation. In equilibrium each forward process will be balanced by a reverse process (as discussed above). In most flames the emissivity of a line or band is usually much below unity. The attainment of electronic equilibrium thus depends on the efficiency of the exciting and de-exciting collisions, such that de-excitation by radiation is only a small fraction of the de-excitation by collisions. At low pressure this will not be the case, even if the gases are otherwise in chemical and vibrational equilibrium. The consequence is the lowering of the population of the electronic state and thus a decrease in line or band emission. This condition has been called 'radiation depletion' or 'collision limiting'. The intensity is then given by (assuming steady state):

$$\frac{I}{I_{\text{equilibrium}}} = \frac{K_+}{K_- + 1/\tau}$$

where I is the intensity of a line or band with radiation depletion, K_+ and K_- are the rates of the excitation and de-excitation process and τ is the lifetime of the electronic transition.

In extreme cases of low pressure or fast flow or both, the assumption of steady state may not even be applicable. For example, the excitation of a metal line by electrons or vibrationally excited nitrogen is further slowed down if electrons or $N_2(v)$ in turn are formed only very slowly. Thus it is clear that the need exists to examine the elementary excitation processes in detail.

Processes of electronic excitation

Much of our information about the temperature, species concentrations and state of equilibrium of flame gases comes from study of the visible and ultra-violet radiation, which results from emission by electronically excited atoms and molecules. For a quantitative interpretation of the data obtained from emission spectroscopy we thus need to know something of the processes leading both to excitation and to quenching.

For electric discharges, excitation is due to impact by fast electrons produced by the discharge and accelerated by the electric field. In flame gases, however, the concentration of electrons is much lower and the proportion of electrons with sufficient energy to cause excitation will be much less. For burnt gases in equilibrium the number of free electrons should certainly be much too small to contribute appreciably to the excitation. For the reaction zone of a flame there is evidence, however, that the ionisation may be abnormally high (*see* Chapter XIII); for organic flames with air the concentration of ions may be as high as 10^{12} per cc. Assuming this figure and a normal gas-kinetic cross-section and making some allowance for the high temperature and greater mobility of free electrons, a molecule should then make about 10^5 collisions per second with free electrons. Thus during a normal radiative lifetime of 10^{-8} sec there would be only one chance in a thousand of collision deactivation by an electron. Unless the effective cross-section is extremely large, collision deactivation by free electrons will therefore be a rare event, and hence the reverse process, excitation by electron collision, will also be infrequent.

The conversion of translational energy of particles of molecular mass into energy of electronic excitation is unlikely to account for appreciable excitation. Classically one can see that it is improbable that a relatively light electron will gain much energy and therefore change its orbit because of collision with a more massive particle. Experimentally it is known that quenching of fluorescence, i.e. collision deactivation of electronically excited atoms, by monatomic gases is extremely inefficient (*see*, for example, Massey, 1949); by the principle of micro-reversibility the conversion of kinetic energy of translation into electronic excitation must also be inefficient. There may be exceptions to this statement in special cases when we have crossing of potential energy curves or surfaces, as indicated below.

There appear to be two remaining methods of producing electronic excitation – conversion of internal molecular energy into electronic excitation by collision, and chemical processes.

The conversion of internal molecular energy (i.e. energy of vibration and rotation) into electronic excitation during a collision is best understood in terms of the potential energy surfaces of the collision complex which is momentarily formed. To begin with, in order to illustrate the physical principles, we shall consider a simple diatomic molecule AB, and then the collision between the two atoms A and B which can form such a molecule.

Fig. 9.2 shows two potential energy curves of a diatomic molecule AB, one dissociating to normal atoms $A + B$, and the other to a normal atom A and an excited atom B^*. The two curves cross at the point X, and we assume that the electronic states to which the curves correspond are of a type between which interaction is possible. Then we have the possibility that a radiationless transition may take place from one electronic state to the other. If the point X lies above the dissociation limit to $A + B$ then we shall have the well-known phenomenon of predissociation, in which the transition from a level of the excited molecule AB^* above X will lead to the formation of a normal molecule AB with more vibrational energy than it can retain, and the molecule will dissociate. If the crossing occurs below the dissociation limit, then in the region of the crossing point X the energy levels of both AB and AB^* will be perturbed, and radiationless transition from state AB^* to AB will only lead to conversion of the electronic excitation energy into vibrational energy.

Now let us consider the collision between a normal atom A and an excited atom B^*. The collision cannot lead to the formation of a stable molecule, because the system must always have sufficient energy for dissociation of the molecule. However, it is possible for the collision to form a molecule in the excited state AB^* momentarily, and for this to undergo radiationless transition to the normal molecule AB which then, having too much vibrational energy, dissociates to normal atoms. The net result is conversion of excitation energy into translational energy with which the atoms fly apart. The reverse process, collision of normal A and B atoms with large amounts of translational energy, could lead to the formation of an excited atom B^*. However, even this process is less simple than pictured here, because usually the collision will not be head-on. If the collision is not head-on, then the system will possess angular momentum, and the potential energy curves must be replaced by effective potential energy curves in which the energy due to the angular momentum, or rotation, is included. The curves will then possess maxima, usually at an internuclear distance of 4 to 5 Å, and this potential barrier must be taken into account (Eyring, Gerschinowitz & Sun, 1935).

For any one pair of dissociation products, there may be several possible electronic states of the molecule; for two atoms in ^1S states only one molecular state, a $^1\Sigma$, is possible, but for two atoms both in ^3P states there

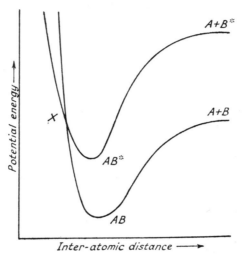

Fig. 9.2.

are eighteen possible molecular states. Thus at first sight there appears a good chance of crossing of potential curves as indicated in Fig. 9.2. However, many of the possible molecular states will be unstable, having repulsive energy curves, and crossings are only important if the selection rules for radiationless transition are fulfilled. For diatomic molecules we have the further restriction that crossing of curves of states of the same electronic species is forbidden (non-crossing rule) (*see* Gaydon, 1968). Thus in practice, conversion of translational energy into electronic excitation by this process will be rare for collisions between free atoms.

For collisions between molecules or between an atom and a molecule, the potential energy curves must be replaced by potential energy surfaces, usually by surfaces in many dimensions. These surfaces may be more numerous than for the diatomic case, and being surfaces instead of lines, there is a better chance of crossing, at least in some regions. Also the non-crossing rule does not apparently apply for polyatomic molecules (Teller, 1937) so that more crossings are allowed.

The high efficiency of interconversion of electronic and vibrational energies was suggested by observation of the efficient quenching of resonance fluorescence by molecular gases (*see* Table 9.3, page 251) but not by atomic gases. Shock-tube studies have, indeed, shown (Gaydon & Hurle, 1963) that the excitation temperature of Na and Cr atoms behind the shock front in some molecular gases (N_2, CO) is initially too low, and its relaxation to the equilibrium value occurs at the same rate as the vibrational relaxation of the molecules. In the quenching of mercury resonance fluorescence by carbon monoxide Polanyi (1963) and Karl, Kraus & Polanyi (1967) have shown that the CO molecules emit in the infra-red and that vibrational levels up to $v = 9$ are excited; a high proportion, although not all, of the electronic energy is converted into vibrational energy. To explain this effect it is necessary to assume the formation of a collision complex with crossing of potential energy surfaces. The reverse effect, the direct electronic excitation of atoms (Na) by vibrationally excited molecules (N_2) has been demonstrated by Starr (1965). For discussions of energy transfer processes see Polanyi (1963) and Callear (1965). It seems that when there are departures from equilibrium the electronic excitation temperature is likely to be closer to the effective vibrational temperature than to the translational temperature.

Chemical processes may also be important in causing electronic excitation. These may fall into two general classes of reaction,

$$A + B + C = AB + C^*$$

and

$$AB + C = A^* + BC \quad \text{or} \quad AB + CD = AC + BD^*,$$

where A, B, C, and D may be either single atoms or molecular groups, and the asterisk, *, always denotes electronic excitation.

The first reaction involves three-body collisions. From gas kinetic values, three-body collisions are about a thousandth as frequent as two-body ones at atmospheric pressure. Thus reactions of this type will have to be very efficient to contribute appreciably to excitation at atmospheric pressure. It should, however, be remembered that effective cross-sections for this type of reaction may be much larger than those from gas-kinetic data, and also for large molecules the formation of collision complexes ('sticky collisions') may increase the proportion of three-body collisions. It would appear, at first sight, that reaction involving three-body collisions must become relatively less important at reduced pressure, compared with other processes;

however, it seems that this may not be so for gases in equilibrium, because the reverse process is dissociation by a two-body collision, and it seems that when all factors are taken into consideration the increase in the concentration of the dissociated fragments A and B at reduced pressure will balance the smaller proportion of three-body collisions.

The second type of reaction requires only two-body collisions and so may appear more likely, but the number of possible reactions of this type is much more limited, and it would, for example, be difficult to find reactions of this type to account for the excitation of free metallic atoms in flame gases. Also, this type of reaction involving a rearrangement of the system may be less efficient because of steric factors.

To summarise this section, we may say that neither collisions with free electrons nor conversion of translational energy of particles of atomic masses into electronic excitation energy appear adequate to account for the emission from flames. Conversion of internal vibrational energy into electronic excitation is probably the main cause of discrete emission from flames. In some cases chemical processes may contribute to the emission, even under equilibrium conditions, if the effective cross-sections for collision are very high.

Conditions for radiative equilibrium

As already stressed, the radiation density in small clear flames will be so low that there is a tendency for excited electronic states of atoms and molecules to be depopulated by radiation. If the collision processes by which they are excited and also deactivated are not rapid compared with this radiation deactivation, then the proportion of excited atoms and molecules will fall below that expected from the Maxwell-Boltzmann distribution law, and any measurement of temperature depending on the determination of the population of excited states, that is, most optical methods, will fail.

The radiative life, that is the average time before an undisturbed excited atom or molecule loses its excitation by radiation, may be determined from quantitative measurements of the strength of the absorption or emission from the gas at known concentration and temperature, or from direct measurement of radiation decay. Table 9.1 gives some illustrative values, taken from the older literature and a few from later collections of data (Bennett & Dalby, 1960; 1962; Gaydon, 1974).

The number of collisions per second which a molecule suffers is rather an indefinable quantity except in special circumstances (Mitchell & Zemansky, 1934), and the gas-kinetic value is not directly related to the value for transfer of excitation energy. However, it may serve as a yard-stick. From the kinetic theory, the number of collisions a molecule suffers per second is equal to

$$4\sigma^2 n \sqrt{\frac{\pi RT}{M}}, \tag{9.5}$$

where σ is the effective diameter of the molecule, n is the number of molecules per cc, R is the gas constant ($= 8\cdot3143$ J deg^{-1} mole^{-1}), T the ab-

TABLE 9.1

Some values of radiative lifetimes

Emitter	Transition	Wavelength, Å	Life, sec
Na	$^2S - {}^2P$	5890, 5896	$1\cdot58 \times 10^{-8}$
K	$^2S - {}^2P$	7699, 7665	$2\cdot59 \times 10^{-8}$
Rb	$^2S - {}^2P$	7947, 7800	$2\cdot79 \times 10^{-8}$
Li	$^2S - {}^2P$	6708	$2\cdot7 \times 10^{-8}$
Mg	$^1S - {}^3P$	4571	4×10^{-3}
Cd	$^1S - {}^3P$	3261	$2\cdot5 \times 10^{-6}$
Cd	$^1S - {}^1P$	2288	$1\cdot8 \times 10^{-9}$
Hg	$^1S - {}^3P$	2537	$1\cdot15 \times 10^{-7}$
Hg	$^1S - {}^1P$	1849	1×10^{-9}
Tl	$^2P_{3/2} - {}^2S_{1/2}$	5350	$1\cdot4 \times 10^{-8}$
C_2	$A^3\Pi - x^3\Pi$	5165	$1\cdot70 \pm 0\cdot2 \times 10^{-7}$
OH	$^2\Sigma^+ \to {}^2\Pi$	3064	$1\cdot5 \times 10^{-6}$
CN	$B^2\Sigma^+ \to x^2\Sigma^+$	3883	$8\cdot3 \pm 1 \times 10^{-8}$
CH	$A^2\Delta \to x^2\Pi$	4315	$5\cdot6 \pm 0\cdot6 \times 10^{-7}$
CH	$B^2\Sigma \to x^2\Pi$	3900	$0\cdot8 \pm 0\cdot2 \times 10^{-6}$
NH	$A^3\Pi \to x^3\Sigma$	3360	$4\cdot55 \pm 0\cdot4 \times 10^{-7}$

solute temperature and M is the molecular weight. A rather more general formula for collisions of molecules of type 1 with molecules of type 2 is:

$$Z = 2n_1 n_2 \sigma^2_{12} \sqrt{2\pi RT \frac{M_1 + M_2}{M_1 M_2}},$$

where n_1 and n_2 are the concentrations, and M_1 and M_2 are the molecular weights.

Values of the effective collision cross-section, σ^2, may be derived from measurements of viscosity, diffusion coefficient, thermal conductivity or van der Waals's equation, and agree within a few per cent. Values from viscosity data (from Kaye & Laby, 1966) for a few typical gases, with collision frequencies calculated from equation (9.5), are given in Table 9.2.

Thus at S.T.P. the number of collisions which a molecule suffers is around 8×10^9 per second. The number will be proportional to the gas pressure, and inversely proportional to the square root of the absolute temperature. Thus in flame gases the high temperature will reduce the collision frequency to around 2×10^9.

TABLE 9.2
Collision cross-sections and collision frequencies from kinetic theory data (viscosity)

Gas	H_2	He	N_2	CO	CO_2	C_2H_4
σ^2	8.8×10^{-16} cm^2	6.6	14.1	13.8	15.2	17.9
Coll. freq. 0°C, 760 torr	15.4×10^9 sec^{-1}	6.9	7.7	7.75	9.3	13.3

The effective cross-sections for deactivation of electronically excited atoms can be derived from quenching of fluorescence and various other methods. Many of the older values quoted in our earlier editions seem to be inaccurate, probably because of self-absorption of the fluorescence emission. Values given in Table 9.3 are from Jenkins (1966; 1968; 1969), who used a flame fluorescence method, and so should be applicable to flame temperatures and conditions; there is still some lack of agreement with other authors but Jenkins's estimates indicate likely errors not exceeding about $\pm 20\%$. Some other values have been quoted in the review by Callear (1965).

TABLE 9.3
Quenching cross-sections, σ^2, in 10^{-16} cm^2, for some atomic resonance lines in various gases, obtained by flame fluorescence

Gas	Atom						
	Li	Na	K	Rb	Cs	Pb	Tl
H_2	5.2	2.87	1.03	0.61	1.7	0.4	0.03
O_2		12.3	15.5	25		15	13.2
N_2	6.75	6.95	5.6	6.1	25	5.7	6.4
CO	12.6	11.9	12.4	11.8		13	13.6
CO_2	9.2	17.0	21.4	24.0		29	32.5
H_2O	1.9	0.5	0.9	1.27	5.5	8	1.75
Ar	<0.3	<0.1	<0.2	<0.3	<0.9	<1.6	<0.1
He		<0.1	<0.08	<0.11	<0.4	<0.6	<0.12

It will be seen that these values vary over a wide range. The cross-sections for quenching for rare gases are limits set by the method of measurement, and actual values are probably almost negligible. The low values for water are rather surprising. Quenching is apparently specific to the excited species and to the quencher and the efficiency may be very small or up to about 2.

Thus in a flame at atmospheric pressure an excited atom may suffer up to perhaps 2×10^9 effective collisions for deactivation per second, although if the flame gases were inert for the particular deactivation, the number might be much less. For a fully allowed transition with a radiative life of 10^{-8} sec, a molecule will thus suffer about twenty collisions during this time. Thus it seems that under normal circumstances, processes of activation and deactivation by collision are likely to be much more important than radiation processes at atmospheric pressure. At low pressure, however, radiation deactivation will tend to become more important. Probably departures may be expected to occur below 1/10 atm, and below 1/100 atm ordinary collision processes will be quite unable to keep pace with radiation depletion.

This appears to be supported by experiment. Thus measurements of flame temperature by the sodium-line reversal method at atmospheric pressure give flame temperatures usually only 50° to 100° below theoretical values, and part of this difference may be due to heat losses from the flame, rather than to any lack of equilibrium. In flames at pressures of around 1/100 atm we have, however, found that temperatures measured in this way for the gases above the reaction zone come several hundred degrees too low. This may again be partly through heat losses, but we suspect it is mainly through lack of radiative equilibrium in the flame gases (*see also* page 282).

More convincing evidence of this failure of radiative equilibrium comes from some experiments involving a predissociation in the spectrum of OH (Gaydon & Wolfhard, 1951b; Gaydon & Kopp, 1971). The upper $^2\Sigma^+$ state is affected by a partially forbidden predissociation, probably to a $^4\Sigma^-$ state. The effect only occurs in vibrational levels with $v' = 2$ or more, and is best seen in the (1, 0) sequence of bands at 2811 Å. In a normal flame the (1, 0) band is the strongest, with the (2, 1) appreciably weaker, and the (3, 2) weaker still. In an oxy-hydrogen flame at about 1/100 atm pressure, however, the (2, 1) band is the strongest and the (3, 2) band is also enhanced, but keeps about the normal ratio of intensity compared with the (2, 1). The effect is shown in the reproductions in Plate 16, strips *a* and *b*. Our interpretation is that at this low gas pressure ordinary collision processes, by which vibrational energy is converted into energy of electronic excitation of OH, are unable to keep pace with radiation deactivation, so that the OH emission is much weaker than for equilibrium. For bands with v' of 2 or more, however, we have the additional process

$$O + H \rightleftharpoons OH\,(^4\Sigma^-) \rightleftharpoons OH\,(^2\Sigma^+).$$

Since the concentrations of O and H atoms in this low-pressure flame are around 10% this process is more important and assists in maintaining equilibrium for those levels affected by the predissociation and its inverse effect. Thus the bands (2, 1) and (3, 2) will be nearer their equilibrium strength and will appear relatively enhanced compared with the (1, 0) band.

Chemiluminescence

When a chemical reaction leads directly to the formation of an atom or a molecule in an electronically excited state, from which radiation may occur, then we may have a light emission out of all proportion to that to be expected from thermal emission. This phenomenon is usually referred to as chemiluminescence. In many cases, however, it is a matter of degree rather than kind. We have seen, page 248, that even in complete equilibrium reversible chemical processes may contribute to the formation of excited atoms and molecules. In some cases, such as the emission of formaldehyde bands from the bluish 'cool flames' of ether, aldehydes and some hydrocarbons at temperatures between 200°C and 400°C there can be no hesitation in calling the phenomenon chemiluminescence. In the case of the OH predissociation mentioned above, the chemical process of recombination of O and H atoms modifies the spectrum of OH, but this would not normally be regarded as a case of chemiluminescence. There is also the possibility of indirect chemiluminescence, in which excited molecules formed by chemical reaction pass on their excitation, by collision, to other species, from which abnormally high emission results; here again there are degrees of the effect according to whether the active molecules are in some unusually excited metastable form, or merely possess rather more than a fair share of vibrational energy.

There has been a lot of discussion about the extent to which the radiation from flames arises from ordinary thermal excitation or is due to chemiluminescence.

For premixed flames, of the Bunsen type, the emission from the interconal gases is probably almost all thermal in origin. For some flames, there is evidence for the persistence of more than equilibrium amounts of atomic hydrogen and oxygen, especially just above the inner cone. Padley & Sugden (1958) have studied effects on metal-atom excitation due to these excess free atoms; in hot flames the thermal emission is dominant, but in cooler near-limit flames the reaction zone is strengthened by chemiluminescence of the type

$$H + H + Na = H_2 + Na^*$$

or

$$H + OH + Na = H_2O + Na^*$$

which depend on the square of the radical concentration; just above the reaction zone thermal processes are dominant, but the full flame temperature is not attained, so there is weaker thermal metal-atom emission than higher in the flame where equilibrium is established.

For the reaction zone, corresponding to the inner cone of the Bunsen flame, there is ample evidence that the emission is much stronger than could be accounted for thermally. For OH in organic flames the radiation is vastly stronger from the reaction zone than from the interconal gases, although the concentration could not possibly be higher to an extent which would account for the effect. Also the effective rotational temperature of

OH (*see* Chapter XI) is very high, rising from around 5400 K at atmospheric pressure to near 10 000 K at very low pressure; this effect is best interpreted as direct chemiluminescence (Gaydon & Wolfhard, 1948). Measurements of excitation temperatures using reversal of Fe or Pb lines also indicate abnormally high electronic excitation, but in this case the effect may be indirect, rather than true chemiluminescence. In organic flames with air there is also emission of the CO Fourth Positive bands below 2400 Å; at these short wavelengths thermal emission from flames with air is quite negligible and the CO emission, although very weak, may be used for fire detection if a special discharge-type detector sensitive only to far ultra-violet radiation is used. The emission of the Swan bands of C_2 and of the CH bands at 4315 and 3900 Å is also very strong in the reaction zone of organic flames and certainly due to chemiluminescence.

The individual reactions responsible for these emissions have been the subject of much research and discussion in the literature and have been fully dealt with in *The Spectroscopy of Flames* (Gaydon, 1974). Here, to avoid repetition, we give only a summary.

For OH there are at least three established processes for chemiluminescence. In hydrocarbon flames the main process is

$$CH + O_2 = CO + OH^*,$$

the electronically excited OH being formed with high rotational temperature. In some hydrogen flames the inverse predissociation already referred to

$$O + H = OH\,(^4\Sigma^-) = OH\,(A^2\Sigma^+)$$

occurs; the strength of this depends on the square of the concentration of free atoms (which are in pseudo-equilibrium with each other and with ground-state OH radicals), and leads to selective excitation to vibrational levels $v' = 2$ and 3. At the base of low-temperature hydrogen flames the recombination

$$H + OH + OH = H_2O + OH^*$$

occurs; this has an intensity proportional to the cube of the concentration of free atoms or radicals but gives a fairly normal distribution of rotational and vibrational energy.

For CH the main reaction is

$$C_2 + OH = CO + CH^*.$$

There is also some other process responsible for CH excitation which is not dependent on C_2. Kinbara & Noda (1971) studied the time history of emission and absorption spectra following flash ignition of

$$C_2H_2 + O^+ + NO_2$$

and discussed as possible reactions

$$C_2H + O_2 = CO_2 + CH\ (A^2\Delta)$$
and $$C_2H_2 + OH = H_2 + CO + CH\ (X^2\Pi)$$

which could be followed by

$$CH + OH + H = CH^* + H_2O.$$

For C_2 it is difficult to find reactions which are sufficiently exothermic to form it in the electronically excited state and with high rotational temperature, as observed. Also, for acetylene/oxygen flames, isotope studies indicate that the two carbon atoms come not from one acetylene molecule but from separate carbon fragments. There is still no generally accepted consensus of opinion about how C_2^* is formed. Gaydon (1974) summarised evidence in favour of reactions involving polyacetylenes. For example, C_8H_2 might decompose

$$C_8H_2 \rightarrow C_6 + C_2 + H_2$$

or reactions might occur such as

$$C_4H_2 + O_2 \rightarrow C_4H_2O_2 \rightarrow C_2 + H_2 + 2CO.$$

Another reaction which has been considered is

$$CH_2 + C = C_2^* + H_2.$$

In a more recent paper, Kinbara & Noda (1975) suggest formation of C_2H, e.g. by $C_2H_2 + O = C_2H + OH$ or $C_2H_2 + H = C_2H + H_2$, with the C_2H polymerising to $(C_2H)_n$ which then decomposes to C_2 and CH fragments; this is really quite similar to the polyacetylene mechanism.

HCO is a feature of the spectrum of some fuel-weak flames and is often associated with the presence of peroxides. Reactions of CH with H_2O_2 or HO_2 may cause the chemiluminescence.

Excited formaldehyde molecules are responsible for the blue light from cool flames, and are formed by reactions of the general type

$$RCH_2O + R' = RR' + HCHO^*$$

where RCH_2O and R' are free radicals; methoxy is a likely radical

$$CH_3O + OH = HCHO^* + H_2O.$$

Suggested reactions for forming electronically excited CO are

$$C_2O + O = CO\ (A^1\Pi) + CO\ (X^1\Sigma)$$
$$CH^* + O = CO\ (A^1\Pi) + H$$
and $$C_2 + O_2 = CO\ (A^1\Pi) + CO\ (X^1\Sigma).$$

Vibrationally excited CO is also formed by

$$O + C_2H_2 = CH_2 + CO_{v=14}$$
and
$$O + CH (x^2\Pi) = H + CO_{v=33}.$$

Departures from equilibrium in flame gases

It seems appropriate to say here a few words about the nature of departures from equilibrium in flame gases. We have already discussed radiation effects. The most obvious type of failure to reach equilibrium will be through incomplete combustion. The main chemical processes are normally completed in a very short time during passage through the reaction zone of the flame, but in some cases the chemical reactions are not complete. This is especially likely to be the case with low-temperature flames, when reactions requiring large activation energies may not take place with sufficient speed. Thus reactions are most likely to be incomplete for flames with inactive diluents such as flames with air, and for near limiting mixtures. Any turbulence which causes quenching of combustion by contact with cold entrained air may result in incomplete combustion, and aldehydes and peroxides and other intermediaries can often be detected by chemical sampling on flame products under these conditions. It is also found in flames with air that nitric oxide may pass through the inner cone largely unchanged; in equilibrium it should be converted mostly to O_2 and N_2, but its decomposition only occurs at very high temperature.

Apart from these major chemical effects, there is evidence for other chemical abnormalities. Many of these may be due to the slowness with which free atoms recombine. The main reactions are probably of the chain type, proceeding by bimolecular collisions; this applies to both chain propagation and chain branching. Chain termination involving removal of free atoms or active radicals usually, however, requires a surface or a three-body reaction and tends to be less rapid. Thus free atoms or radicals may persist rather above the main reaction zone. This subject has already been discussed in connection with the effect of excess free radicals on flame propagation (page 100). Smith (1940*b*) has attributed candoluminescence of solids in flames containing hydrogen to recombination of excess hydrogen atoms at the surface.

In low-temperature hydrogen flames this excess of free atoms and OH radicals leads to chemiluminescent excitation of metal atoms (Padley & Sugden, 1959) by reactions such as

$$Pb + H + OH = H_2O + Pb^*$$
$$Tl + H + H = H_2 + Tl^*.$$

On the whole the effect of these chemical abnormalities on the radiation is not likely to be very great, although there is always the possibility that recombination of free atoms on a third body may lead to excitation of this third body. The abnormally high intensity (Plate 16c) of the (2, 1) and (3, 2) bands of OH in H_2/air flames at atmospheric pressure (Gaydon & Wolfhard, 1949a; 1951b; Charton & Gaydon, 1958) is due to an excess population of free atoms; in this case, H and O atoms recombine directly through the inverse predissociation to form OH in the vibrational levels with $v' = 2$ and 3. Recombination of free atoms could affect the temperature, and therefore the radiation, from hot particles.

There is a marked excess of electronic excitation energy in the reaction zone, as will be discussed in the next chapter, but we think it unlikely that this will persist appreciably into the interconal gases. While quenching cross-sections are very variable and are apparently specific to the type of excited atom and the quenching molecule, it seems generally that they are several orders too high to allow electronically excited atoms or molecules to persist for long; we do not believe that the 'latent energy' of flame gases can be explained in this way.

There has also been a good deal of discussion about persistence of vibrational and rotational energy during combustion. Observations on the dispersion of sound do clearly show that the conversion of vibrational energy into translational energy may in some cases be difficult and require many thousand collisions. However, in ordinary flame gases, containing a fair amount of water vapour, this persistence is not very important. If we assume that it takes about 10^3 collisions and that there are about 10^9 collisions per second, the departure from equilibrium should not persist more than something of the order of a microsecond. While the persistence of vibrational energy in cold gases may be fairly long, the relaxation time decreases at high temperature. Thus, the curve by Glick & Wurster (1957) indicates that the vibrational relaxation time for O_2 is about 7 μsec at 1000 K, 0·7 μsec at 2000 K and only 0·07 μsec at 3000 K. The vibrational relaxation time for nitrogen is about 100 μsec at 2400 K (Clouston, Gaydon & Glass, 1958). Only in very high velocity gas streams, such as in expansion through a nozzle from a rocket motor, are such effects likely to be detectable. From the failure to detect the right type of ultrasonic dispersion it seems that there is very little persistence of rotational energy. Shock-tube experiments (Greene, Cowan & Hornig, 1951) indicate that rotational relaxation in N_2 requires only about 20 collisions.

Radiation and scattering from solid particles

Small particles are formed in many flames, and they may profoundly affect the radiation. The most important case is the formation of soot in many organic flames, but in some other flames (and in photoflash lamps) particles of metallic oxides such as MgO or Al_2O_3 are formed. The spectrum is then always mainly continuous, instead of consisting of discrete bands, and is more like that from a black body. However, it is wrong to assume that it is identical with that of a black, or even a grey body at the flame temperature. There are two reasons for this. First, the emissivity of the material of the particle may vary with wavelength; this is particularly true for metallic oxides, though carbon is fairly black. The second reason is associated with the scattering of light by small particles; Mie (1908) has shown that particles which are smaller than a wavelength of light will not only scatter according to the well-known law that the intensity of the scattered light is proportional to the inverse fourth power of the wavelength, but will absorb with an absorption coefficient which also depends on wavelength, even if the material of the particle is black or grey. Thus by Kirchhoff's law, the emissivity of small particles will also vary with wavelength. The emissivity may be calculated from Mie's equations if the complex refractive index is known; for very small particles the Rayleigh fourth-power scattering law dominates; for large particles it will approach that for the material of the particle, and for particles comparable in size to the wavelength, the emissivity will change with wavelength in a more complex manner.

It has long been thought that the colour temperature of a candle flame is about 100°C higher than the true flame temperature. Senftleben & Benedict (1919) were able to show that this was caused by a change in emissivity with wavelength due to the small size of the particles. However, later measurements on the colour temperature of soot particles in benzene and acetylene flames (Behrens & Rössler, 1957) indicate substantial agreement with the gas temperature.

The sizes of carbon particles formed in flames vary rather widely according to the type of flame, usually between the limits 50 to 2000 Å. Thus the emissivity is likely to vary rather markedly with wavelength. It must be remembered, however, that the emissivity of a flame and of the particles which it contains may differ; as the flame becomes larger, self-absorption will become important, and the emission will approach closer to that of a black body. Stull & Plass (1960) have shown curves of emissivity against wavelength for various sizes and size distributions for a range of particle concentrations; for a very high concentration the emissivity will approach unity in the visible region, but remains low in the infra-red. However, nearly all laboratory flames are optically thin, and the emissivity is mainly determined by the character of the particles.

The absorption coefficient of soot may be measured either directly in the flame, or after collection on a quartz plate. Many measurements have been made but there is a great diversity in the values ; it is now realised (Wolfhard & Parker, 1949a) that this may be because of the different particle size distribution in various flames; even for the same fuel it may depend on whether a diffusion flame or a rich premixed flame is used, whether the supporting atmosphere is air or oxygen, and on the size of the flame, which may affect the time factor. It is convenient to express the absorption coefficient K_λ in the form

$$K_\lambda = k/\lambda^\alpha$$

The following are some typical values for the constant α from Rössler and Behrens (1950).

Flame	α
Acetone	1·43
Amyl acetate (candle)	1·39
Coal-gas/air	1·29
Benzene/air	1·23
Nitrocellulose	1·14
Benzene/NO	1·05
Acetylene/air	0·66 to 0·75

Rather different values, mostly around $\alpha = 1·0$, are obtained by Siddall & McGrath (1963), e.g. amyl acetate 0·89 to 1·04, benzene 0·94.

The effect of this varying emissivity on measurements of flame temperature using the radiation from particles will be discussed again in Chapter XI.

We should also consider to what extent the temperature of particles in a flame is likely to differ from that of the flame gases. Particles lose heat by radiation and gain heat from the flame gases by various heat transfer processes. The temperature of the particle will adjust itself so that the temperature gradient is sufficient for the heat gain to equal the heat lost by radiation. For large surfaces, the temperature difference between gas and solid may be quite large, but heat transfer theory shows that for very small particles the heat transfer is very efficient. For small particles the heat transfer number tends to be nearly proportional to the reciprocal of the particle diameter; the physical reason for this is that we are dealing with a three-dimensional problem, the temperature gradient near the surface of a small particle being greater than for a large one. Thus a smaller temperature difference will ensure sufficient heat transfer to balance the radiation loss. However, calculations are difficult because usual heat transfer formulae involve the velocity of the particle relative to the gas; for a stationary particle there is no true equilibrium. Some old estimates by Schack (1925) indicated that for soot particles the temperature difference between particle and flame gases caused by radiation loss was only about 1°C; it seems to be generally accepted that the temperature difference is indeed very small, but improved quantitative data would be welcome.

This discussion, showing that small particles closely approach the temperature of the flame gases, only holds as long as there is no catalytic heating. Energy transport and mass transport tend to be proportional and heating effects due to surface combustion, e.g. by reaction of solid carbon surfaces with O_2, will increase in importance, relative to radiation losses, for small particles. Recombination effects, of the type

$$O + O + \text{surface} = O_2 + \text{heated surface},$$

will follow a different law but, for small particles, will still dominate radiation.

Light scattering by particles is of interest in two ways – it may give information about particle size, and it may, if large, interfere with optical measurements of gas temperature. The subject has been discussed in books by Green & Lane (1965), Van der Hulst (1957), Hottel & Sarofim (1957), Fabelinskii (1968) and Kerker (1969) and in reviews by Hawksley (1952).

For particles much smaller than the wavelength of the scattered light we have the Rayleigh law. There are two components, i_1 and i_2, polarised perpendicular and parallel to the plane of observation. If θ is the angle, measured from the forward direction, and m is the complex refractive index ($= n(1 - ik)$), then for a particle of radius r and volume V

$$i_1(\theta) = \frac{9\pi^2}{2r^2} \frac{V^2}{\lambda^4} \left(\frac{m^2 - 1}{m^2 + 2}\right)$$

and

$$i_2(\theta) = \frac{9\pi^2}{2r^2} \frac{V^2}{\lambda^4} \left(\frac{m^2 - 1}{m^2 + 2}\right) \cos^2\theta.$$

Thus the i_1 component is uniform, independent of direction, while i_2 falls to zero at $\theta = 90°$.

For larger particles we have to use the Mie equations, which involve complex mathematical series, including Bessel functions. These are best solved by modern computer methods. When $2\pi r > \lambda$ scattering is much stronger in the forward direction and has, for monodisperse particles, maxima at one or more values of the scattering angle θ; when white light is scattered coloured halos may be produced.

Scattering by soot in flames has been studied experimentally by Erickson, Williams & Hottel (1964) and Dalzell, Williams & Hottel (1970). Fig. 9.3 shows the best fit between the experimental scattering and calculations, assuming monodisperse spherical particles. A better fit (Fig. 9.4) is obtained by assuming chains of spherical particles. The agreement is very close for i_1, the perpendicularly polarised component, and that for i_2 is quite good except close to $\theta = 90°$. Dalzell concludes that for this propane/air flame the particles are about 350 Å in diameter and form clusters about

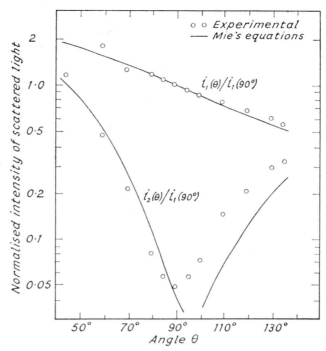

Fig. 9.3. Comparison of experimental light scattering with Mie equations assuming spherical particles with $2\pi r = 1\cdot 20\,\lambda$ for $\lambda = 4358$ Å using the complex refractive index $m = (1\cdot 60 - 0\cdot 50\,i)$. This is for a propane/air flame. [*Redrawn from Dalzell*

3700 Å long. Other work on light scattering by soot particles has already been discussed in Chapter VIII. Medalia & Heckman (1969) and A. R. Jones (1973) have extended the calculations on emissivity from spherical to elongated ellipsoidal particles. Chippet & Gray (1978) have used a number of methods to measure particle size and conclude that measurements of the light transmission are more sensitive to the size distribution than are those of angular scattering; they find a complex refractive index $m = (1\cdot 9 - 0\cdot 35\,i)$ which is rather different from the value used by Dalzell.

Calculations of radiative heat transfer

The bulk of the energy radiated by flames lies in the infra-red. For clear (i.e. non-sooty) flames the radiation is due to transitions between vibrational energy levels of molecules. Selection rules prevent radiation from homonuclear diatomic molecules like O_2 and N_2 and from symmetrical modes of molecules like CO_2, but the vibrations of heteropolar diatomic molecules like CO and NO and the asymmetric modes of polyatomic molecules are infra-red active. For hot flames the bulk of the energy is radiated in the CO_2 bands at 2·7 and 4·4 μm and for water vapour at 2·8 μm.

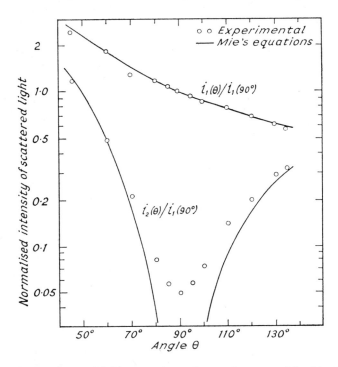

Fig. 9.4. Comparison of light scattering using a cluster model with chains of length 3704 Å (i.e. $r/\lambda = 0.85$); other data as for Fig. 9.3.

For less hot gases, and in atmospheric radiation processes, bands at 6·3 μm (H_2O) and 15 μm (CO_2) are also important.

These infra-red bands all have a complex rotational fine structure, with many lines overlapping in some regions. The relative strengths of the lines change with temperature and the intensity contours of individual lines depend on temperature and pressure and on the nature of any diluent gas. Thus detailed calculation of gas emissivity is very complicated. Theoretical calculations usually involve assumptions about the form of the line contours and sometimes assume some value for the average separation between lines. The subject has been discussed in the book by Penner (1959). Experimental work to supplement the theoretical calculations is made difficult by the need to use high resolving power so as completely to resolve the fine structure of the bands (e.g. Benedict, 1951). Alternatively, a semi-empirical method may be used in which the emission or absorption of individual bands is studied under smaller resolving power as a function of temperature and pressure. As an example, some values for bands of CO_2 have been given by Tourin (1961), Ferriso & Ludwig (1964) and Ferriso, Ludwig & Acton (1965).

A still more practical approach has been to measure the total thermal emission from hot gases under specified conditions and to draw up graphs

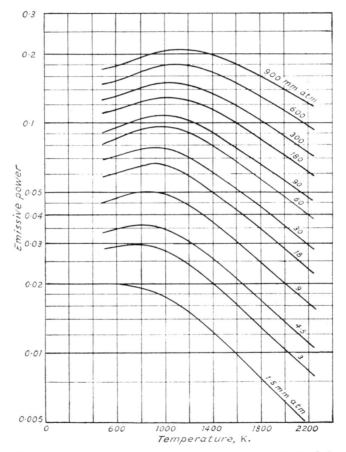

Fig. 9.5. Emissive power for carbon dioxide for various values of the optical path pl, where p is the partial pressure of CO_2 in atm and l is the length in millimetres.

[*Data of Hottel & Mangelsdorf, redrawn from Lander* (1942)]

of emissive power or total radiation against gas temperature. A good deal of work of this type was reported in the older literature, especially by Hottel & Mangelsdorf (1935) and Lander (1942), and this has been well

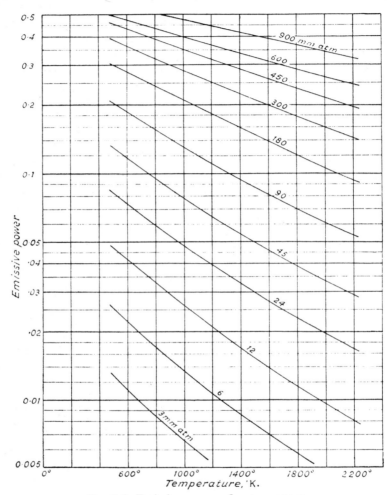

Fig. 9.6. Emissive power of water vapour.

[*Redrawn from Lander* (1942)

summarised in the books by Thring (1962) and Hottel & Sarofim (1967). Most of the results have been expressed in engineering units, such as degrees Fahrenheit and B.T.U. or C.H.U. per square foot of surface. These graphs are therefore rather inconvenient for scientific use. We have redrawn some of the graphs from Hottel & Mangelsdorf or Lander in Figs. 9.5, 9.6, 9.7 and 9.8 using K for the temperature and cal per square cm per sec for amount of radiation. The emissive power is defined as the ratio

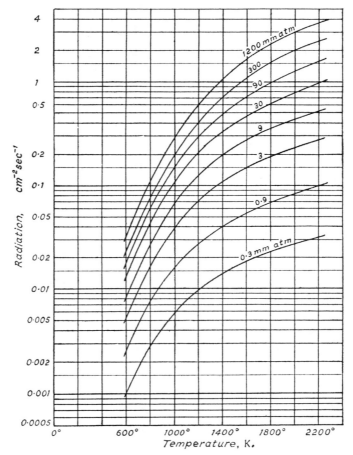

Fig. 9.7. Radiation from carbon dioxide for various values of the optical path pl.
[*Redrawn from Hottel & Mangelsdorf* (1935)]

of total emission from the gas to that from a black body at the same temperature. The amount of radiation is expressed as total emission in all directions from an element of area at the centre of the flat side of a hemisphere of radius l, so that the path length in all directions has this length l. Measurements have usually been made of mixtures of the emitting gas with air at atmospheric pressure. The emission is then a function of the product of the path length *l* and the partial pressure of the emitting gas, p; values of pl, originally in ft atm have been converted to mm atm for our figures. Because of the dependence of line contour on pressure, there is a correction

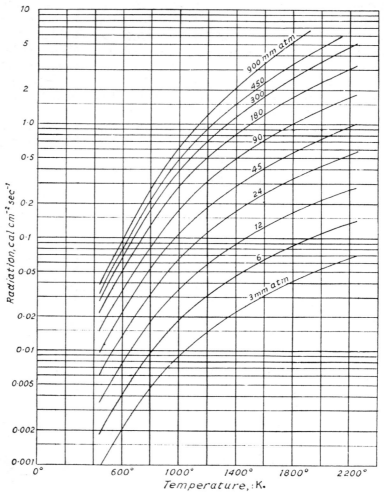

Fig. 9.8. Radiation from water vapour.
[*Redrawn from Hottel & Mangelsdorf* (1935)]

if the emissive power is required at a pressure differing from 1 atm; this correction is fairly important in the case of water vapour.

For actual flame gases, containing both water vapour and carbon dioxide, the radiation is slightly less than that obtained from the sum of

H_2O and CO_2 because there is appreciable overlapping of the bands around 2·8 μm. Some data for producer gas, coke-oven gas and blast furnace gas are tabulated by Thring (1962). For luminous flames we also have the contribution from the soot particles; the emissivity of soot varies with wavelength because of the effect of particle size discussed in the previous section. Many measurements on large industrial-scale flames have been made at IJmuiden under the auspices of the International Flame Research Foundation; the results have been published in a series of papers from 1951 in the Journal of the Institute of Fuel. For luminous flames a grey body approximation is usually employed and complex flames may be studied by a zone method (Johnson & Beer, 1973). Markstein (1975) found that for highly luminous diffusion flames the grey body approximation was satisfactory, but for less luminous flames the weighted mean of two grey bodies was better.

Chapter X

Flame Temperature.
I. Measurement by the Spectrum-line Reversal Method

Flame temperature?
The possibility that flame gases are not in equilibrium, and the consequent difficulty in defining the temperature is now generally realised. Many of the causes and results of departures from equilibrium are discussed in the previous chapter. However, it seems that departures from equilibrium are slight for the interconal or burnt gases for most small, steadily burning flames of the Bunsen type, and there is a real need for many purposes to measure the temperature. There is also, as we shall see, ample evidence that in the reaction zone or flame front, there are major departures from equilibrium. In this respect flames resemble in many ways the high-temperature gases in hypersonic flow, where reaction times are greater than stay times at any particular region of temperature and pressure; thus, as in the reaction zone of a flame, no strict definition of temperature is possible and the partitioning of energy into the electronic and vibrational states is of great interest for the understanding of the energy transfer processes in the gas. The extent and type of these departures from equilibrium may be of great interest for a full understanding of detailed processes in both flames and hypersonic flow. Studies of the energy distribution or effective temperature in the various forms the newly released energy may take can thus be of great value even when equilibrium does not exist, and may contribute to the solution of problems such as the mechanism of formation of C_2 and CH radicals and the abnormal ionisation in flame gases and lead to a better understanding of the mechanism of flame propagation.

In this and the following chapter it is proposed to give both methods and results for temperature measurements by a variety of means. Obviously, since this is a book on flames and not on temperature measurements as such, the discussion of the methods must be to some extent limited. Of the basic methods of measurement, optical ones have, in general, an advantage over wire ones in that they do not disturb the flames gases. On the other hand they often have the disadvantage that it is not possible to make a point-by-point study of the temperature distribution through the flame, the methods only giving a mean value along the path of the light beam.

One of the most convenient and most frequently used methods of measuring flame temperature is by the spectrum-line reversal method, which forms the subject of the whole of this chapter. The method has become so generally used that it has become quite orthodox. It is not always realised that the result of the measurement is the effective electronic excitation temperature for the particular element used, and it is not necessarily more likely to give a true estimate of the temperature of the flame gases than do some other methods such as those for effective rotational temperature or effective vibrational temperature, which are often frowned on as meaningless.

The sodium-line reversal method

The spectrum-line reversal method is most frequently carried out in practice using sodium, and for the sake of simplicity in description we shall describe first the method as used with sodium. It is well known that when sodium is introduced into a flame it emits the two yellow D lines at 5890 and 5896 Å. Also when light from a bright background source is passed through sodium vapour, these same two lines appear dark, in absorption, against the continuous spectrum from the background. It can be shown from Kirchhoff's law that when light from a bright background source giving a continuous spectrum is passed through a flame containing sodium vapour, the sodium lines will appear either in absorption, as dark lines against the continuum, or as bright lines, standing out brighter than the continuum, according to whether the brightness temperature of the background source is higher or lower than the flame temperature. When the brightness temperature of the background and the flame temperature are exactly the same, then the lines are invisible, having the same brightness as the background. Thus if we can vary the brightness of the background until the sodium lines just disappear and then measure the brightness temperature of this background with an optical pyrometer, we can determine the flame temperature.

The principles of the method date back to Kirchhoff's law and early work by Bunsen, the first practical use for measurement of flame temperature being by Kurlbaum (1902) and Féry in 1903, who used the method with sodium, and the spectrum-line reversal method is sometimes called after them. The number of papers in which the method is described or used, with various modifications, is very great; among the most important early ones are those by Henning & Tingwaldt (1928), Griffiths & Awbery (1929), Jones, Lewis, Friauf & Perrott (1931), Bundy & Strong (1949) and Strong, Bundy & Larson (1949).

Light from a suitable background source B, such as a strip-filament tungsten lamp, is focused with a lens L_1 to give an image of the lamp in the flame F. A second lens L_2 then forms an image of both flame and lamp on the slit S of the spectroscope C. An aperture stop A is used to restrict the aperture so that the solid angle of light taken from the flame is the same as that from the image of the lamp; the correct position for this aperture stop is at the position of the image of the lens L_1 formed by L_2 (Thomas, 1968b); the stop is often placed close to the lens L_2, and was shown in this position

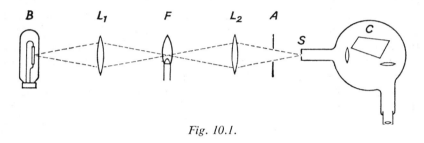

Fig. 10.1.

in our earlier editions; however, this may lead to some error for off-axis rays if a large image of the flame is thrown on the slit, unless the aperture of L_1 considerably exceeds that of L_2.

The lenses L_1 and L_2 must be of adequate quality to give a good image of the background source on the slit; any loss of light by bad lens quality or poor focusing will tend to weaken the image of the small background source more than that of the larger flame, and thus will tend to require a higher brightness for the background for reversal, so the error will cause the temperature reading to be too high. For the same reason, it is necessary to be very careful to see that the system is correctly lined up and that the cone of light from the image of the background source in the flame completely and uniformly fills the aperture determined by the stop A. The lens L_1 should, however, be a single lens, as complex lenses tend to lose light by reflection (*see* below).

The spectroscope must be of sufficient resolving power so that it is possible to see easily the sodium lines in absorption when the flame is cooler than the background. Instruments of very large dispersion are not, however, so sensitive, as the light is not bright enough. The ordinary constant-deviation spectrometer is very suitable, and a good direct-vision spectroscope is adequate. Small pocket spectroscopes are not usually satisfactory, nor are large grating instruments.

There are a variety of ways by which the sodium may be introduced into the flame. The most frequently used method is by spraying; a small quantity of sodium chloride solution is atomised into the air stream, and the small droplets are carried into the flame; the method is fairly easy and gives a steadily and controllably coloured flame, but the introduction of water may modify the flame itself a little (in the combustion of carbon monoxide, cyanogen, etc., water is a strong catalyst). The simplest method is to introduce the salt as a solid, either on a strip of platinum, or if a Méker-type burner is used for the flame, by placing it on the grid of the burner; the supply of sodium is not so steady or controllable, but it is possible to some extent to choose the region of the flame being coloured. Another method is to pass the air stream over a suitable salt which is strongly heated in a quartz tube. For low-pressure flames, we have managed by passing the gas through a small bottle containing very finely powdered NaCl, the bottle being tapped gently. Kaveler & Lewis (1937) passed the air through an electric arc between electrodes of NaCl packed round a copper wire. In some cases it has been possible to restrict the coloration to the centre of the flame by using a burner of concentric tubes, only the gases in the inner tube being fed with sodium (Padley & Sugden, 1958); for a long rectangular burner the coloration may be restricted to the centre.

Corrections for reflection loss and for change of brightness temperature with wavelength

If, as is common practice, a tungsten strip lamp is used as background source, and the temperature of this is determined from a calibration against a black body using red light, then there are two corrections which must be made to this temperature. The calibration is for the brightness temperature in the red, usually for a mean wavelength of 6550 Å; i.e. the temperature a black body would have to have to give the same light intensity at that wavelength. Now the true temperature of the tungsten strip is appreciably higher than its brightness temperature, and when we come to use the yellow sodium line the brightness temperature for this wavelength is rather closer to the true temperature.

This correction can be made quite easily by the use of Wien's law (page 239) if the emissivity of tungsten is known. This emissivity changes slightly with both wavelength and temperature. Values for this emissivity are listed in convenient form by de Vos (1954); the following table gives some values taken from his curves.

TABLE 10.1

Emissivity of tungsten

Temp. (true) K	1600	2000	2400	2600	2800
λ					
2500 Å				0·417	0·410
2750		0·465	0·456	0·450	0·446
3000	0·482	0·473	0·465	0·461	0·456
3500	0·479	0·473	0·467	0·464	0·461
4000	0·481	0·474	0·467	0·464	0·461
4500	0·474	0·467	0·460	0·457	0·454
5000	0·469	0·462	0·455	0·451	0·448
5500	0·464	0·456	0·450	0·446	0·443
6000	0·455	0·448	0·441	0·438	0·434
6500	0·449	0·442	0·434	0·430	0·427
7000	0·444	0·436	0·428	0·423	0·419
8000	0·431	0·420	0·409	0·404	0·400

Using these values we have calculated the true temperature and the brightness temperature for sodium light for a few values of the brightness temperature in the red. These are listed in Table 10.2.

TABLE 10.2

Tungsten temperatures, K

Brightness temp. at 6550 Å	True temp.	Brightness temp. at 5893 Å	Effective brightness temp. allowing 10% loss
1500	1586	1514	1499
1800	1929	1816	1802
2100	2276	2119	2098
2400	2641	2426	2402
2700	3014	2735	2703

The second correction is because the light from the background source is weakened by reflection losses in passing through the lens L_1, whereas that from the flame is not. The proper way to correct for this is to determine the effective brightness temperature of the source with an optical pyrometer viewing it through the same lens. In practice it is common to make an approximate correction by assuming a 10% loss (i.e. 5% at each surface); the calculation is then easily made by assuming an effective emissivity of 10% lower. It so happens that in the case of sodium this reflection loss almost exactly cancels the brightness temperature correction. This is shown in Table 10.2, in which the effective brightness temperatures of a tungsten lamp for the sodium wavelength, after allowing 10% reflection loss, are given in the last column. This cancellation of errors only holds, of course, for sodium with a tungsten background source; for the Li red line the brightness temperature correction is negligible, but the reflection loss is still about the same.

Effect of complex flame structure

For most flames there are appreciable temperature gradients as we pass through the flame, and especially the outer zone of the flame may be cooled by entrainment of excess air, or for premixed flames of rich mixtures may be hotter due to secondary combustion (*see* page 286, Fig. 10.3). Unless the sodium vapour can be restricted to some selected region of the flame, we shall obtain some sort of mean temperature.

There are two general effects. First, the concentration of free sodium atoms tends to be highest in the hottest part of the flame, and also with higher temperature the sodium lines are wider, so that in taking this mean temperature there is some tendency for the hottest parts of the flame to dominate. Secondly, if a large amount of sodium is present, there is a tendency for the gases nearest the spectroscope to determine the effective brightness of Na emission and therefore the apparent mean temperature, light emitted by the hot centre often being absorbed in the cooler outer layer.

Thus Henning & Tingwaldt (1928) found that when sodium was introduced, on a strip of platinum, to the back of a flame, they obtained a temperature of 2019 K, but when the sodium was introduced into either the front or both sides of the flame the temperature only appeared to be 2001 K. Griffiths & Awbery (1929) used two flames of different temperatures, one in front of the other, and both coloured with sodium; in one of their experiments the separate flame temperatures were 1736°C and 1625°C; with the hotter flame nearer the spectroscope, the mean value was 1728°C, and with the flames reversed only 1678°C; these values illustrate the two effects discussed above.

This subject of reversal in complex or multiple flames has been examined in great detail by Strong, Bundy & Larson (1949) and Strong & Bundy (1954). They have made a study of the contours of the sodium lines under a variety of conditions, including an experimental study of the large flame from a rocket motor using a special interferometer. The reversal point for the flanges of the lines is mainly determined by the hottest part of the flame and can be used to measure the temperature of the core. The centre of the line usually has lower intensity because of self-absorption by the cooler outer layer of gases and gives a lower temperature. The actual contour of the emission line usually shows a 'dimple' in the centre due to this self-reversal. The amount of sodium added does not greatly affect the mean result because the lower intensity in the centre due to strong self-absorption when a lot of sodium is present is offset by the greater strength of the flanges, which have a higher effective temperature. Quantitative treatment (e.g. by the Abel inversion method) of the temperature distribution, even in circularly symmetric flames, is very difficult because of self-absorption of radiation and non-uniformity of the sodium concentration in the flame.

Bundy & Strong (1949) have used the method for studying the temperature distribution in the shock 'diamond' of a rocket motor flame; in their experiments the temperature is about 400° higher in the shock diamond. They have also made estimates of the pressure from the broadening of the sodium lines, and of gas velocity from the Doppler shift of the lines.

Some detailed theoretical calculations on sodium line contours in flames with cool boundary-layers have been made more recently by Vasilieva, Deputatova & Nefedov (1974) and Daily & Kruger (1977). The results are very sensitive to the thickness assumed for the cool boundary. At relatively high sodium concentrations the most intense parts of the line contour move away from the line centre and the reversal point at these intensity maxima gives the maximum temperature in the flame.

Use of other lines for reversal work

The sodium lines are most frequently used because they lie in a region where the eye is very sensitive. Sodium tends, however, to be less sensitive for flames of weak mixtures, apparently because of its tendency to form an oxide (Kaskan, 1965). The red line of lithium and the green line of thallium have frequently been tried. The former usually gives results agreeing with that for Na; from our results for the reaction zones of flames (*see* later) we should expect this line to be less affected by departures from thermal equilibrium and therefore perhaps preferable to sodium, but the eye is less sensitive in the red, so though the absolute result may tend to be more reliable, the accuracy is lower. The thallium line is not a true resonance line, and although in true equilibrium this should not matter it seems to make this line more sensitive to any departure from equilibrium; Jones, Lewis, Friauf & Perrott (1931) found difficulty in getting a reversal with Tl and obtained values from 100° to 140° lower than with Na; Kohn (1914) used Na, K, Li, Rb and Tl lines and obtained substantial agreement. However, we are of the opinion that the Tl line is unsuitable. We are not aware of any work on flames using the indium blue or violet lines, but the blue line has been successfully used by one of us for temperature measurements behind shock waves. It is quite easy to introduce Pb, as lead tetra-ethyl, or Fe as iron carbonyl, into a flame, and we have found these metals suitable for studying departures from equilibrium in low-pressure flames, but we would not recommend them for the determination of true flame temperatures. For time-resolved studies of shock-heated gases and detonations, the chromium lines at 4254, 4274 and 4289 Å have been used (Gaydon & Hurle, 1963); the chromium is introduced as the carbonyl. For lines which are not in the visible region, it is of course necessary to use photographic or other means of recording. Since the reversal method is a null method this does not in itself lead to any loss in accuracy or any great difficulty, but in practice the calibration of the background source for the ultra-violet may lead to less accurate results. At short wavelengths the radiant intensity changes more rapidly with temperature, however, and this tends to increase the sensitivity.

While it is usual to introduce sodium or some other metal into the flame, it is also possible to use emission from molecules present in the combustion products. In the ultra-violet the OH band at 3064 Å can be used; since band lines are sharper than atomic lines it is necessary to use fairly high dispersion, and there is the practical difficulty mentioned above of getting an accurate temperature calibration for the background source in this region. Wolfhard & Parker (1952) have obtained reversal with the Schumann–Runge bands of O_2 at the limit of the quartz ultra-violet; here again there is considerable difficulty with accurate calibration of the background, but results have been sufficient to establish the thermal nature of the excitation for O_2 in diffusion flames. In the infra-red the strong bands due to the vibration–rotation spectrum of CO_2 and H_2O may be used. Henning & Tingwaldt (1928) used the 4·4 μm band of CO_2; for this region it is necessary to use an infra-red spectrometer with a thermopile, or more modern photoelectric detector, such as Pb Se, and to use mirrors instead of lenses for the optical system; it seems to us that loss of light by poor reflection or poor definition through astigmatism or other defects might easily lead to error (we note Henning & Tingwaldt used an effective temperature of only 3400 K for the carbon arc, instead of the usual 3800 K for the visible). Also for these infra-red bands there may be self-absorption by the large amounts of cool CO_2 and water vapour surrounding the flame. Unfortunately the usual type of flame gases do not give any useful emission in the visible region. The C_2 and CH bands are strong in the reaction zone, but their excitation is abnormal and it is difficult to obtain reversal, so they are not suitable for temperature measurements.

Background sources

For general use the best background source is a strip-filament tungsten lamp. In Britain gas-filled lamps of this type are supplied by the General Electric Co., and are calibrated by the National Physical Laboratory up to a temperature of about 2300°C against a standard black body, using the gold point and Planck's law; temperatures in this range are accurate to $\pm$ 10°. It is possible to run such lamps much hotter, up to a brightness temperature of about 2800 K, but they are not reliable as permanent standards. Thus for work above 2600 K it is best to have two lamps, one for use, and one for standard, the temperature of the one in use being determined with a disappearing filament-type optical pyrometer, which is calibrated with filters of known transmission against the standard.

The tungsten lamp has two limitations. It can only be used in the region where the glass is transparent, i.e. only the visible, not the ultra-violet or infra-red. The highest temperature available, around 2800 K, is inadequate for dealing with very hot flames such as those of hydrocarbons with oxygen. By using a lamp with a tubular filament drilled with a small hole, which serves as a black body, an effective emissivity of nearly unity, can be achieved, this giving a brightness temperature up to 300° higher; the small size of the hole makes the optical alignment difficult. Tungsten lamps with a quartz window may be obtained from Philips Research Laboratory. Stair, Schneider & Jackson (1963) have described a coiled-coil tungsten-iodine lamp which has a brightness temperature of 3000 K. A zirconia lamp may give a slightly higher temperature than a tungsten lamp, but we have not tried this. An electrically heated strip of tantalum carbide may be useful. In favoured localities, the sun can be used as background source (Wilson, 1953); this has a temperature of about 6000°C, but variable atmospheric conditions and the Fraunhofer lines may cause some difficulty.

A carbon arc is the most readily available background source for higher temperatures. Henning & Tingwaldt used this, and found a value of 3818 K for the visible region, but much lower for the infra-red. Details of a suitable arc have been given by Krygsman (1938). The positive crater of the arc is used and he found a positive electrode of 8 mm diameter and a negative one of 6 mm, both of pure carbon, satisfactory. He used 440 V D.C., with a series resistance and inductance, the current being adjusted to give the area of the anode spot the same size as that of the electrode. The current was 10·2 A, and voltage drop across the arc 60 V. The brightness appeared independent of current, but increased slightly with the voltage. He found a temperature of 3800 K for the visible and near ultra-violet, but below 2700 Å the continuous radiation from the gas contributes appreciably and the brightness temperature rises a little. The arc has some limitations. Usually it contains some sodium as impurity, which makes work with the Na lines difficult. The violet region is masked by CN bands. The far ultra-violet is partially interfered with by CO Fourth Positive bands, but some brightness temperature determinations of the arc in the middle ultra-violet have been made by Euler (1954). For later work on a standard carbon arc *see* Madgeburg & Schley (1966).

It is usual to reduce the intensity of the arc with either a rotating sector or with a neutral filter, until its brightness matches that of the spectrum lines in the flame. We have found some difficulty with the arc as source, but this may be due either to poor quality carbon or to a ripple on the direct current supply. The best procedure is probably to use either the lithium red line or the indium blue line, as these elements are not usually present as impurity, and to measure the effective temperature of the arc at the required wavelength. This can be done with a monochromator and photomultiplier using Planck's law and a standard tungsten-strip lamp, the brightness temperature of this being corrected to the required wavelength (*see* page 271).

As an alternative to a carbon arc, an electrically heated carbon tube has the advantage that emission from Na, CN, CO, etc. may be avoided. Anacker & Mannkopff (1959) have described such a source, in which the tube has a small slit cut in it, light from the slit area giving black body radiation at a temperature of up to the sublimation point of carbon at 4000 ± 7 K. It is necessary to avoid accumulation of carbon vapour near the slit, and this is done by sweeping the vapour away with a gas flame.

The xenon high-pressure arc lamps, which are now commercially available, have a brightness temperature around 5000 K in the brightest point of the arc, but the rapid variation of brightness with position in the arc causes difficulty. The arcs also tend to fluctuate with time, but this unsteadiness can be reduced by using a direct-current xenon arc at maximum current. Faizullov *et al.* (1960) and Hurle (1964) have successfully employed xenon arcs for time-resolved studies of shock waves; light from the arc is passed through the shock tube and a focused image of the arc is thrown on to the slit of a monochromator, and a portion of this image selected with a stop at the slit position; the mean brightness of the lamp for this portion of the image is then determined under actual conditions of use, with a photomultiplier, calibrated against a standard lamp.

In the past, hydrogen discharge lamps have been used for the ultra-violet region, but these are seldom bright enough to reverse hot flames and have been replaced by Xenon lamps or flash discharge lamps for absorption spectroscopy. Flash lamps with a long-duration pulse have also been used for time-resolved studies (Fairbairn, 1962).

Adaptations for non-visual recording and for time-resolved studies

Although, for steady flames, sodium-line reversal measurements are most often made visually, it is sometimes desirable to use instrumental recording methods. These are especially necessary for time-resolved studies of explosion flames, for large flames where for safety reasons it is necessary to use remote control, for automatic control, or for use of spectrum lines not in the visible region. Photographic methods are cumbersome, and it is now practice to use photomultipliers with suitable optical and electronic systems.

For study of transient phenomena such as shock waves and detonations, Clouston, Gaydon & Glass (1958) used the reversal method with a monochromator photomultiplier and oscilloscope. The oscilloscope circuit was arranged so that it responded only to changes in light intensity; thus as the shock front or detonation passed the observation window the oscilloscope trace was deflected upwards for increased emission or downwards for absorption by the sodium line. With a single beam system of this type it was possible to say whether, at any selected time, the gas was hotter or cooler than the background source. Later (Gaydon & Hurle, 1963) a double-beam system was devised in which two photomultipliers were sighted through the detonation on to background sources at effectively different brightness temperatures T_1 and T_2; the sodium wavelength was in each case selected either with monochromators or narrow-band interference filters, and the photomultiplier amplification systems were adjusted to give the same sensitivity.

In the simple case where T_1 was around 200 K higher than T_2 and the gas temperature T_g was between T_1 and T_2 it was possible to make a simple linear interpolation to measure T_g. Let e be the deflection of the C.R.O. in emission against the background at T_2, and a be the deflection in absorption against T_1, then since we have similar beams, of equal path length and sodium concentration, to a first approximation

$$T_g = T_2 + e(T_1 - T_2)/(a + e).$$

This method has been used to study the temperature history of shock-heated gases and shock-ignited explosions, and can give an accuracy of about $\pm$ 30 K and a time-resolution of 2 μsec.

When there were large temperature fluctuations behind the front, as in detonations or when the gas temperature slightly exceeded that of the available background source, Gaydon & Hurle (1963) used a modification, which really approached the brightness and emissivity method (*see* Chapter XI), employing Wien's law. Using the double-beam method, the photomultipliers were adjusted for equal sensitivity, as before, and then the background light in one beam was cut off (i.e. $T_2 = 0$). If ΔI_1 is the emission deflection of the C.R.O. for the hot gas alone at T_g, and ΔI_2 is the increased emission in the beam with background source at temperature T_B, then

$$\Delta I_1 = K \exp(-c_2/\lambda T_g)$$

and

$$\Delta I_2 = K \exp(-c_2/\lambda T_g) - K \exp(-c_2/\lambda T_B)$$

where c_2 is the second radiation constant, K is a constant depending on instrumental factors and on the sodium concentration and λ is the wavelength. If $\Delta I_1/\Delta I_2 = R$, then it can be shown that

$$\exp\left(-c_2/\lambda T_g\right) = \frac{R}{(R-1)} \exp\left(-c_2/\lambda T_B\right)$$

If $T_g > T_B$ then R is positive and greater than 1. If $T_g < T_B$, then ΔI_2 becomes negative (i.e. absorption, not emission) and R is also negative, but the method still applies. If $T_g \gg T_B$ then R approaches 1 and $R/(R-1) \to \infty$ so the method becomes insensitive. The method, as used by Gaydon & Hurle, enabled gas temperatures to be measured with reasonable accuracy up to about 300 K above that of the background source. A similar method, but with a short-duration flash tube as background, has been used by Carnevale et al. (1967) to measure temperatures of shock-heated gases up to 6700 K; agreement with values obtained by an ultra-sonic method was found.

In a method used for large pulverised coal flames (English & Dingle, 1966), a monochromator with vibrating exit slit is employed; the magnetically driven slit is adjusted to oscillate over a distance equivalent to about 4 Å so that it scans backwards and forwards over one of the sodium lines at mains frequency. The output from a photomultiplier placed behind the vibrating slit may be displayed on an oscilloscope; a tuned amplifier system which would make the system adaptable for a servo-mechanism has been described.

A rather different system for a similar purpose (Thomas, 1968b) uses two narrow band (10 Å) interference filters, one set for the mean sodium wavelength (5893 Å) and the other at a near-by wavelength just clear of the sodium lines. A beam splitter is used to divide the light beam, after passage through the flame, so that the divided light passes through the filters on to two photomultipliers. These are adjusted to the same sensitivity with the flame extinguished. With the flame running, the difference ΔI between the photomultiplier signals then depends on the departure from the reversal condition. Two neutral filters of different optical density are switched alternately in front of the background source (a carbon arc) so that two different effective background temperatures, denoted, T_1 and T_2 are produced. If the signal-differences are denoted ΔI_1 and ΔI_2, then it may be shown that

$$\frac{\Delta I_1}{\Delta I_2} = \frac{\exp(-c_2/\lambda T_g) - \exp(-c_2/\lambda T_1)}{\exp(-c_2/\lambda T_g) - \exp(-c_2/\lambda T_2)}.$$

The method is somewhat similar to that used by Gaydon & Hurle, except that one monochromator is set just off the sodium wavelength, and the time resolution is limited to the frequency with which the neutral filters are switched, which is usually about once a second. It is useful for steady flames, whereas the Gaydon & Hurle method as initially used is only suitable for transient phenomena.

A variation of the method which enables reversal to be obtained at flame temperatures well above the brightness temperature of the background source has been developed by Snelleman (1967). Two adjustable rotating-sector light choppers are used, one placed between the background and the flame and the other between the flame and monochromator; with suitable electronic arrangements and adjustment of the relative sector apertures the advantages of the null method are retained and temperatures up to 300 K above that of the background have been measured, and the method might possibly be used up to a temperature excess of 800 K.

Accuracy and reliability

For temperatures below about 1500°C it is possible with care to adjust the temperature of the background source to within about 5° for reversal, and the background temperature can then itself be calibrated within another 5°, so that an overall accuracy of within 10° may be attained. This very high sensitivity of the method is due to the rapid variation of light intensity with temperature; at 2500 K and 5890 Å the intensity varies as about the tenth power of the temperature! Snelleman (1967) has made some calculations on the signal-to-noise ratio for photoelectric detection, and concludes that random errors need not exceed ± 1 K; photographic recording is less good, and for visual estimates he estimates that the random error might reach ± 30 K, which is greater than we find experimentally. For higher temperatures, up to 2300°C, the calibration of the background source is accurate to only $\pm$ 10° and the accuracy must be correspondingly less. For really high-temperature flames it is probably possible to set the reversal point to $\pm$ 10°, but an error of at least 20° must be accepted as probable for the calibration of the background source.

In practice, the most likely errors are systematic ones through faults in the optical system, and these nearly always result in the temperature being read too high. It is most important to see that the alignment is correct and that light from the background fills the aperture properly not only for the centre of the slit, but for the top and bottom of the image as well. Dust or dirt on L_1 reduces the brightness of the background. Dust on L_2 may also cause trouble (Snelleman, 1967) as it may scatter sodium light from the flame (which is closer to L_2 than the background source is) towards the spectroscope and thus apparently strengthen the sodium emission; deliberately powdering L_2 produced an error of 35°.

The method is entirely dependent on Kirchhoff's law, which states that, in equilibrium, for any wavelength the emissivity is equal to the absorption coefficient. There seems little reason to doubt that for the measurement of the temperature of hot gases in equilibrium the spectrum-line reversal method is reliable and accurate. The main difficulties arise when the gases are not in equilibrium or when the flame contains solid particles.

For Kirchhoff's law to hold there must be no loss of light from the flame by reflection or scattering. For clear flames there is little doubt that reflection is negligible. When the flame contains carbon particles (i.e. soot) then Plate 15a clearly shows that some scattering of light from the background will occur; this will weaken the background continuous spectrum. However, it will also weaken the sodium emission from the back of the flame, but not that from the front of the flame, so that the Na emission will not be weakened as much as the background. Thus again the tendency will be to read the flame temperature too high. In small laboratory-scale flames the optical absorption and scattering by soot is quite small and much work in the older literature shows that the particles are fairly black and the errors are very small. Kuhn & Tankin (1968) have used the sodium-line method to study the temperature distribution through the sooting zone of a propane diffusion flame; they find good general agreement between the reversal temperature and the particle temperature, although the soot particles tended to be slightly too hot on the oxygen edge and slightly too cool on the fuel edge of the carbon zone of the flame. D. L. Thomas (1968a) has discussed these scattering errors in connection with studies of large hot coal-dust flames which have a high emissivity; with some assumptions he shows that for a flame at 3000 K a correction of the order 40 K may be necessary. Gaydon (1966) has considered the use of a combined reversal measurement and light-chopper study of the scattering to make the correction. If a flame contains liquid droplets or solid transparent crystals, then obviously light from the background source will be lost by reflection and scattering, and not replaced by emission from the droplets or particles.

Another similar possibility of error is due to refraction of the light beam on entering the flame. For normal use this will be quite immaterial, but if attempts are made to study a thin reaction zone using a narrow-angle beam of light, then errors could result; we have encountered difficulty for this refraction effect in studying absorption spectra in the preheating zone of flames.

In all flame gases in which the amount of added metal is small so that the flame is optically thin for the line under examination, there will be some departure from equilibrium due to emission of radiation, as discussed in Chapter IX. For a flame with air, having a temperature of 2000 K, and assuming that nitrogen is the most important constituent, we can use the data of Table 9.3 to show that at 1 atm pressure the collision life for deactivation of an excited Na atom will be around 1.0×10^{-9} sec; the radiative life (Table 9.1) is 1.58×10^{-8} sec. Thus the population of the excited state may be reduced by 6% by radiation depletion. Using the Maxwell-Boltzmann law, it can then be shown that this causes the temperature to be 10 K too low. For a flame at 1/10 atm pressure, the error would be 78 K, and for 1/100 atm 280 K. The errors will be less than this if plenty of sodium is added to the flame, but even then the wings of the line will show the same effect, and, even for the centre, the gases nearest the observer can only receive balancing radiation over one hemisphere, while they radiate to a whole sphere, so an effect of at least half the magnitude may still be expected. Direct measurements of the fluorescence of Na in a flame have been made by Boers, Alkemade & Smit (1956); results in general agreement with our estimates have been obtained, and they find a discrepancy of up to 8 K at 1 atm between true and reversal temperatures for an optically thin flame.

In Chapter IX it was shown that electronic energy most readily interchanges with the vibrational energy of molecules. In the reaction zones of flames there may initially be excess vibrational energy in the bonds of the newly formed molecules, and we expect the electronic excitation, which controls the reversal temperature, to relax with the vibrational energy of these molecules. In some gases at room temperature vibrational relaxation times may be several milliseconds; at high temperatures they are shorter, but may still be of the order 100 μs for some pure gases such as N_2, CO and CO_2 (Gaydon & Hurle, 1963). However, in flame gases, with water vapour present, relaxation times will only be of the order of a few μs so that the disequilibrium cannot persist for more than a small fraction of a millimetre above the reaction zone. A possible exception is the dry cyanogen flame (see page 308) for which the products, N_2 and CO, have a rather long vibrational relaxation time.

In the reaction zone, atoms and molecules are formed in electronically excited states by chemical reactions; this chemiluminescence has been discussed in Chapter IX. If the electronic energy is transferred to the species used for reversal-temperature measurements then the electronic excitation temperature of this species will be raised above the translation temperature of the gas molecules; this may cause major departures from equilibrium in the reaction zone and lead to anomalously high reversal temperatures, as discussed on page 289.

Just above the main chemical reaction zone, we often have an 'afterburning' zone in which there is an excess concentration of free atoms and radicals. Reactions of the type

$$H + H + Na = H_2 + Na^*$$

then tend to raise the population of electronically excited species. However, these three-body collision processes are less frequent than the normal bi-molecular processes responsible for maintenance of thermal equilibrium; thus these effects do not usually produce a major difference between reversal and translational temperature. In near-limit flames when thermal excitation is weak because of the low temperature, these departures from equilibrium due to excess population of free H or free O atoms may cause the reversal temperature to be slightly high. Early criticism of the reversal method by David and colleagues (e.g. David, 1934) probably resulted from their study of near-limit flames which showed strong afterburning or 'latent energy', associated with an excess population of free atoms.

Thus the reversal method has some limitations, especially for non-equilibrium conditions. Care with the optical arrangements is essential; most errors lead to the reversal temperature being too high. However for hot flames we are practically forced to use an optical method, and this one has great advantages in simplicity, especially because it is a null method.

From the experimental results to be discussed later, it appears that there are major departures from equilibrium in the reaction zones of flames, and reversal temperatures are not related to the mean gas temperatures. Above the reaction zone, however, the reversal temperature seems in general to give useful results. For flames supported by air most investigations (e.g. Griffiths & Awbery, 1929) indicate good agreement between maximum flame temperature by the reversal method and by calculation or from measurement by hot-wire methods (assuming proper radiation-loss corrections to the latter). Some measurements on oxy-acetylene and oxy-hydrogen flames by Lurie & Sherman (1933) using a carbon arc as background indicate temperatures only a little below the theoretical maximum flame temperature, and our observations also tend to show that for the interconal gases there is agreement between calculated temperatures and those obtained by the reversal method, using either sodium or lines of the OH band. In shock waves through molecular gases, there is good agreement between measurements of reversal temperature and the equilibrium gas temperature calculated from measurement of the shock speed (Gaydon & Hurle, 1963), but for monatomic gases such as argon the reversal temperature is low because of radiation depletion.

The Kurlbaum method

The first use of a reversal technique was by Kurlbaum (in 1902), who did not use any added metals, but worked on luminous flames containing carbon particles. The principle of this method of measuring the temperature of luminous flames is essentially the same as for the sodium-line reversal method. It is only sensitive if the flame contains enough particles to absorb a fair proportion of the light from the background source. There may be a slight tendency for the results to come a little high because of loss of light by scattering and reflection, but errors on this account are probably less than those from the same cause when the colour temperature of the flame (two-colour method, *see* Chapter XI) is made the basis of measurement.

In this method no spectroscope is necessary. An optical pyrometer, usually with a red filter, is sighted alternately on a suitable background source directly and on the same source viewed through the flame; the brightness of the background is adjusted until the two values are the same and the brightness temperature is then read from the optical pyrometer.

Results for interconal gases

For premixed flames with *air* there seems to be a general consensus of results that the spectrum-line reversal method gives the correct maximum flame temperature, except where the flame itself is cooled by radiation losses and heat transfer to the surroundings, when the results may be a little low; thus Jones *et al.* (1931) found values from 50° to 100° low, while the results of Melaerts *et al.* (1960) for small flames of hydrogen/nitrogen/oxygen and acetylene/nitrogen/oxygen mixtures are around 200° to 300° low at atmospheric pressure and up to 500° low at 50 torr pressure.

Lewis & von Elbe (1943) have made a most valuable study of the temperature distribution across a section of a flame. They used a rectangular burner 21.9×7.55 mm and coloured only the central portion with sodium and viewed the burner end-on. The results for slightly lean and rich mixtures of natural gas with air are reproduced, by their kind permission, in Figs. 10.2 and 10.3.

The interesting result for the weak mixture is that the temperature does not attain its maximum value, which is equal to the theoretical flame temperature, immediately above the reaction zone, but only some 6 to 10 mm higher, corresponding to a time delay of about 0·0025 sec. Results in general agreement with this have been obtained by other investigators who have studied the temperature as a function of height above the inner cone. The reason for this presents an interesting problem. It seems unlikely to us that it could be accounted for by vibrational energy lag, because we should expect this to last only a few microseconds. It is possible that this delayed attainment of maximum temperature especially in lean flames is connected with the observed delayed afterburning of carbon monoxide as observed in low-pressure flames by Friedman & Cyphers (1955) and Fristrom & Westenberg (1957).

For the rich mixture there is also another interesting effect; the outer regions of the flame give a higher temperature, due to secondary combustion in the surrounding air.

For flames with oxygen there is more doubt whether the reversal method gives reliable results, and there are some reports of temperatures exceeding the theoretical values (e.g. David, 1934). This effect is also shown in Lewis

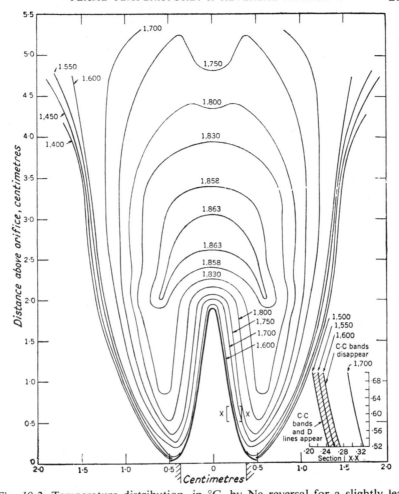

Fig. 10.2. Temperature distribution, in °C, by Na reversal for a slightly lean (near stoichiometric) flame (8·7% natural gas in air).

[*Lewis & von Elbe, U.S. Bureau of Mines*

& von Elbe's mapping of the temperature distribution, reproduced in Fig. 10.4. In this the temperature just above the reaction zone is 120° above the theoretical value. Lewis & von Elbe initially attributed this to vibrational excitation lag, but we have pointed out that vibrational relaxation times in flames are very short.

There seem to be three possible causes, (i) continued chemiluminescent reactions of the type occurring in the reaction zone, (ii) reactions of the type $O + O + Na \rightarrow O_2 + Na^*$ or $CO + O + Na \rightarrow CO_2 + Na^*$ in the

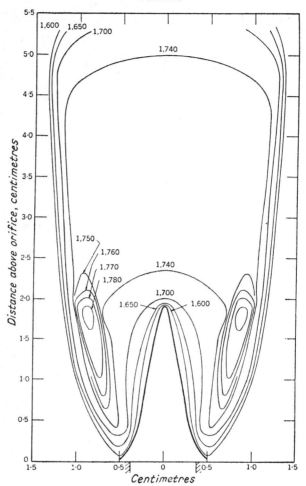

Fig. 10.3. Temperature distribution by Na reversal for a rich flame (10·96% natural gas in air).

afterburning zone, or (iii), a real temperature excess because diffusion gradients in the flame change the mixture strength towards the stoichiometric value. Lewis & von Elbe comment on the brightness of the Na emission from the reaction zone in this flame, but their measurements of temperature in the zone do not appear to us to be trustworthy because of the zone being so thin (to examine a zone of the order 0·1 mm thick over a length of even 5 mm requires keeping the flame very flat and steady and restricting the aperture of the light beam to less than $f/50$).

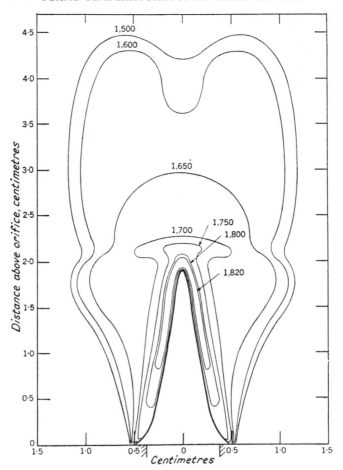

Fig. 10.4. Temperature distribution by Na reversal in a weak natural gas/oxygen flame.

Results for reaction zones

As indicated immediately above there is considerable experimental difficulty in studying very thin reaction zones. At atmospheric pressure the reaction zone for an oxy-acetylene flame is only 1/50 mm thick, and for butane/air about 1/5 mm. We (Gaydon & Wolfhard, 1951a) have therefore made measurements in the much thicker reaction zones of flames at reduced pressure. We have used several elements (Na, Fe, Pb, Tl) and find that in general non-resonance lines cannot be reversed, always appearing in emission against even the full carbon arc as background, but that resonance lines can be reversed. This difference between resonance and non-resonance lines is itself a clear indication of departure from equilibrium, and is illustrated in Plate 16f, in which the Pb line 2833 Å shows in absorption against a carbon arc, while other lines are shown in emission.

For resonance lines the reversal temperature tends to increase for lines of shorter wavelength, greatly exceeding the theoretical flame temperature for lines in the ultra-violet. A typical set of results for a stoichiometric acetylene/air flame at 24 torr pressure is given in Table 10.3.

TABLE 10.3

Reversal temperatures for C_2H_2/air at 24 torr

Line	Reversal temp. K	E/kT	Line	Reversal temp. K	E/kT
Na 5890·6	2010	12·1	Fe 2983·5	3050	15·7
Fe 3859·9	2565	14·4	Fe 2966·9	3068	15·7
Fe 3824·4	2580	14·45	Fe 2936·9	3100	15·7
Tl 3775·7	2350	16·2	Pb 2833·0	2970	17·0
Fe 3719·9	2650	14·5	Fe 2719·0	3310	15·9
Fe 3679·9	2630	14·8	Fe 2522·8	3365	16·8
Fe 3440·6	2880	14·4	Fe 2501·1	3410	16·75
Fe 3020·6	3058	15·5	Fe 2483·3	3450	16·7

For this flame the theoretical temperature is 2340 K, but the actual temperature is probably rather lower, due to heat loss. For the interconal gases the reversal temperatures are around 2000 K, but some correction for radiation depletion may be necessary (*see* page 281).

The result, from a reversal temperature, is really the temperature corresponding to the actual ratio of populations of the normal and excited atoms. In equilibrium this would be proportional to the Maxwell-Boltzmann value $e^{-E/kT}$, where E is the excitation energy. Thus if T is the observed reversal temperature $-E/kT$ is equal to the natural logarithm of the ratio of populations of the excited and ground states. In equilibrium a plot of E/kT against E should give a straight line through the origin, the slope giving the temperature. The results from Table 10.3 are plotted in Fig. 10.5. The points for Fe lie on a fairly good straight line, but this does not pass through the origin, and its slope corresponds to 8000 K.

Clearly there is some major departure from equilibrium in the reaction zone, lines of high excitation energy being emitted much too strongly. The effect apparently varies according to the element being excited, and for low energies, below about 18,000 cm^{-1}, thermal excitation becomes the more important. The results are discussed in detail in the original paper. This abnormal excitation occurs for all organic flames studied, except formaldehyde, but is not observed in reasonably hot flames of hydrogen or carbon monoxide; there is a slight abnormality for relatively cool hydrogen/air flames.

This abnormal excitation is not greatly influenced by pressure, the reversal temperatures rising a little from low pressure towards atmospheric, and falling at very low pressure. The effect is not very sensitive to mixture strength, but rich mixtures tend to show it rather more than weak mixtures. While the abnormally high excitation is difficult to measure quantitatively in flames at atmospheric pressure, its existence can be demonstrated conclusively and simply by adding a little iron carbonyl to a flame; it is then found that the interconal gases only emit the resonance lines of Fe, but that in the inner cone all the lines of neutral Fe, including some requiring as much as 174 kcal per mole for their excitation, appear, and that the intensity distribution among the Fe lines in the inner cone is very similar to that in a very hot electric arc (e.g. the ordinary iron arc comparison). This is shown in Plate 16g and h.

The cause of this abnormal and very marked high electronic excitation in the reaction zones of flames is unknown, and is one of the major outstanding problems in flame research. It has been discussed by Gaydon (1974); there appears to be a general failure of the Maxwell-Boltzmann distribution in very fast reacting systems, especially when bimolecular exo-

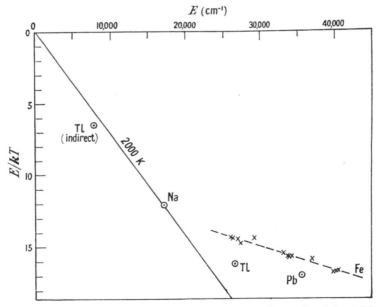

Fig. 10.5. Plot of E/kT against E for reaction zone of acetylene/air at 24 mm pressure.

thermic reactions occur. In systems such as burning hydrogen, where the main exothermic steps involve three-body collisions, the chemical disequilibrium may persist longer, but the departures from thermal equilibrium are then less marked.

It seems that some other flame abnormalities, such as the high ionisation in the reaction zone (*see* Chapter XIII) may be linked with this abnormal electronic excitation. We should also take into account the possibility of the flame propagation itself depending on the presence of energy-rich molecules; judging from the actual concentration of electronically excited atoms which are formed, we think it unlikely that these are numerous enough to affect the propagation, but there remains the possibility that the underlying cause of the abnormal excitation may itself be of a type which could influence the propagation.

Chapter XI

Flame Temperature
II. Other Methods of Measurement

There is no perfect way of measuring flame temperature. All methods are subject to both practical and theoretical limitations, and the choice of the most convenient and reliable method will depend on the type of flame being studied. Here we shall not be able to discuss all methods in great detail; indeed for many of the methods described we have had no practical experience of their use. We shall, however, endeavour to indicate the principles of the various methods which have been suggested, and to point out their advantages and limitations.

For flames below about 1750°C (i.e. below the melting-point of platinum) hot-wire methods can usually be employed. The main problems, with which we are most interested, concern hotter flames. For these, optical methods, which we shall discuss first, are most suitable. Of these the sodium-line reversal method discussed in the previous chapter is the most generally suitable, although it cannot easily be used for highly luminous flames* or for extremely hot flames, and its application to a study of temperature distribution within a flame is limited. For luminous flames we have the Kurlbaum method already described, or the two-colour method. Some optical methods, such as the line-ratio method, or the measurement of rotational temperatures from band spectra are not limited to a comparison with any standard or background source and can therefore be used for the highest temperatures, although these methods are most subject to any departures from equilibrium in the flame gases.

The advent of the laser has given us two valuable new methods, still rather in the development stage, which enable a point-by-point determination in the flame of the temperature of ground-state molecules; laser-Raman scattering gives their vibrational temperature, while Doppler broadening of the Rayleigh scattering from a fine laser line may give the translational temperature.

---

* It may be used provided the luminosity of the flame is not so high that its emissivity is comparable with that of the spectrum line; in this case the sensitivity falls due to lack of contrast between the line and the continuum.

One of the favourite methods for determining the temperature of gaseous explosions is, of course, from the rise in pressure. The method is not applicable to steady flames, and is indirect and limited in scope; if the gas composition before and after explosion is known, then the temperature can be calculated theoretically with precision and there is little point in measuring it; if the gas composition is not known, perhaps due to departures from equilibrium, then the pressure rise does not enable the temperature to be deduced accurately. We have no experience of the method and shall not describe it any further.

Apart from their use in determining actual flame temperatures, some of the methods are of particular use in getting information about detailed chemical and physical processes. Thus it is often possible to recognise the occurrence of chemiluminescence from the existence of abnormally high effective rotational temperatures. Such methods will therefore be examined as well, even though not always likely to be of practical use in the determination of mean flame temperatures.

Brightness and emissivity methods

If for a flame we know the emissivity and the brightness temperature at any wavelength, then we can apply either Planck's or Wien's law (page 239) to derive the gas temperature; the brightness temperature can be obtained by isolating a certain wavelength with a monochromator or spectroscope and comparing the brightness with that of a standard source such as a tungsten strip-filament lamp; to determine the emissivity we again use Kirchhoff's law that emissivity equals the absorptivity for gases in thermodynamic equilibrium.* If the brightness temperature is determined visually it is necessary to devise an optical system in front of the slit of the spectroscope to bring the images of the flame and comparison lamp together. Photographic photometry also is tedious and subject to errors through Eberhard effect and other causes. Photoelectric measurements of the brightness have advantages of direct reading and speed; for the far infra-red a thermopile is satisfactory.

The absorption may be determined by using a background source which has a much higher brightness temperature than the flame, so that the flame emission may be neglected. When the flame emission is small but not negligible it may be measured separately and subtracted from the measurement of the transmitted intensity. Carbon arcs and xenon high-pressure arcs may be used as background, or a flash discharge-tube can be employed if a synchronised shutter is used to exclude flame emission except at the moment when the flash occurs.

* It has been shown previously, page 281, that small flames are not in radiative equilibrium and that Kirchhoff's law does not hold strictly. The errors due to radiation depletion, however, are small for flames at atmospheric pressure.

As an alternative to using a monochromator or spectroscope it is possible to use a colour filter and to sight on the flame and on the comparison source directly with an optical pyrometer through the filter. This method is, of course, only satisfactory if the spectrum of the flame is continuous over the range transmitted by the filter. Schmidt (1909) made measurements of flame temperature from brightness and emissivity, using infra-red radiation, and this is sometimes referred to as the Schmidt method.

A simplification of this method is to assume unit emissivity. Thus Kandyba (1948) saturated a flame with sodium vapour and studied the absolute intensity photoelectrically for the centre of the Na D lines. It would seem that one can only attain unit emissivity by using a large amount of sodium and by restricting observation to the centre of the line, which requires high resolving power and therefore probably loses the advantage of speed. It would seem to us that errors due to inadequate resolving power and to absorption by the cooler outer regions of a flame would lead to serious inaccuracies.

Another way of measuring the absorption is by the two-path method. In this, the light received directly from the flame is compared in intensity with the light from the flame plus that reflected back through the flame by a mirror of known reflection coefficient. From the ratio of the intensities, the absorption by the flame can be deduced. Alternatively, two or more similar flames in line may be used. The method is particularly useful for luminous flames giving a continuous spectrum and for studying temperature variation with time as in the flame from a pulse-jet motor. It has been used (e.g. Quinn, 1950) with the sodium D lines, but in this case an average absorption, and therefore emissivity, over the contour of the line is obtained, and we think this is likely to lead to error; Penner (1951) has discussed this and shown how it is possible to calculate the contour of a line and make allowance for this type of effect, but the procedure is rather elaborate.

A third method for determining the absorptivity is to use a light-chopper between a background lamp and the flame, with a photoelectric detector with amplifier circuit tuned to the chopping frequency. The output then depends only on the fluctuating signal, and the steady emission from the flame is not recorded, so that the absorption can be measured by observing the signal from the background with and without the flame interposed (Fig. 11.1). For turbulent flames, random fluctuations in light emission may give a rather high noise level; a correction for this may be made by observing the signal from the flame alone when the background is cut off. As an alternative to the tuned amplifier, a phase-sensitive detector may be employed, this giving a better signal-to-noise ratio.

This light-chopper method was first used by Silverman (1949), who studied CO_2 bands in the infra-red, and received the radiation on a bolometer with tuned 10-cycle amplifier and chopper device. He compared the readings: (a) of the flame alone, with the chopper in front of the flame; (b) with a globar alone as standard; and (c) with the globar behind the flame and with the chopper between globar and flame. This last record (c) together with (b) gave the absorption by the flame, since the tuned bolometer was only sensitive to the chopped light from the globar, the strength of which was however reduced by later passage through the flame. The method is again subject to the limitations that a mean emissivity is obtained,

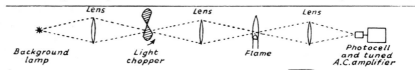

Fig. 11.1. Light-chopper system for measuring absorptivity of a flame.

and that the cooler outer layers of the flame make an appreciable contribution to the result. However, Silverman studied the results for different parts of the CO_2 band, and was able to show that on the low-frequency side of the band maximum the emission came mainly from higher vibrational levels which were only populated in the hot part of the flame, and that this region of the spectrum gave the maximum flame temperature all right. The method has been discussed fully by Tourin (1966) who has used it in the infra-red for the study of luminous detonations.

In the infra-red the bands of H_2O, CO_2 or CO may be used for combined emission-absorption temperature measurements. It is generally not feasible to use isolated rotational lines as such high resolving powers are not readily available. Band models are used in which the lines are smeared out so that the band appears continuous. The emissivity or absorptivity is measured or calculated for a given temperature as a function of optical depth (in cm atm). Most calculations of band shapes are based on the assumption of collision-broadened lines and care should be used in applying them to flames at low pressure. We do not intend to treat the subject of band models here and the reader is referred to books by Penner (1959) and Goody (1964). In using these infra-red bands, troubles due to absorption in the cooler outer layers of a flame are reduced if the measurements are made on a part of the band corresponding to transitions between relatively high vibrational and rotational levels.

A modified brightness and emissivity method may also be used with discrete spectrum lines if due consideration is given to effects of line contour and instrumental slit function. An example of this type of application is that by Fissan (1971) who used both Na lines and OH bands to determine detailed temperature contours for an oxy-methane flame. He used the

Abel inversion method to build up information about the various zones of this circularly symmetrical flame and found that the maximum temperature on the flame axis by both the Na and OH methods was very close (10 to 20 K low) to the adiabatic flame temperature of about 3060 K. Without using the Abel inversion, the temperature, assumed isothermal through the flame, was 120 K below the adiabatic value when using OH and 200 K low for Na.

The methods dependent on measurement of brightness and emissivity are not necessarily limited, like the reversal method, to temperatures below that of the comparison source. Also if the measurements are made in the infra-red there is perhaps less likelihood of small departures from equilibrium causing major changes in temperature. The method is, however, likely to give lower precision than the reversal method, and unless very high resolving power is available is only accurate for luminous flames. It is, of course, only applicable to flames of reasonably high emissivity; many small luminous flames are optically rather thin and the emissivity is too low to be measured accurately by the two-path or other methods. It has the same limitation as the reversal method in that it is not suitable for flames containing particles which reflect or scatter light.

The colour-temperature method

For luminous flames, efforts have been made to determine the flame temperature from the colour temperature. The method requires some assumption about the emissivity or at least about the variation of emissivity with wavelength. If the emissivity is known, and the brightness can be compared with that of a standard lamp at any single wavelength, then the flame temperature can be determined at once, without need to work at two wavelengths. Similarly if we assume that the flame is a black body, then measurement of the brightness at a single wavelength is sufficient. If, however, we assume some law for the dependence of emissivity on wavelength, then we can determine the flame temperature from measurement of the relative brightness at two wavelengths, or with sufficient accuracy, through two colour filters.

A simple assumption would be that the flame was grey, consisting of black particles with regions of complete transparency between the particles. However (*see* page 259) it is found in practice that for thin flames the emissivity increases markedly to shorter wavelengths.

The method has been used by Hottel & Broughton (1932). They used Kirchhoff's law that the monochromatic emissivity E_λ was equal to the absorptivity A_λ where

$$A_\lambda = 1 - e^{-K_\lambda L},$$

K_λ being the monochromatic absorption coefficient and L the path length.

They assumed that the absorption coefficient varied as a simple power, α, of the wavelength
$$K_\lambda = k/\lambda^\alpha.$$
For a flame of length L, we then have
$$E_\lambda = 1 - e^{-kL/\lambda^\alpha}.$$

If the brightness temperature, T_r, is measured for red light of mean wavelength λ_r, it can then be shown from Wien's law that the true flame temperature T is given by the equation
$$\frac{1}{T_r} - \frac{1}{T} = \frac{\lambda_r}{c_2} \log_e(1 - e^{-kL/\lambda_r^\alpha}),$$
where c_2 is the second radiation constant (page 238).

If a similar determination of the brightness temperature is made for green light we have two equations from which kL may be eliminated and T determined, provided a value for α is known. Hottel & Broughton studied diffusion flames of amyl acetate, and used the two-path method (*see* page 293) to determine α. They found a value of 1·39; we have seen however (page 259) that α is different for different fuels (Rössler & Behrens, 1950; Wolfhard & Parker, 1949*a*). Sanderson, Curcio & Estes (1948) found that the two-colour method gave temperatures from 10% to 15% lower than did the sodium-line reversal method; this was probably because the Na method gave the temperature for the hottest part of the flame, while the two-colour method gave a mean value. Sobolev & Shchetinin (1950) however found agreement between the two methods, and rather surprisingly also found that the absorption coefficient was independent of wavelength, i.e. the flame behaved as a grey body.

If the flame is not optically fairly thick, then a determination of emissivity, and hence of α, is likely to be rather inaccurate. In this case, it should be possible to measure the brightness temperature at three wavelengths and then to eliminate α between three equations; we then assume only the form of the relationship $K_\lambda = k/\lambda^\alpha$ and not the actual values of K_λ or α. Siddall & McGrath (1963) studied flames of several fuels and found that this form of relationship was satisfactory in the visible part of the spectrum, but that in the infra-red α was not constant but took the form $\alpha = a + b \ln \lambda$; for many fuels $\alpha = 0·906 + 0·283 \ln \lambda$ (with λ in μm).

It should be possible to determine the colour temperature absolutely with a bolometer and monochromator of known transmission factor at various wavelengths. In practice it is usually more convenient to make relative measurements against a standard comparison lamp. This again can be done visually, photographically or photoelectrically. A simple visual method is to balance first the flame and then the comparison lamp against a disappearing filament optical pyrometer, using colour filters; if ordinary gelatin or glass filters are used, considerable care is necessary in determining the effective mean wavelength of the filters; the modern interference filters, covering a narrow wavelength band are especially suitable for this type of work. Photoelectric recording may be used either with a monochromator or with filters, readings being taken on the flame and on the comparison lamp for each colour; this method has the advantage that it is quick and can be used for transitory flames.

The colour temperature method is limited to luminous flames. It has the advantage of simplicity, but, as often carried out with two colour filters and uncertain assumptions about the variation of emissivity with wavelength, we fear that it is not very reliable. Providing the flame has a reasonably high emissivity, the two-path or Kurlbaum methods are better. For small, optically thin luminous flames, the colour method may have some advantage, provided care is taken to study the dependence of emissivity on wavelength. Care must then be taken to avoid wavelengths at which banded (e.g. C_2 or CH) emission may occur. Any errors due to surface combustion or catalysis will affect the colour temperature, brightness and emissivity and Kurlbaum methods similarly. The method has been used to study the temperature variation of large individual particles in flames, high-speed photography through colour filters being employed.

The line-ratio method
The optical methods of measuring temperature discussed so far have all required fairly high emissivity and absorption for their successful use. The next four methods to be discussed (line-ratio, Doppler broadening, rotational intensity distribution and vibrational intensity distribution) all require the reverse, that the strength of the spectrum lines being used shall be so small that self-absorption is negligible. As the concentration of emitters increases, the centre of the line becomes increasingly affected by self-absorption, the intensity in the centre being ultimately limited to that of a black body at the flame temperature. On the other hand the wings of the line continue to strengthen long after the centre has reached practically the maximum intensity. Thus as the concentration of emitters increases, the strength of the spectrum line increases, at first linearly, but as self-absorption becomes important the increase in intensity continues more slowly, and the so-called 'curve of growth' is complex and does not rigorously obey any simple law as it depends on the contour of the spectrum line; in many practical cases the intensity varies as the square root of the concentration. With ordinary spectrographs it is not possible adequately to resolve the contour of the spectrum line, and hence it is not usually practical, even by the two-path method, to measure this effect of self-absorption so as to correct for it. It is best just to make sure that the effect is negligible.

If several spectrum lines of the same element, with frequencies v_1, v_2, v_3, ... and different excitation energies $E_1, E_2, \ldots$, have transition probabilities $P_1, P_2, \ldots$, then from the Boltzmann distribution law their emission intensities will be proportional to

$$P_1 \, v_1^4 \, e^{-E_1/kT}, \quad P_2 \, v_2^4 \, e^{-E_2/kT}, \text{ etc.}$$

When measurements can be made on several lines, it is best to plot $\log_e$ (intensity) $- \log_e (Pv^4)$ against E and deduce T from the slope, as is done for rotational temperatures (page 301). The method has the advantage that one is not limited by the maximum temperature of a comparison lamp, as in the reversal method. Although not of much value for the study of simple flames, it has been used successfully for hotter sources such as plasma jets, electrically augmented flames and shock-heated gases. It is likely to be particularly sensitive to any departures from thermal equilibrium, since it is not usually used with resonance lines because of self-absorption, and non-resonance lines are most sensitive to such errors. There is also difficulty in getting up these non-resonance lines in flames – they usually require very high temperature. The errors due to self-absorption are probably the most serious source of error, and the method has been criticised on this account (Sobolev, 1949). The method requires determination of relative intensities of lines which are often at appreciably different wavelengths; photographically this introduces some difficulty, and for steady flames a photomultiplier is best; this can be calibrated against a standard lamp.

For flames, the best metal to use is probably iron. Dieke & Crosswhite (Crosswhite, 1958) have made determinations of the relative transition probabilities of a number of lines using a flame of known temperature. Broida & Shuler (1957) have used their results for an actual study of flames, showing marked departures from equilibrium in the reaction zones of organic flames, and some smaller abnormality even for hydrogen flames.

An advantage of the line-ratio method is that it may be used to study temperature distribution through a non-uniform source. It is most suitable for the study of flames, plasma-jets, etc. which have a radially symmetrical temperature distribution. The source is considered in a series of annular zones; the emission of the outermost zone is determined first; the second zone, inside it, is then studied and the now known contribution from the outermost zone is subtracted to get that from this second zone; a third zone is then studied, and so on to the centre. The Abel integral inversion technique enables the method to be treated mathematically (Tourin, 1966; Griem, 1964).

The line-ratio method generally needs the transition probabilities, or at least the ratio of the transition probabilities of the lines used. Great care is needed in taking such values from the literature as measurements made in complex sources such as electric arcs are often unreliable. Thus Corliss & Bozman (1962) have given a valuable list of transition probabilities of lines of seventy elements but take care to point out in their preface that their own trials have shown that the values are often unsuitable for temperature determination by the Boltzmann-plot method.

A modification of the line-ratio method has been used for arc plasma by Garton & Rajaratnam (1957) in which the emission strengths of a number of lines were compared quantitatively, using a standardised carbon arc as reference, and then the absorption strength of the lines was measured with a flash-tube as background. If J is the intensity in emission and W is the equivalent breadth of the line in absorption, it can be shown that for a line of frequency, v

$$\log \frac{J}{Wv^3} = \text{const.} - \frac{n}{k}\frac{v}{T},$$

so that a graph of $\log J/W v^3$ against v will have a slope from which T can be determined. This method does not require an independent knowledge of the transition probabilities and is applicable to strong self-absorbed lines, such as resonance lines; it seems that it might be used successfully for high-temperature flames.

Effective translational temperature by Doppler broadening

The position of a spectrum line is displaced to longer or shorter wavelengths according to whether the emitter is moving away from or towards the observer. This is known as the Doppler effect. For a hot gas the random movement of the molecules causes a broadening of the spectrum lines. The half-breadth, b, is given by

$$b = 2\sqrt{(\log 2)} \sqrt{(2RT/Mc^2)} \cdot v = 0.71 \times 10^{-6} \sqrt{(T/M)} \cdot v \text{ cm}^{-1},$$

where M is the molecular weight, R the gas constant and c the velocity of light and v the wave-number of the spectrum line.

Under ordinary conditions broadening or structure due to other causes masks the Doppler broadening. Pressure broadening and self-absorption are the main difficulties, but hyperfine structure of the spectrum lines due to nuclear spin is also troublesome. We have explored the possibility of the method at low pressure (Gaydon & Wolfhard, 1949c); for lines of Pb and Fe and for the OH band lines, about the expected flame temperature was obtained, but the method was insensitive. For CH we obtained definite evidence of an abnormally high effective translational temperature, the result being about 4000 K compared with a maximum theoretical flame temperature of 2700 K; this was for the reaction zone of a flame and serves again to illustrate the lack of equilibrium. It also suggests the occurrence of some exothermic process in the formation or excitation of the CH radical. In oxy-acetylene at atmospheric pressure the OH lines are pressure broadened and although the CH lines approximate to a Doppler width there is evidence (Harned & Ginsburg, 1958) of some pressure broadening.

For our measurements we employed a Fabry–Perot interferometer to study the emission-line contours. We used photographic recording and this involved problems of plate calibration, and also we had to make allowance for the influence of the intensity contours of the interference fringes themselves in determining the true line contour. The method as used in this way is unsuitable for normal measurement of flame temperatures, although the observation of high effective translational temperature for electronically excited CH was interesting.

A possible modification is to use absorption lines, and Lück & Müller (1977) have suggested the use of a tunable dye laser to study the line contours of OH. This would have the considerable advantage of giving the translational temperature of the unexcited molecules but collision broadening effects would still contribute to the line width and detailed calculations of the Voigt profile would be necessary to interpret the observations.

An entirely different way of using Doppler broadening to get translational temperatures is from the study of Rayleigh scattering of a fine laser line (Pitz et al., 1976). The Doppler broadening of the scattered light may originate from three causes, gas movements due to the flow or to turbulence, Brillouin components due to density fluctuations travelling with the speed of sound, and kinetic effects due to thermal motion of the scattering molecules. In the high-density unheated gas mixture the Brillouin components are not negligible, but in the hot burnt mixture the kinetic effects dominate and may be used to measure the temperature. If θ is the angle between the incident laser beam and the scattered beam an additional factor of $2\sin(\frac{1}{2}\theta)$ is introduced into the expression for the line half-breadth. The method has the special advantage of enabling measurement of translational temperature at a particular point in the flame, instead of giving an averaged value over a long light path. The breadth of the scattered line is not affected by collision broadening, as is a spectrum line in ordinary emission or absorption, but it does depend on the molecular weight of the scattering species and thus on the composition of the flame gases.

Pitz et al. used the 4880 Å line from an argon-ion laser operating in a single longitudinal mode. The line contour was measured photoelectrically by means of a Fabry–Perot interferometer with piezo-electric scanning; an interference filter was used to isolate the laser line. They studied small hydrogen/air flames and compared observed and calculated line contours.

Good fits were found with the adiabatic flame temperature for near stoichiometric mixtures, but for lean flames ($\phi = 0\cdot6$ and $0\cdot7$) the measured temperatures were about 200 K high; this might be a real effect due to poor mixing in the flame gases. The method appears useful and the absolute value should be free from errors due to lack of equilibrium in the flame gases, but the error range tends to be rather large because of the square-root dependence of line breadth on temperature. Pitz *et al.* express the view that the accuracy could be improved by adjustments to the optical system. Dust particles in the air are a source of trouble, and it is necessary to know species concentrations in the flame.

Rotational temperatures

It is possible to determine the temperature of a gas from the distribution of intensities among the lines of the rotational fine structure of a band spectrum. The principle of the method is the same as that of the line-ratio method, but in this case the relative transition probabilities can be derived from theory instead of being determined experimentally.*

The experimental technique and methods of calculation of rotational temperatures have been discussed in *The Spectroscopy of Flames* (Gaydon, 1974) and will not be repeated here in detail. The intensity I of a line of the rotational fine structure of a band is given by

$$I = C.P.v^4 e^{-E_r/kT}, \qquad (11.1)$$

where C is a constant which is the same for all lines of the same band, P is the rotational transition probability (including the statistical weight factor), v is the wave number (i.e. frequency), E_r is the rotational energy of the initial state, k is Boltzmann's constant and T the temperature. The transition probability, P, can be calculated if the type of electronic transition, molecular constants and rotational quantum number K are known. The rotation energy, E_r, is usually given with sufficient accuracy by $E_r = B'.K'.(K'+1)$ where B' is the rotational energy constant and K' is the rotational quantum number for the upper electronic state.

Knowing, P, v and E_r, it is possible, in theory, to determine T from measurement of the intensity of two lines. In practice, there are three methods of deriving the temperature: (1) from the position of the maximum

---

* Learner & Gaydon (1959) and Learner (1962) have shown that the usual practice of assuming that the rotational and vibrational intensity factors may be treated independently may lead to errors which are not negligible. For a rotating molecule the potential energy curve has an extra term in $K(K+1)/r$ and this alteration to the potential energy curves modifies the vibrational intensity distribution. The effect is likely to be greatest for weak vibrational transitions, the vibrational transition probability changing appreciably at high rotational energies. Exact calculations are difficult, but preliminary work shows that with one assumption about variation of electric dipole moment with internuclear distance there may be an error of as much as 10% in rotational temperature even for the strong (0, 0) band of OH.

intensity in a single branch of the band; (2) by the so-called iso-intensity method, in which lines of equal intensity are found, one near the beginning and the other in the tail of a branch; and (3) by the plotting method, in which we rewrite (11.1) as

$$\log_e I = \log_e C + \log_e (P \cdot v^4) - E_r/kT$$

so that if we plot $\log_e I - \log_e (Pv^4)$ against E_r, we should obtain a straight line of slope $-1/kT$.

The intensity-maximum method (1) is of low accuracy and is rarely used. The iso-intensity method (2) has the advantage that quantitative measurement of intensity is avoided, and the method is fairly free from errors due to self-absorption, but in practice it needs high dispersion to isolate lines near the head of a band, and the method does not readily give information whether or not a Maxwell-Boltzmann distribution exists. The plotting method (3) is subject to errors due to self-absorption, which tends to weaken strong lines and causes the curve to dip in the middle; it does, however, show at once any departures from equilibrium and, as it makes use of a number of lines, is not seriously affected by a chance blending of one or two lines.

The most suitable band which is emitted from equilibrium hot flame gases is that of OH at 3064 Å. Dieke & Crosswhite (1962) have published full data, including transition probabilities, and shown that it can be used fairly successfully. Penner has stressed that even the iso-intensity method is not immune from errors due to self-absorption in flames of uneven temperature distribution, i.e. with cooler outer layers.

Hot flames of hydrogen do not show any clearly defined reaction zone in OH light, and the rotational temperature is always around or rather below the theoretical flame temperature. Cooler hydrogen flames do show some chemiluminescence of OH in the reaction zone (*see* page 257), but the rotational temperature for the (0, 0) band is still in fair agreement with the expected flame temperature.

For moist carbon monoxide there appears, visually, to be a luminous reaction zone, but the OH bands again give a reasonable temperature. For flames of methyl alcohol, formaldehyde and formic acid with oxygen, there is a well-defined reaction zone in which the OH emission is much stronger than that from the hot gases above the zone, but the effective temperature from OH is near the expected flame temperature and the points for a graph of $(\log I - \log Pv^4)$ against E_r (or against $K(K + 1)$) lie on a straight line.

For most organic fuels (hydrocarbons, higher aldehydes, higher alcohols, ethers) the OH bands emitted from the reaction zone give, however, a very high effective rotational temperature and in some cases (e.g. methane, ethyl alcohol) the graph is far from straight, points at low rotational energy giving a slope corresponding to a fairly normal temperature, while those for high rotational energy indicate a very high temperature. Some typical graphs for flames of stoichiometric mixtures are shown in Fig. 11.2; in these the pressure at which the flame was run and the value for the effective rotational temperature from the slope are indicated beside each graph.

We have also studied the effect of pressure on the OH rotational temperature in the reaction zone. For oxy-acetylene there is a marked rise of temperature at low pressure,

Pressure of flame	OH rotational temperature, K
1 atm	5400
13 torr	5700
5·5 torr	6200
2·5 torr	7000
1·5 torr	8750

The highest temperature, about 10 000 K, is obtained with a low-pressure flame of oxy-hydrogen containing a trace of acetylene. The effect of adding the acetylene in strengthening the OH in the reaction zone and in raising the rotational temperature is very obvious and can be seen from Plate 16d and e.

The effect of pressure can best be explained in terms of a highly exothermic chemical process forming the OH radicals in the electronically excited state and at the same time with a rotational energy distribution initially equivalent to a temperature of at least 10 000 K. Collisions with other molecules will most frequently lead to a loss of rotational energy but may also lead to loss of electronic excitation. At very low pressure the OH radicals tend to emit while still at high rotational temperature. At high pressure a limiting effective rotational temperature is reached corresponding to the temperature to which the radicals are on the average reduced before they are electronically deactivated (Gaydon & Wolfhard, 1949a). Initial work on quenching of OH fluorescence (Broida & Carrington, 1955) appeared difficult to reconcile with the above interpretation, but later measurements (Carrington, 1959) seem to agree.

OH rotational temperature measurements have been discussed more fully by Gaydon (1974) and it has been shown that the exothermic chemical reaction is fairly definitely

$$CH + O_2 = OH^* + CO.$$

The non-thermal excitation is most obvious when the normal thermal radiation is reduced. Thus in flames at atmospheric pressure cooling by dilution with inert gases relatively enhances the OH rotational anomaly.

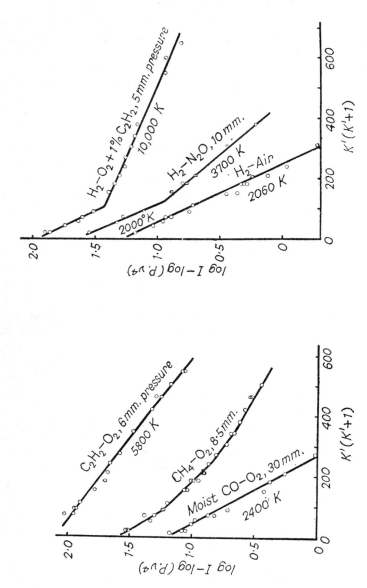

Fig. 11.2. Graphs for effective rotational temperature from OH bands, for the reaction zones of various flames.

This was noted by Wolfhard (1939) and has been studied quantitatively by Broida.

In the case of OH it is also possible to study the rotational intensity distribution in absorption, using a flash-tube as background. This gives the rotational temperature for the ground state and largely overcomes the non-equilibrium difficulties. Kostkowski & Broida (1956) have discussed this method and shown how a modification of the iso-intensity method may be used even for strong absorption. Kistiakowsky & Tabbutt (1959) have studied rotational temperatures of detonations in absorption; near the front the values are still anomalously high.

For the Swan bands of C_2, we again find very high rotation temperatures in the reaction zones of organic flames (Gaydon & Wolfhard, 1950a). For oxy-acetylene we find about 4700 K at 1 atm falling to about 3800 at very low pressure. Typical graphs are shown in Fig. 11.3. In the mantle of rich oxy-acetylene flames, away from the reaction zone, the C_2 rotational temperature in both absorption and emission agrees with the adiabatic temperature of about 3200 K (Jessen & Gaydon, 1969).

The effect of pressure on the rotational temperature in the reaction zone is here more difficult to explain. The C_2 may be formed in the ground electronic state but with high rotational energy and then subsequently excited before it has lost its rotational energy excess; alternatively the C_2 may, like OH, be formed in the excited state by an exothermic reaction, but with initial thermal energy of the reacting species being carried over into the product C_2.

For CH bands at 3900 and 4315 Å the rotational temperatures in the reaction zones are around or only slightly above the adiabatic flame temperatures (Gaydon & Wolfhard, 1949c). This is a little surprising in view of the high translational temperatures of CH in low-pressure flames.

The CN Violet bands in a diffusion flame of cyanogen with nitric oxide have been used to study the rotational temperature of this very hot flame (Cros, Bouvier & Chevaleyre, 1971); the maximum value of 4750 ± 100 K is close to the calculated adiabatic value of 4850 K.

For the NH band at 3360 Å, temperatures appear to be high. For the reaction zone of an acetylene/nitrous oxide flame the rotational temperature is 3760 K at 1 atm and 3500 K at 11 torr. Even the hydrogen/nitrous oxide flame has a well-defined reaction zone (unlike the flame with O_2) and the NH bands have a high rotational temperature, about 3500 K at 1 atm. For ammonia/oxygen and ammonia/nitrous oxide, the NH band gives a temperature only slightly above the adiabatic value.

From these results it is clear that rotational temperature measurement are useful for indicating the state of equilibrium and giving pointers to the particular chemical reactions occurring, but cannot give a measure of true temperature in a reacting system, despite the fact that rotational energy is usually equilibrated in less than 100 collisions. In equilibrium hot gases, the OH bands will give a measurement of temperature, but in large flames errors due to self-absorption are difficult to avoid. The only other bands normally emitted by the interconal gases are those of the Schumann–Runge system of O_2 and the NO γ system; these lie at rather short wavelengths and are usually weak and not very suitable for measurements.

Reliable rotational measurements can be derived from infra-red rotation–vibration bands. H_2O and CO_2 are not usually suitable as they have such a large number of overlapping rotational lines. In some cases it is

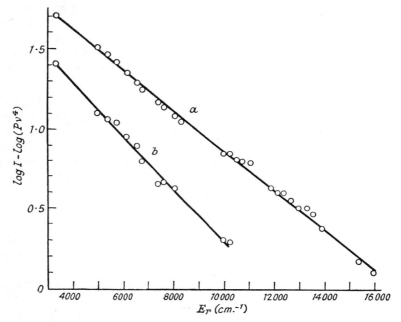

Fig. 11.3. Rotational temperatures for C_2. Graphs of $\log I - \log Pv^4$ against rotational energy term E_r for stoichiometric oxy-acetylene (a) at 1 atm with a welding burner, giving $T = 4880 \pm 200$ K; (b) at 8·5 torr with 21 mm burner, giving $T = 3700 \pm 400$ K.

possible to introduce a small amount of HF into the flame. The HF band stretching from 2·35 μm (R-head) to 3·4 μm has a very open rotational structure, with the distance between adjacent lines varying from 20 to 50 cm^{-1}, so that the use of high resolving power is unnecessary; the implications of measurements using HF are treated by Simmons (1967). The CO bands at 4·7 and 2·35 μm are also suitable for rotational temperature measurement, although somewhat higher resolving power is needed.

An interesting method of measuring gas temperatures and densities is due to Muntz (1962). He injected a narrow beam of fast electrons (10 to 100 keV) into the gas or flame and studied the excitation of the nitrogen First Negative bands (N_2^+, $B^2\Sigma_u^+ \to X^2\Sigma_g^+$). The overall intensity of the bands is proportional to the density of nitrogen and the rotational temperature is found to be the nitrogen ground-state temperature (gas temperature). Although the overall accuracy of the method may be not much better than 5% it has the advantage of being a point measurement and is thus not influenced by hotter or colder surrounding layers. The temperatures can be measured with two narrow-wavelength interference filters and detectors and this allows the measurement of rapid density and temperature changes. Marrone (1966) has also used the method to measure the rotational temperature in supersonic nitrogen jets expanding from room temperature and has been able to measure temperatures down to the extraordinarily low value of 9 K.

Vibrational temperatures

The measurement of effective vibrational temperatures from the vibrational intensity distribution of bands within a system is very similar to the measurement of rotational temperatures. On the practical side the method may have the advantage that it is usually adequate to use small dispersion, measuring either the maximum intensity near the head of each band or the integrated intensity of the whole band. Thus exposure times may be shorter than for rotational temperatures. On the other hand there is more likelihood of difficulty with change of plate sensitivity with wavelength, and it may be necessary to examine the structure of the bands in some detail to see how the closeness of the lines of the rotational fine structure and the value of the rotational quantum number at which the head occurs will affect the intensity contour of the bands.

For diatomic molecules, vibrational transition probabilities may readily be calculated if the molecule is assumed to behave as a harmonic oscillator with constant dipole moment. In many practical cases, however, corrections for anharmonic factors, for vibration–rotation interaction, and for variation of dipole moment with nuclear separation are not negligible. The first two corrections may be made by more refined calculations if sufficient data about the spectrum are available, but little is known theoretically about the variation of dipole moment. Thus we are often still dependent on measurements of relative intensities of bands in a source at known temperature to obtain the vibrational transition probabilities.

One of the first measurements of vibrational temperature was that for the CN Violet bands, in an arc, by Ornstein & Brinkman (1931). These CN bands have also been used for one special flame study by Thomas, Gaydon

& Brewer (1952). For cyanogen/oxygen the expected temperature is extremely high, the actual value obtained by calculation being either 4850 K or 4350 K according to whether the dissociation energy of nitrogen is taken as 9·76 or 7·38 eV. This flame gives some emission of the CN Violet bands from the main body of the flame gases. For this work the vibrational transition probabilities were based on an experimental study of CN emission from a King furnace at known temperature. The vibrational temperature of these bands in the flame was found to be 4800 ± 200 K, thus supporting the high value for the dissociation energy of nitrogen, which was at that time slightly uncertain. This flame of $C_2N_2 + O_2$ is indeed one of the hottest natural flames; this is because of the low dissociation of the product gases CO and N_2. In the reaction zone of the flame the CN vibrational temperature was only around 3500 but rose to the final value a couple of millimetres above the zone. In this case we therefore have evidence of either a vibrational energy lag, or a lag in reaching chemical equilibrium.

For OH, values of the vibrational transition probabilities are available (Nicholls, 1956), with corrections for vibration–rotation interaction (Learner, 1962; Watson & Ferguson, 1965). A number of flames, detonations and shock-tube sources have been studied. A difficulty is that there is a weak predissociation in the upper $A^2\Sigma^+$ electronic state for levels with v greater than 1, so that an inverse effect may cause enhancement of bands with $v' = 2$ and 3 if the gases contain an excess population of free O or H atoms (Gaydon & Wolfhard, 1951b). Thus observations in emission tend to be unreliable and serve mainly to detect departures from chemical equilibrium.

For C_2, data for transition probabilities have also been determined with sufficient accuracy (Jain, 1964; Mentall & Nicholls, 1965). Pillow (1952) found that the C_2 vibrational temperature in the inner cone of a Bunsen flame was very high, around 3700 K. Marr (1957) found high temperature in a rich oxy-acetylene flame, and recently Jessen & Gaydon (1969) have confirmed the high temperature in the reaction zone and shown that even in the mantle region the vibrational temperature is still 300° to 600° too high. For a low-pressure oxy-acetylene flame Jones & Padley (1975) quote vibrational temperatures of 6000 K for C_2 ($A^3\Pi$), 5400 K for OH ($A^2\Sigma^+$), 2600 K for CH ($A^2\Delta$) and 4020 ± 150 K for CO ($A^1\Pi$); for the last value, on the CO Fourth Positive bands, it was necessary to sweep away flame products with nitrogen to prevent self-absorption; the result confirms the chemiluminescent excitation of CO.

Laser-Raman scattering

The advent of the laser has given us a potentially powerful tool for point-by-point temperature measurement in flames. When monochromatic light is scattered by a molecular gas the main, though very faint, Rayleigh scattering is of the same wavelength as the incident light apart from small Doppler shifts, but there is also much weaker scattering in which the light frequency is altered by subtraction (Stokes) or addition (anti-Stokes) of the characteristic frequencies of the vibrational and rotational modes of the molecule. This is known as the Raman effect. The rotational Raman effect, in which the rotational quantum number changes by ± 2, leads to frequency alteration of the order 10 to 100 cm^{-1} and so the lines are rather close to the central Rayleigh line. The vibrational Raman effect is, however, of the order 1000 cm^{-1} and so the bands are well separated from the exciting line; typical Raman frequencies are N_2 2326, O_2 1556, CO 2145, H_2 4158 and CO_2 1285 and 1388 cm^{-1}. In the normal Stokes bands, absorption of energy from the incident light quantum raises the vibrational quantum number in the scattering molecule by one unit, e.g. from 0 to 1, from 1 to 2 or from 2 to 3 etc. In anti-Stokes bands the molecule gives up energy to the incident quantum and the observed bands correspond to changes in vibrational quantum number from 1 to 0, 2 to 1, etc. The equilibrium population in each of the various vibrational energy levels is given by the familiar Boltzmann distribution law that the number of molecules in a level with energy E is proportional to $\exp(-E/kT)$. Thus measurement of the relative intensity of the $(0 \rightarrow 1)$ Stokes band to the $(1 \rightarrow 0)$ anti-Stokes band will depend on the population in the $v = 0$ and $v = 1$ levels and could be used to give the temperature. Alternatively, as mostly used so far, Stokes bands $(0 \rightarrow 1)$, $(1 \rightarrow 2)$, $(2 \rightarrow 3)$ etc. may be compared; these bands overlap slightly but being close together can more readily be scanned as a group, without the need to calibrate the photomultiplier for variation of sensitivity with wavelength.

The practical problems are mainly due to the weakness of the scattered light. Jessen (1971) gave a useful assessment of the practical technique and points out that even the Rayleigh scattering is so weak that it requires passage through about 100 miles of atmosphere to reduce green light to half its initial intensity, and that the Raman scattering is of the order 1000 times weaker. The Raman light too is distributed among a number of vibrational and rotational transitions. Scattering is relatively strongest for short wavelengths, varying as $1/\lambda^4$, and so it is an advantage to use a laser line at the blue end of the spectrum; so far the 4880 Å line from an argon-ion laser has been favoured. The Raman spectrum must be studied with a monochromator of adequate dispersion, usually a grating, and a double monochromator is probably necessary to cut down stray light. Recording is usually made with a very sensitive photomultiplier as this is faster than a photographic plate and also has a linear intensity response. Stray light from the Rayleigh line, which is strongly polarised, may be reduced by using a suitably oriented piece of polaroid. By careful focusing of the laser beam and studying scattering at 90° it should be possible to obtain a spatial resolution in the flame of about 1 mm.

Most investigators so far have aimed at proving the technique. Papers of interest are those by Widhopf & Lederman (1971), Lapp, Goldman & Penney (1972), Vear, Hendra & Macfarlane (1972) and Stricker (1976), who studied the temperature distribution in a propane flame; there are also a number of papers in *Laser-Raman gas diagnostics* (Lapp & Penney, 1974). It is usually necessary to allow for the finite resolving power of the monochromator and to compare computed and observed band profiles. Thus Lapp *et al.* (1972) assumed a triangular slit function and compared band profiles for O_2 and N_2 in hydrogen flames; their curves are shown in Fig. 11.4.

Results computed in this way involve both the effective vibrational and rotational temperatures of the ground-state O_2 and N_2 molecules; the vibrational contribution would probably be the more important though. Both these temperatures should be in close equilibrium with the translational temperature under flame conditions, even in the reaction zone. The active Raman frequencies are different from those for the infra-red, and symmetrical vibrations such as those of homonuclear molecules like O_2, H_2, and N_2 are observed. The laser-Raman method is thus ideally excellent, giving information about the ground states of the main stable molecular species, with good spatial resolution. It is also possible to measure concentrations of these species. Practically, though, the very low light intensity necessitates the use of refined expensive apparatus, and some spectroscopic know-how in computing band profiles is required. The technique has now been proved successful, but as yet has not been applied much in practical cases.

Another method now being explored (Beattie *et al.*, 1978; Stenhouse & Williams, 1978) is coherent anti-Stokes Raman spectroscopy ('CARS'). This involves irradiating the flame with two laser beams, one of frequency v_1 and the other of frequency v_2, where $v_2 = v_1 - v_m$, v_m being the vibrational frequency of the molecule being studied. v_2 is obtained from a tuned dye laser. The two beams are inclined at a small angle and the CARS radiation occurs at a frequency $2v_1 - v_2 = v_1 + v_m$. The total light emission, proportional to the cube of the irradiating intensity, is extremely weak, much weaker even than normal Raman emission, but because it occurs in a narrow coherent beam its brightness in this direction is relatively high. Troubles from fluorescence, which can interfere with normal Raman spectroscopy, are avoided. The spatial resolution is poorer but still adequate. Preliminary studies of the rotational temperature of N_2 in flames are promising and it should be possible also to measure species concentrations.

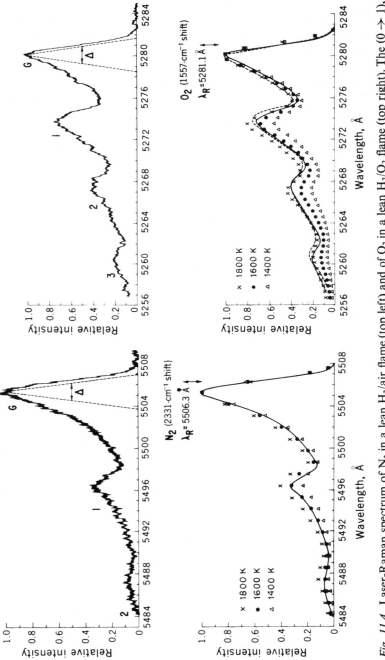

Fig. 11.4. Laser-Raman spectrum of N_2 in a lean H_2/air flame (top left) and of O_2 in a lean H_2/O_2 flame (top right). The $(0 \to 1)$, $(1 \to 2)$, $(2 \to 3)$ and, for O_2, the $(3 \to 4)$ Stokes bands are observed and are labelled G, 1, 2, and 3. The slit function is indicated by the Δ sign. Below are given the smoothed band contours and computed points for 1400, 1600 and 1800 K. For O_2 (bottom right) two contours from the different runs indicate the experimental scatter. Excitation by 4880 Å. [*Redrawn from Lapp* et al. (1972)]

Refractive-index methods

The methods of measuring flame temperature described in the next three sections (refractive index, α-particle and velocity of sound) all depend essentially on the density of the gas. They thus all have the great advantage that the result gives the mean translational temperature of the gas molecules, and is not therefore sensitive to slight departures from equilibrium. They have the disadvantage that a knowledge of the gas composition is required and are rather insensitive because a small thickness of gas at room temperature is equivalent to a much greater thickness at flame temperature, so that any error in locating the edge of the flame will produce a magnified error in the determination of the flame temperature.

Direct measurement of the refractive index through the flame gases is best made interferometrically, comparing the optical path through the flame with that through air. The simple Rayleigh and Jamin refractometers may be used, but for study of the distribution of temperature through a flame, the Mach-Zehnder interferometer (page 52) is better; the displacement of the parallel fringes may be recorded photographically. For a rectangular flame with path length l, the fringe shift should be equal to $(l/\lambda)(n - n_0)$, where λ is the wavelength and n is the refractive index of the flame gases and n_0 that of the air in the reference beam.

The refractive index depends mainly on temperature but also on gas composition and wavelength. At constant pressure, for monochromatic light we may use

$$n - 1 = (n_0 - 1) T_0/T \cdot (n_p - 1)/(n_r - 1)$$

where n_p and n_r are the refractive indices of the product gas mixture and of the initial reactant gases at the initial temperature T_0; T is the flame temperature being measured. Obvious difficulties are the accurate determination of the path length l through the flame and the refractive index n_p of the products; above about 2000 K dissociation of molecules to form OH and free atoms begins to be significant and it is difficult to obtain the refractive indices of these, although data from shock-tube measurements are now becoming available. Early studies using a Mach-Zehnder interferometer were made by Olsen (1949), and the method has been fully reviewed by Weinberg (1963), who has also used the Weinberg-Wood four-grating interferometer (page 53). A laser may now be used for illumination, and this greatly facilitates setting up these interferometer systems.

Variations of refractive index can also be studied by shadow or schlieren photography and by ray deflections. Dixon-Lewis & Wilson (1951) analysed schlieren photographs to get information about the temperature distribution in the preheating zone. The best method, however, for these studies is that developed by Weinberg and colleagues using an inclined slit

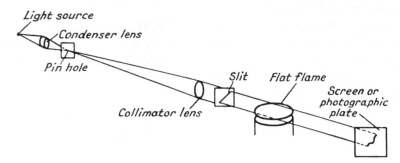

Fig. 11.5. Layout of inclined-slit arrangement for studying variation of refractive index. [*From Weinberg*

(Burgoyne & Weinberg, 1954; Weinberg, 1963). The set-up is illustrated in Fig. 11.5. The downward deflection of the light distorts the slit image, as in Fig. 11.6. For a flat flame, the deflection at any point is given to a close approximation by $D.l.dn/dx$, where D is the distance from the centre of the flame to the photographic plate, l is the path-length in the flame, and dn/dx is the vertical refractive index gradient.

In the preheating zone, before there is any change in composition, we may take $n - 1 = (n_0' - 1) T_0/T$, and for the burnt gas region we again have
$$n - 1 = (n_0 - 1) T_0/T \cdot (n_p - 1)/(n_r - 1).$$
It is possible to integrate the change of refractive index over any region to get the refractive index and hence the temperature. This analysis of tem-

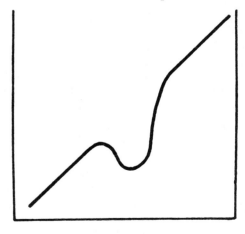

Fig. 11.6. Image of inclined slit.
[*From Weinberg*

perature distribution through the preheating and reaction zones has been referred to already (page 97). It can also be extended, in a simpler way, to give the final flame temperature. The area contained by the deflected and un-deflected slit images is, with sufficient accuracy, related to the final refractive index, n_f, by

$$A = \frac{D \cdot l}{m} (n_0 - n_f)$$

or to the final temperature by

$$T_f = \frac{T_0}{1 - Am/Dl\,(n_0 - 1)} \left(\frac{n_p - 1}{n_r - 1}\right)$$

where m is the slope of the slit.

This method of measurement of final flame temperature has been used mostly for flat Egerton-Powling type flames with lean mixtures. It is, however, also applicable to curved flames. The various approximations and limitations have been considered in detail. In hot flames of fast-burning mixtures, it would be difficult, because of diffraction effects and optical limitations, to measure the deflection of the slit image with sufficient accuracy, and dissociation effects in the burnt gases would lead to some uncertainty in the value of the refractive index of the products. The main use of the method is undoubtedly for determining the temperature course through the preheating and reaction zones. By using a pulsed laser as light source it is possible to take instantaneous photographs to study transient or fluctuating flames, or the method may be applied to holograms of the flame.

α-particle and X-ray methods

Shirodkar (1933) has determined the temperature of the interconal gases of a coal-gas/air flame by measuring the range of α particles. A polonium source was used and the ionisation produced by the particles was measured as a function of distance using an electroscope. Curves were plotted for air and with the flame interposed between the source and electroscope. The curves were of similar form but displaced horizontally, in the actual experiment by a distance of 5·55 mm. The thickness of the flame was determined by plotting an approximate temperature distribution curve with a thermocouple, the flame used for the experiment having a thickness of 6·7 mm. Thus 6·7 mm of flame gases have a stopping power equivalent to 6·7 − 5·55 mm of air = 1·15 mm of air. It was found that the cooled flame gases had the same stopping power as air. Hence

$$\frac{\text{flame temperature}}{\text{room temperature}} = \frac{6\cdot 7}{1\cdot 15}.$$

With room temperature at 290 K, this gave the flame temperature as 1690 K. It was found necessary to put water-cooled shields round the flame to prevent heating of the surrounding air. In later work a counting device was used to replace the electroscope, and multiple flames were studied.

As stressed in the previous section, one difficulty is to determine the effective thickness of the flame accurately; if a refractory radioactive source could be placed on a probe in the flame gases, it might be possible to eliminate end effects. For very hot flames it might be difficult to correct the results for reduction in density of gases due to dissociation.

A similar method using X-ray absorption has been discussed by Weltmann & Kuhns (1951), and Mullaney (1958) has used a radioactive source, Fe^{55}, of monochromatic X-rays of about 2 Å wavelength. The scattering of a beam of neutrons has been studied by Delaney & Weber (1962), who made some trial measurements on the temperature of an oxy-hydrogen flame, but concluded that the method was more suitable for plasma at higher temperature.

Sonic methods

The velocity of sound, c, depends on the temperature, T, the ratio of specific heats at constant pressure and volume, γ, the molecular weight M and the gas constant R

$$c = (\gamma\, RT/M)^{\frac{1}{2}}.$$

Hence if we can determine the velocity of sound in a gas and also know the mean molecular weight and γ, then we can calculate the temperature. Care must be taken not to use very high frequency sound because the ratio of specific heats may alter because of delay in the molecules taking up vibrational energy. There are various possible ways of generating the sound wave and measuring its velocity. Marlow, Nisewanger & Cady (1949) have used a quartz oscillator of known frequency as source and measured the wavelength by shadow photography using an instantaneous spark flash. The velocity is then of course the product of frequency and wavelength. In this way the end errors are absent; exact knowledge of the chemical composition is not required, but the effect of dissociation on molecular weight and γ would have to be known or assumed. The photographic method could not be applied to turbulent flames, and we think that in some cases change of refractive index near the flame edges and in the reaction zone could distort the photographs. However, this is a method worthy of more attention, and the development of the laser should assist experimental work. Determination of the local sound velocity by laser–schlieren–Doppler measurements has possibilities (Schwar & Weinberg, 1969). Transit times of ultra-sonic pulses have been used by Carnevale et al. (1967) to measure temperatures in shock tubes, and a similar method should be useful for detonation flames in tubes.

Hot-wire methods

The temperature of a wire immersed in flame gases may be measured in three ways:

(a) from its electrical resistance;
(b) from thermo-electric e.m.f. at a junction (i.e. thermocouple);
(c) from its brightness and emissivity, with an optical pyrometer.

The wire will normally be cooler than the flame gases owing to heat losses by radiation, and perhaps by conduction down the leads. A large number of methods have been used to reduce or correct for this error.

The convective heat transfer coefficient increases for thinner wires so that as we use thinner and thinner wires we get nearer the true flame temperature. There is obviously a limit set by the mechanical strength of the wire, but it is possible to extrapolate results for wires of different thickness to that for an infinitely thin wire. We do not think that very accurate results can be obtained in this way, especially for hot flames where the radiation losses are very great. We note a recent paper by Sato *et al.* (1975) in which the thermocouple is traversed through the flame in the direction of the leads and an iterative calculation enables correction for conductivity and radiation losses.

The heat transfer also depends on the velocity of flame gases past the wire, so that if the flame gases are sucked over a thermocouple then the temperature of the thermocouple will rise with speed of sucking. Hottel & Broughton (1932), for example, have used a suction pyrometer on this principle. They also attached the thermocouple to an electrically heated ribbon and adjusted the heating current so that the temperature of the thermocouple did not vary with the speed of sucking of gases past it, at which condition the couple and gases must be at the same temperature; with the couple and ribbon hotter than the gases, increased sucking rate would lead to a cooling of the couple, and conversely with the couple below gas temperature increased sucking would lead to a rise.

The heat transfer to a wire or thermocouple depends on the gas velocity V and the wire diameter d, and the heat transfer coefficient may be expressed as

$$h_c = CV^n d^{-m}$$

where $n = 0.5$ to 0.7 and $m = 0.3$ to 0.5. In practice the maximum velocity, and thus heat transfer, for a suction pyrometer is obtained when the pressure drop is sufficient to accelerate the gases to the velocity of sound. Small sonic thermocouples can be obtained in which this sound velocity can be achieved in a suitably designed throat (Lalos, 1951). A good review of temperature measurements in industrial flames is given by Chedaille & Braud (1972).

The heat loss by radiation can also be reduced by the use of shields around the couple; shields of polished metal such as platinum are most effective, but for relatively cool flames it is possible to use steel shields.

The radiation losses may also be corrected for in the case of the resistance-wire method by heating the wire electrically. In a method due to Schmidt the radiation from the wire is measured with a thermopile for various heating currents with the wire in air; knowing the total heat loss by radiation at any temperature it is then possible to compensate for this loss by supplying the required amount of heat electrically. A simpler method has been used by Griffiths & Awbery (1929); they measured the temperature of the wire with an optical pyrometer and plotted temperature against heating current for the wire *in vacuo* and for the wire in the flame; it can then easily be shown that the point of intersection of these two curves is at the true flame temperature, because at this temperature there must be no heat transfer to or from the flame gases, the condition which obviously holds *in vacuo*.*

The chief limitations of hot-wire methods are due to the melting-point of the wire and to catalytic effects on the surface.

Platinum, with a melting point of 1769°C, is the most suitable metal and is most generally used. For thermocouples it may be alloyed with other platinum-group metals. Rhodium melts at 1966°C but because it forms an oxide around red heat is used alloyed with platinum; platinum/platinum + 10% rhodium or platinum/platinum + 13% rhodium thermocouples are in common use and may be used to 1760°C; platinum + 24% rhodium/platinum + 40% rhodium may be used to 1880°C. Iridium melts at 2410°C and iridium/iridium + 40% rhodium thermocouples are stated to reach 2300°C and to be commercially available (Land, 1963). Osmium melts around 3000°C but is more difficult to process and has a volatile oxide. The tungsten-group metals have high melting points but oxidise rapidly at high temperature and can only be used in a reducing or inert atmosphere. Morgan & Danforth (1950) have experimented with bare tungsten/tungsten + 50% molybdenum to 2900°C; tungsten/tantalum will reach about 3000°C but the e.m.f. has a maximum around 2000°C. Davies (1960) has used tungsten/tungsten + 25% rhodium to 2800°C. Chedaille & Braud (1972) also include a curve to about 2500°C for tungsten/tungsten + 26% rhenium.

Wires may be protected by refractory shields, usually of metal oxides; the maximum useful temperatures, as quoted by Chedaille & Braud are MgO 2200°C, Al_2O_3 1850°C, ZrO_2 1000°C, BeO 2200°C and ThO_2 2800°C. There is some tendency for the oxide to dissolve in the metal of the wire

* Prof. Weinberg has pointed out that this is only true as long as radiation from the flame to the wire is negligible.

and, usually over a time of the order 100 hours, to cause a change in electrical properties and some embrittlement.

Thermocouples must be calibrated by heating in a furnace whose temperature is measured optically using the radiation laws. We note the description of a suitable furnace for calibrating thermocouples to 2200°C by D.B. Thomas (1962). The cold junction of a thermocouple is normally immersed in an ice/water mixture; if the junction is left at room temperature this only causes an error of about 0·5°C; one should never just add the room temperature, in °C, to the reading.

A suction pyrometer may be protected by a surrounding water-cooled jacket, so that the thermocouple junction is not heated beyond the melting-point of the metals (Matton & Fouré, 1957). Although the temperature of the junction may then be much lower than the flame-gas temperature, it is uniquely determined by this gas temperature. Such a pyrometer can be calibrated by placing it in a Méker-burner flame whose temperature is determined by the sodium-line reversal method. After calibration it may be used for point-by-point determinations of temperature. A cooled probe of this type is likely to disturb the gas flow and local temperature somewhat but the instrument might be very useful for studying temperature distribution in large hot flames, such as gas turbines.

A venturi pneumatic pyrometer (*see* Chedaille & Braud, 1972) uses two venturi throats with a resistance thermometer in the relatively cool gas beyond the second throat. Pressure differences measured across the two throats enable the gas temperature to be calculated. Probes of this type developed by Land–BCURA have been successfully tested at IJmuiden.

If the flame gases are in true equilibrium, then there is little reason to doubt that, within the limits set by the melting-point, a hot-wire method can be made to give an accurate value for the temperature. If there is any departure from chemical equilibrium, then catalytic heating on the surface may cause error. The most extreme case is the well-known effect of palladium in igniting a mixture of air and alcohol, but at high temperatures it is likely that most metallic surfaces are more or less active in promoting combustion. Marked differences between the temperatures of bare wires and those coated with quartz have been found by the late Prof. David at Leeds. This effect is probably mainly due to recombination of free oxygen atoms, but may also be caused by surface oxidation of carbon monoxide. The temperature rise due to such effects is not, of course, reduced by using smaller wires or altering the flow of gas over the wire, as with radiation errors. Surface catalysis may be reduced by covering the wire with a protective coating of quartz; this is usually done by spraying hexamethyldisiloxane into a propane/air flame at 1600°C, and passing the wire through the flame a few times. However, when used in a reducing flame above 1100°C the silica is reduced to silicon which dissolves in the platinum and weakens it and also alters its electrical properties. To overcome this, Kent (1970) has recommended a paste of 10 to 15% beryllium oxide with yttrium oxide, fired on at 1600°C.

The heat transfer from flames to various solid surfaces has been investigated by Kilham (1949) and by Kilham & Garside (1949). They used a large flame from a rectangular burner, and measured the radiation from a refractory tube coated with various metallic oxides, and also the temperature of the tube by a thermocouple in it, and the temperature of the flame by the sodium-line reversal method. The tube was rotated, to ensure uniform heating. The heat transfer to the tube must be equal to that lost by radiation, and can therefore be obtained from the measured amount of radiation. Knowing the temperature of the flame and of the tube it is possible to compare the calculated heat transfer, using known coefficients for heat transfer by forced convection, with the measured heat transfer. For carbon monoxide/air flames the measured heat transfer agrees with that expected from forced convection and there is no evidence for specific catalytic effects. For a hydrogen flame, however, the experimental heat transfer coefficients are higher than those calculated for forced convection for the region just above the inner cone, but higher in the flame there is again agreement. This appears to indicate some lack of equilibrium just above the inner cone in agreement with other observations.

Hot-wire methods are not generally suitable for measurements in the reaction zone of a flame because of catalytic effects and because the wire is large or at least comparable in size with the thickness of the reaction zone and disturbs the flame itself. Using low-pressure flames, with a much thicker reaction zone, Klaukens & Wolfhard (1948) have used a thermocouple to measure the rise of temperature in the preheating zone and have shown that exothermic reaction begins at a temperature well below that expected on a thermal theory of flame propagation (*see* page 96).

In fast gas streams any obstacle becomes heated. We may define two temperatures for the gas, the static temperature, T_s, taken up by a body moving with the gas stream, and the total temperature T_t, corresponding to the total energy of the gas. It can be shown that

$$T_t = T_s + v^2/(2Jc_p),$$

where v is the velocity, J is the mechanical equivalent of heat and c_p is the specific heat at constant pressure.* For a wire in a gas stream, the part of the wire normal to the stream is heated by impact of the gas molecules and tends to take up the full temperature, T_t, but the sides of the wire are

---

* If we use v in cm/sec and c_p in cal/gm then $J = 4\cdot18 \times 10^7$ ergs; alternative if v is in m/sec and c_p in cal/kg then $J = 4\cdot18$.

heated less by friction, and the shielded side is unheated. In practice we may define a recovery factor r as

$$r = (T_w - T_s)/(T_t - T_s),$$

where T_w is the temperature taken up by the wire (after allowance for radiation and other losses). The recovery factor is about 0.68 ± 0.07 for a wire perpendicular to the flow direction and 0.86 ± 0.09 for a wire parallel to the flow. For ordinary thin wires the temperature rise due to this heating effect is only about 2 K for a velocity of 75 m/s, rising to 10 K at 180 m/s. In a sonic suction pyrometer the gas will cool, by around 200 K, as it passes through the nozzle, but presumably most of this is recovered by the thermocouple wires.

Although thermocouples and resistance wires have an appreciable response time, often of the order of tenths of a second, it is possible by combining them with special electrical circuits to use them for studying fast temperature fluctuations, such as occur in turbulent flames. If T_w is the instantaneous temperature of the wire, T_g is the gas temperature and τ is the time constant, then

$$\tau \left(\frac{dT_w}{dt}\right) = T_g - T_w \quad \text{or} \quad \tau \left(\frac{dT_w}{dt}\right) + T_w = T_g.$$

An electrical circuit can be devised which effectively differentiates the electrical signal E and then adds this value back to the original signal to give

$$\tau \left(\frac{dE}{dt}\right) + E$$

which is of the form given above. The method, used by Shepard & Warshawsky in 1952, has frequently been employed for the study of turbulence in flames. Kunugi & Jinno (1959) used a bare thermocouple of wires 0.06 mm thick, with electrical compensation of this type, and oscilloscope display and were able to follow temperature fluctuations due to turbulence up to 1000 Hz. Lockwood & Ododi (1973) achieved a response time of 10^{-4} sec using 40 μm wires of 6% Rh/Pt vs 30% Rh/Pt for a thermocouple. Ho, Jakus & Parker (1976) used an iridium resistance wire 12.5 μm diameter to study hot and cold temperature spikes in turbulent flow.

In summary, hot-wire methods are particularly suitable for large industrial flames and furnaces (Chedaille & Braud, 1972; Thring, 1962). In these there is less risk of departures from chemical and thermodynamic equilibrium, and since such flames are usually with air, the temperatures are not too high. In these cases it is often important to know the temperature distribution and for this point-by-point plotting of temperature, the hot-wire methods are obviously preferable to optical methods, most of which give a mean temperature along an optical path right through the flame. Hot wires are less suitable for studying very hot flames, small flames which may be disturbed by the probe, and for giving details of processes in reaction zones.

Ionisation and electron temperatures

The degree of ionisation and the velocity distribution of ions and electrons are temperature dependent and provide a potential method of temperature measurement.

The concentrations of ions and free electrons are, in equilibrium, given by the Saha equation (*see* Chapter XIII). Electron concentrations can be measured by micro-wave absorption, by conductivity measurements with probes, or, at high temperatures, by Stark-effect broadening of spectrum lines (e.g. Holtsmark broadening of hydrogen lines for temperatures around 6000–12 000 K) while positive ion concentrations may be studied by observation of spectrum lines of neutral and ionised atoms, e.g. of Ba and Ba^+. In practice the persistence of ions and electrons produced in very high concentration in the reaction zones of organic flames by chemi-ionisation processes, and also the sensitivity of ionisation to the presence of trace impurities such as alkali and alkaline-earth metals, makes concentration studies unsuitable for realistic temperature measurements in flames.

Electron temperatures (i.e. their random velocity distribution) may be studied with Langmuir probes or double probes (*see* page 345). Such measurements are usually made on low-pressure flames where the electron mean free path is longer than the probe dimensions; at higher pressures probe theory is very complex. In recent years a large number of papers have been published on measurements of electron temperatures in various low-pressure flames. Values are usually above, sometimes well above, adiabatic flame temperatures and there has been considerable debate on the interpretation of the results. Table 11.1, based on one given by Bradley & Ibrahim (1975), shows some values obtained by double-probe studies on low-pressure hydrocarbon flames. From study of papers such as those by Bell & Bradley (1970), Porter (1970) Silla & Dougherty (1972) and Brule, Michaud & Barassin (1973) it seems that very high values of T_e are likely to be spurious because of errors in interpretation of probe characteristics; this is particularly so where large currents are drawn from the probe near the condition of electron saturation. Most reliable results are obtained with very small double probes, and the effect of probe size is illustrated by the last two entries in Table 11.1. Water cooling of a probe may also produce errors due to negative ion attachment (Sahni, 1969), and probes must be kept out of the secondary reaction zone (i.e. the outer cone). Temperature excesses of around 200 K are, however, likely to be real.

TABLE 11.1

Probe measurements of maximum electron temperatures (T_e) compared with gas temperatures (T_g) for hydrocarbon/oxygen/nitrogen flames

Flame	Probe size* mm	Pressure torr	Burner diam. mm	T_e K	T_g K	$T_e - T_g$ K	Authors
C_3H_8/air diffusion	0·5	50	48	4620	1310	3310	Kinbara et al.
$C_2H_2 + 1·58\ O_2 +$ 10·28 N_2	0·25 × 10	18	250	1850	1615	235	Porter (1970)
$C_2H_2 + 3·25\ O_2 +$ 10·17 N_2	0·25 × 10	18	250	1800	1630	170	Porter (1970)
CH_4/air	1·9	40	76	2900	1320	1580	Bradley et al.
C_2H_4/O_2 diffusion	0·5	2·8	4·8	2600	1200	1400	Brule et al.
C_2H_4/air diffusion	0·4 × 33	98	65	1940	1595	345	Taran et al.
CH_4/air	1·04	38	76	2400	1740	660	Bradley et al.
CH_4/air	0·508	38	76	1940	1805	135	Bradley et al.

* Diameter of spherical probe or diameter and length of cylindrical probe.

Although the electron temperature is a measurement of the translational temperature of the electrons, this may not equilibrate very quickly with the translational temperature of the gas molecules because of the big mass difference between electrons and molecules, this preventing appreciable energy transfers during elastic collisions. Inelastic collisions with electronically or vibrationally excited species in the flame may raise the effective translational temperature of the electrons. Bradley & Sheppard (1970) have shown that the relaxation time should be only of the order 10^{-8} sec but that T_e could be raised slightly above T_g by energy transfer from vibrationally excited species such as CO, CO_2 or OH and the position of temperature excess should follow, point by point, any chemiluminescence. Bradley & Jesch (1972) showed that T_e followed electronically excited hydroxyl, OH*, rather than C_2*; they calculated that the concentration of OH* would not be high enough to account for the observed excess and suggested that vibrationally excited OH ($x^2\Pi$) might be responsible; however, remembering that the reaction $CH + O_2 = CO + OH$* also produces CO molecules which probably have high vibrational energy, we think that these CO molecules are more likely to be the source of the excess energy.

For carbon monoxide flames (von Engel & Cozens, 1963; Bradley & Matthews, 1967) very high values of electron temperatures have been reported, even up to 30 000 K in one very narrow zone of the flame. These have been attributed to energy transfer from vibrationally excited CO_2 molecules, but there is considerable doubt about their reliability. The subject of flame ionisation is discussed more fully in Chapter XIII.

Chapter XII

Flame Temperature
III. Calculated Values

The temperature to which a reactive mixture may rise after combustion can be calculated from thermochemical data. Many flames, such as those of coal-gas/air or of hydrogen with either excess air or excess hydrogen, have only moderate temperatures, and dissociation of the product gases will not be appreciable. In these cases calculation of flame temperature is fairly easy. We may consider the combustion in two stages; for the first, we consider the reaction taking place at room temperature and determine the heat liberated in formation of water and carbon dioxide, etc.; in the second stage we heat the product gases to such a temperature that their enthalpy is equal to that for the heat of reaction. Data for this type of calculation can be found in Tables 12.3 and 12.4.

If the initial temperature of the gas mixture is not standard room temperature, we have to calculate the heat of reaction for the required temperature with the aid of Kirchhoff's law

$$W_{T_2} = W_{T_1} + \int_{T_1}^{T_2} (c_p' - c_p'') \, dT,$$

where W_{T_1} and W_{T_2} are the heats of reaction at temperatures T_1 and T_2 and c_p' and c_p'' are the specific heats of the unburnt and burnt mixtures at constant pressure. For the heat balance calculation, the enthalpy has, of course, to be taken from T_2 as starting-point.

Flame temperatures are usually required for flames at constant pressure, but it is equally possible to calculate values for constant volume, in which case we have to use the difference in internal energy instead of the enthalpy difference. In practice, however, in explosions in closed vessels, there are temperature differences within the vessel, because adiabatic compression may convey energy from that part of the charge which burns first to that which burns later. The exact calculation, if we allow for this type of effect, is therefore very difficult and beyond the scope of this book.

Theoretical flame temperatures are only attained if there is no heat loss by radiation, thermal conduction or diffusion to the walls. The maximum temperature is therefore only reached in the centres of rather large premixed flames in the region just above the inner cone. Diffusion flames, small premixed flames and turbulent flames of all kinds will normally fail to reach the full adiabatic temperature as the heat loss is appreciable. It is, of course, possible to assume a heat loss of say 20% and then calculate the resulting temperature, but usually accurate data for radiation and other losses are not available, and these will vary from flame to flame.

The method always assumes chemical equilibrium in the burnt gases. The state of equilibrium in flame gases has been discussed in the preceding three chapters; in some cases there is evidence for delay in reaching equilibrium. Thus temperature measurements using the sodium-line reversal method in some cases indicate departures from equilibrium for some millimetres above the reaction zone, and temperature differences between bare wires and quartz-coated wires have been commented on.

The simple method of calculation outlined so far is satisfactory for flames which are at fairly low temperature, but not for very hot flames. Thus if we try to calculate the temperature for a stoichiometric hydrogen/oxygen flame, we find that the heat of reaction is 57·80 kcal/mole; the enthalpy of 1 mole of water vapour reaches this value, however, only at a temperature of about 5000°C. Similarly, a stoichiometric oxy-acetylene flame might be expected to reach about 7000°C. Such high temperatures are not, however, actually attained because of the dissociation of the burnt gases. At high temperature large amounts of CO, H_2, O_2, OH, O and H are formed. This dissociation uses up an enormous amount of energy and limits the flame temperature. A curious example of an unusually high flame temperature is the cyanogen/oxygen flame. The temperature of this may reach 4850 K because there is so little dissociation of the very stable product gases, CO and N_2 (*see* pages 308, 339).

The occurrence of dissociation at high temperature complicates the calculations. It is necessary to know the temperature to be able to calculate the amount of dissociation to get the gas composition, but on the other hand the temperature itself depends on the composition. We have therefore to solve the problem in stages by successive approximation. These reiteration methods are rather laborious for individual calculations, but may readily be adapted to modern computer techniques.

Systematic calculation of composition for a given temperature

The first step is to calculate the composition of the flame gases for a given temperature. The case most frequently met with is for a mixture containing carbon, oxygen, hydrogen and nitrogen. The dissociation equilibria which then have to be considered are:

$$CO_2 \rightleftharpoons CO + \tfrac{1}{2}O_2$$
$$H_2O \rightleftharpoons H_2 + \tfrac{1}{2}O_2$$
$$H_2O \rightleftharpoons \tfrac{1}{2}H_2 + OH$$
$$\tfrac{1}{2}H_2 \rightleftharpoons H$$
$$\tfrac{1}{2}O_2 \rightleftharpoons O$$
$$\tfrac{1}{2}N_2 + \tfrac{1}{2}O_2 \rightleftharpoons NO$$

The assumption of equilibrium between N_2, O_2 and NO, is somewhat doubtful, because the formation of NO in flames is a slow process which is not complete just above the reaction zone. It has, however, been shown that NO is formed and does eventually reach its equilibrium concentration.

In chemical equilibrium, the following equations must be obeyed:

$$K_1 = \frac{p_{CO} \cdot \sqrt{(p_{O_2})}}{p_{CO_2}}, \qquad K_2 = \frac{p_{H_2} \cdot \sqrt{(p_{O_2})}}{p_{H_2O}},$$

$$K_3 = \frac{p_{OH} \cdot \sqrt{(p_{H_2})}}{p_{H_2O}}, \qquad K_4 = \frac{p_H}{\sqrt{(p_{H_2})}},$$

$$K_5 = \frac{p_O}{\sqrt{(p_{O_2})}}, \qquad K_6 = \frac{p_{NO}}{\sqrt{(p_{N_2} \cdot p_{O_2})}};$$

where p_{CO}, p_{O_2}, etc. are the partial pressures of CO, O_2, etc., in atmospheres, and K_1 to K_6 are the equilibrium constants, which depend on temperature. Values for these are given in tables of chemical data and in Table 12.2, page 334. The illustrative calculations we have made all use constants whose values are close to those of Tables 12.2, 12.3 and 12.4 although not necessarily quite identical.

We can also set up a number of equations by considering the number of gram atoms of each element. For example, for the initial mixture $C_2H_2 + 2\tfrac{1}{2}O_2 + 5N_2$ we know that the number of gram atoms of carbon and hydrogen are the same, or

$$n_C/n_H = 1.$$

Also in this example

$$n_O/n_H = 2\cdot 5 \text{ and } n_N/n_H = 5.$$

In general we have

(1) $n_C = p_{CO_2} + p_{CO}$,
(2) $n_H = 2p_{H_2O} + 2p_{H_2} + p_H + p_{OH}$,
(3) $n_O = 2p_{CO_2} + p_{CO} + p_{H_2O} + 2p_{O_2} + p_{OH} + p_O + p_{NO}$
(4) $n_N = 2p_{N_2} + p_{NO}$.

Furthermore the total pressure is

(5) $p = p_{CO_2} + p_{CO} + p_{H_2O} + p_{O_2} + p_{H_2} + p_{OH} + p_O + p_H + p_{N_2} + p_{NO}.$

We have ten unknown partial pressures and ten equations, these equations being the six equations with the equilibrium constants, the three ratios of the numbers of gram atoms, and the total pressure equation.

Many systems have been devised for solving these equations, and tables have been published to facilitate the calculations. We have found the method of Damköhler & Edse (1943) especially suitable for making an occasional calculation of flame temperature and composition.

The method is a trial and error calculation where two partial pressures are first assumed. If no carbon is present we assume values for p_{H_2O} and p_{O_2}. If carbon is present we assume values for p_{H_2O} and p_{CO_2}/p_{CO}.

We shall explain the method in some detail, this being especially necessary as this wartime publication is now difficult to obtain. The best way of explaining the method is to work through a definite example, and for this we shall choose a stoichiometric ethylene/oxygen flame

$$C_2H_4 + 3O_2$$

and as an initial assumption we shall assume a temperature of 2700°C. At this temperature (2973 K) the equilibrium constants have the values

$$K_1 = 0.297, K_2 = 0.0435, K_3 = 0.0465,$$
$$K_4 = 0.154, \text{ and } K_5 = 0.110.$$

The last figure in each case is in some doubt as they have been obtained b graphical interpolation.

The ratios of the numbers of gram atoms are

$$n_C/n_H = 0.5 \text{ and } n_O/n_H = 1.5.$$

As initial trial assumptions we take $p_{CO_2}/p_{CO} = 1.500$ and $p_{H_2O} = 0.350$. The scheme of Fig. 12.1 shows how all the other partial pressures may be derived. K_1 and p_{CO_2}/p_{CO} determine $\sqrt{(p_{O_2})}$, and from this we may calculate p_{O_2} and then using K_5 we get p_O. Then p_{H_2}, p_H and p_{OH} can be calculated in turn, as indicated. All partial pressures involving n_H are now evaluated and n_H can therefore be calculated using equation (2) above. In our example $n_C = n_H/2$ and with the identity $p_{CO} = n_C \bigg/ \left(1 + \dfrac{p_{CO_2}}{p_{CO}}\right)$, and the assumed value for p_{CO_2}/p_{CO}, we derive p_{CO} and then p_{CO_2}.

We now have to test out trial assumptions. We calculate n_O using equation (3) above and see whether n_O/n_H comes to 1.5 as it should, and whether the total pressure, obtained with the aid of equation (5), comes to 1. Actually we find $n_O/n_H = 1\cdot800$ and $p = 1\cdot191$. The results for this and succeeding approximations are set out in Table 12.1.

We shall see later that n_O/n_H depends mainly on our assumption for p_{CO_2}/p_{CO}, and the total pressure mainly on p_{H_2O}. We should therefore reduce

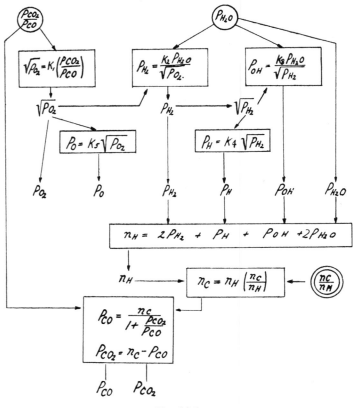

Fig. 12.1.

both assumptions, but it may be more useful, in explaining the method, to alter only one parameter at a time. We therefore reduce p_{CO_2}/p_{CO} drastically from 1·5 to half this value, 0·750. The second column of Table 12.1 shows the result; n_O/n_H is now below the true value, but the total pressure is affected as well, though to a smaller extent. For our third approximation we keep p_{CO_2}/p_{CO} the same and change p_{H_2O}; this brings the total pressure down without affecting n_O/n_H too much.

We now have values of p and n_O/n_H for three assumptions and we will show how the next assumptions should be chosen by graphical interpolation. We plot our first three assumptions as points in a plane with p_{H_2O} as x-axis and p_{CO_2}/p_{CO} as y-axis. We now imagine the pressure p obtained from the calculations to be plotted along the z-axis. The values of p will then give a surface above the plane. This surface will cut the plane $z = 1$ atm in a line. We know three points on the surface and the task is to find this line. As first approximation we can assume that the values of p along the line 1–3 (Fig. 12.2) change linearly, and we use a linear interpolation to find the point c over which $p = 1$ atm. Between the points 2 and 3, for one of which p is too high and for the other too low, we adopt the same procedure and find a point d. The straight line between c and d is then an

TABLE 12.1

Calculation of composition of $C_2H_4 + 3O_2$ flame, assuming $T = 2700°C$

Approxn. No.	1	2	3	4	5
p_{CO_2}/p_{CO}	1·500	0·750	0·750	1·030	1·040
p_{H_2O}	0·350	0·350	0·250	0·307	0·315
p_{O_2}	0·199	0·050	0·050	0·094	0·096
p_O	0·049	0·025	0·025	0·034	0·034
p_{H_2}	0·034	0·068	0·049	0·044	0·044
p_H	0·029	0·040	0·034	0·032	0·032
p_{OH}	0·088	0·062	0·053	0·068	0·070
n_H	0·885	0·939	0·684	0·803	0·820
n_C	0·443	0·470	0·342	0·401	0·410
p_{CO}	0·177	0·268	0·196	0·198	0·201
p_{CO_2}	0·265	0·201	0·147	0·204	0·209
n_O	1·593	1·207	0·915	1·202	1·229
n_O/n_H	1·800	1·285	1·337	1·497	1·499
p	1·191	1·065	0·802	0·980	1·001

approximation for the contour line for $p = 1$ atm. We then adopt a similar procedure to find the contour line for $n_O/n_H = 1·5$ by assuming that the z-axis represents the ratio n_O/n_H; the points a and b are the linear interpolations between the points 1–3 and 1–2, and the straight line between a and b is the required approximation for the contour line $n_O/n_H = 1·5$. We then take as our next, fourth assumption, the values of p_{CO_2}/p_{CO} and p_{H_2O} given by the intersection of the contour lines a–b and c–d, i.e. those values corresponding to the point 4 in Fig. 12.2. The results of this fourth pair of assumptions are shown in Table 12.1. We see that we have got quite near the required values for p and n_O/n_H. We then try to find more accurate contour lines. Between points 1 and 4 as well as between 2 and 4, the value of p must pass through $p = 1$ atm, and we can interpolate the points e and f as a better approximation than the line c–d for the $p = 1$ contour. Between points 1 and 4 the n_O/n_H ratio must take the value 1·5; the interpolated point g is very near point 4 and must be very near the true $n_O/n_H = 1·5$

contour line. We could try to get a further point on this line, but a line through g which is parallel to a–b will serve equally well. The point of intersection, 5, between the new contour lines provides the values of p_{CO_2}/p_{CO} and p_{H_2O} for our next, fifth, approximation. The result of our calculations using these fifth assumptions shows that we have accomplished our task, namely to find values of p_{CO_2}/p_{CO} and p_{H_2O} which will give the required values for n_O/n_H and p. Thus column 5 of the table gives the correct composition of the burnt gases if they are at a temperature of 2700°C.

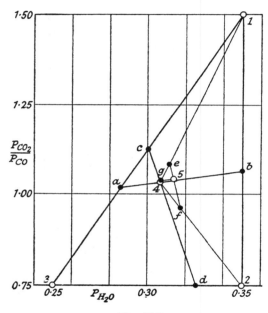

Fig. 12.2.

Calculation of temperature

In order to find the theoretical flame temperature we have to make a heat balance. First, however, we have to find the mole number of the burnt gases. Without dissociation we should, in our example, burn four moles of the ethylene/oxygen mixture to four moles of CO_2 and H_2O, but because of the dissociation the true mole number of the burnt gases will be higher. We conducted our calculations in such a way that the sum of all the partial pressures is 1 atm. Now in any mixture the partial pressures and mole numbers of the constituents will be proportional to each other, and therefore the partial pressures in column 5 of Table 12.1 will represent relative mole numbers. We can get the true mole numbers by multiplying these relative mole numbers by $n_C^*/n_C = n_H^*/n_H = n_O^*/n_O$, where n_C^*, n_H^* and n_O^* are the numbers of gram atoms for the simplified reaction

$$C_2H_4 + 3O_2 = 2CO_2 + 2H_2O,$$

and have the values $n_C^* = 2$, $n_H^* = 4$ and $n_O^* = 6$. The values of n_C, etc. are given in column 5 of the table, and we find that, using n_C, the true mole number is $2/0.401 = 4.88$, instead of 4.

We now have to calculate how much heat would be required to heat the burnt gases from room temperature to 2700°C. Values for the heat contents of all atoms, molecules and radicals can be found in Table 12.4. The heat content of the total 4·88 moles of burnt gases is thus calculated to be 131 kcal.

Next we have to see how much heat is produced in the flame. We again give these calculations in detail for the same example. The heats of formation, at 25°C from the elements in their standard states, are shown in column 2 of Table 12.3. The main source of heat liberated is in the formation of the stable molecules, H_2O, CO and CO_2, and is proportional to the product of their partial pressures and heats of formation.

$$\begin{array}{ll} H_2O, & 0.315 \times 57.8 = +18.21 \text{ kcal} \\ CO, & 0.201 \times 26.4 = + 5.31 \\ CO_2 & 0.209 \times 94.0 = +19.66 \\ & +43.18. \end{array}$$

The total heat liberated in this way for the whole 4·88 moles of burnt gas is therefore $43.18 \times 4.88 = 210.7$ kcal. We must add to this the heat liberated in the decomposition of ethylene to its elements, which is 12·6 kcal, so the total heat liberated is 223·3 kcal.

The next step is to allow for the heat used to form the free atoms and radicals. We have

$$\begin{array}{ll} O, & 0.034 \times 59.56 = 2.02 \\ H, & 0.032 \times 52.12 = 1.67 \\ OH, & 0.070 \times 9.1 = 0.64 \\ & 4.33 \text{ kcal} \end{array}$$

or 21·2 kcal for the 4·88 moles. The net heat production in the flame is therefore $223.3 - 21.2 = 202.1$ kcal. This compares with only 131 kcal used to heat the burnt gases from room temperature to 2700°C. Thus the true temperature of the flame, assuming no heat loss, will be much higher than 2700°C, and we must repeat the whole calculation for composition for a higher temperature, say 2900°C, and then repeat the heat balance test and so on until a temperature is found for which the heat used to bring the gases to that temperature is equal to the heat liberated in forming a mixture with the composition at that temperature.

The true values of composition and temperatures have been calculated for ethylene/oxygen mixtures over a whole range of mixture strength, and the results are shown in Fig. 12.3. We see how strongly the dissociation

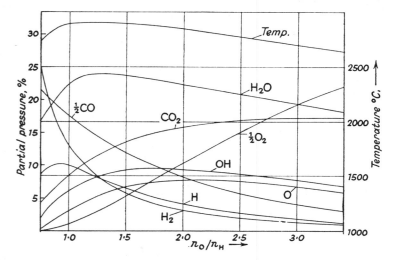

Fig. 12.3. Temperature and composition of C_2H_4/O_2 flame against mixture strength. $n_O/n_H = 1.5$ is stoichiometric.

increases in the range between 2700°C and 2900°C. In this range more heat is used for increasing the dissociation than in increasing the normal heat content of the gas.

For flames containing nitrogen, we make the calculations in the same way as indicated in the example, as far as the calculation of n_H and n_C, including the derivation of p_{CO} and p_{CO_2}, but then use the method indicated in Fig. 12.4 to obtain p_{NO} and p_{N_2}. This diagram should be self-explanatory. It is not possible directly to calculate p_{NO}, and we obtain first an approximate value $p_{NO}{}^*$, which is usually sufficiently accurate, but the scheme indicates how to obtain an accurate value of p_{NO} as well, if necessary. The value for p_{NO} has, of course, to be included in the sum for n_O, and both p_{NO} and p_{N_2} in the sum for the total pressure p.

These methods of calculating flame temperature and composition are restricted to mixtures which are not very rich. If there is insufficient oxygen to burn the carbon at least to carbon monoxide, then solid carbon will be liberated, and the calculations become more difficult. The value of n_C/n_O can no longer be derived from the initial composition of the mixture;

values for the vapour pressure of carbon, p_C, are available, but the best substitute for the equation for n_C/n_O is the Boudouard equation; the calculation is discussed in Chapter VIII, page 216, and values of the Boudouard equilibrium constant are listed.

For flames in which no carbon is present (e.g. an ammonia flame) we start the calculation with a value of p_{O_2} as initial assumption, and plot the graph with p_{H_2O} along the x-axis and p_{O_2} along the y-axis, and then determine contour lines for n_O/n_H and for p again.

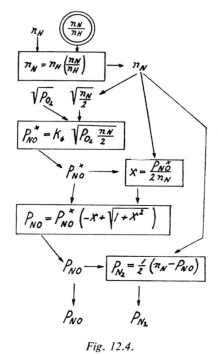

Fig. 12.4.

The method of calculation which we have outlined, that based on Damköhler & Edse's system, is convenient for the occasional calculation of a single flame temperature. When a large range of temperatures and compositions have to be calculated for a similar set of fuel mixtures, it is often a saving in time to prepare systematic graphs and tables or use computer methods. Many reports and papers on the subject were published in the immediate post-war period 1948–51, but these are mostly outdated because they used old inaccurate thermochemical data (*see* below). The little book on combustion calculations by Goodger (1977) is useful for students but is restricted to flames with air and all values of flame temperatures are

20 K to 40 K high because only dissociation to H_2, O_2 and CO is considered, not that to O, H and OH. Weinberg (1956b) has developed a modified method of successive approximation, in which explicit equations are used to derive corrections to the original assumptions, enabling more rapid convergence to the correct values. When using computers there are difficulties in using only tabulated data, and continuous functions are easier to programme. Early attempts to use simple polynomial expressions for enthalpies and equilibrium constants gave results of rather limited accuracy. Prothero (1969) has used data from JANAF (1965–67) tables to derive sixth-order polynomials and has tabulated these; to obtain adequate accuracy it has still been necessary to give these for two temperature ranges, 300 to 2000 K and 2000 to 6000 K.

Values for equilibrium constants, enthalpies, etc.

At the time of writing the first and second editions of this book we had to draw on a number of sources for data for equilibrium constants. For a long while the heat of sublimation of carbon, the dissociation energy of carbon monoxide and the dissociation energy of nitrogen were the subject of some controversy, and although we were in no doubt over the correct values of about $169 \cdot 4 \pm 0 \cdot 7$, $255 \cdot 6$ and $225 \cdot 1$ kcal/mole, respectively, use of lower values in many compilations of data was still occurring. For the third and this edition we have recalculated all constants from the tables of Gurvich et al. (1962), as the values of the latent heat of carbon and dissociation energies of CO and N_2 which they use agree almost exactly with those recommended in a critical survey by Gaydon (1968). We understand that a revised edition of the tables by Gurvich et al. is now being prepared. The JANAF (1965–7, 1971) tables are also very valuable and in substantial agreement with the values we have used.

There is a general tendency to quote the various equilibrium constants to four or five significant figures. There seems little justification for this; first, the values on which the calculations are based are not accurate enough; secondly, the errors in flame temperature due to using slightly different values of equilibrium constants are not serious. Our standards of temperature around 3000°C are not better than to at least $\pm 10°$, and there is therefore little point in attempting calculations to better than this.

We include for completeness, tables of equilibrium constants, heats of formation and enthalpies which are in substantial agreement with values used in our calculations. Values for the Boudouard equilibrium constant, $K_p{}^B = p_{CO}{}^2/p_{CO_2}$, have been given on page 216. The water–gas equilibrium $K_w = p_{CO_2} \cdot p_{H_2}/p_{CO} \cdot p_{H_2O} \equiv K_2/K_1$.

TABLE 12.2
Values of equilibrium constants

T (K)	K_1 $\dfrac{p_{CO} \cdot p_{O_2}^{\frac{1}{2}}}{p_{CO_2}}$	K_2 $\dfrac{p_{H_2} \cdot p_{O_2}^{\frac{1}{2}}}{p_{H_2O}}$	K_3 $\dfrac{p_{OH} \cdot p_{H_2}^{\frac{1}{2}}}{p_{H_2O}}$	K_4 $\dfrac{p_H}{p_{H_2}^{\frac{1}{2}}}$	K_5 $\dfrac{p_O}{p_{O_2}^{\frac{1}{2}}}$	K_6 $\dfrac{p_{NO}}{p_{N_2}^{\frac{1}{2}} \cdot p_{O_2}^{\frac{1}{2}}}$	p_C
1000	5.69×10^{-11}	8.12×10^{-11}	5.19×10^{-12}	2.26×10^{-9}	1.66×10^{-10}	8.83×10^{-5}	6×10^{-30}
1200	1.73×10^{-8}	1.25×10^{-8}	1.62×10^{-9}	1.96×10^{-7}	2.49×10^{-8}	5.23×10^{-4}	1.1×10^{-23}
1400	9.74×10^{-7}	4.45×10^{-7}	9.86×10^{-8}	4.84×10^{-6}	9.40×10^{-7}	1.91×10^{-3}	3.3×10^{-19}
1600	1.98×10^{-5}	6.54×10^{-6}	2.16×10^{-6}	5.41×10^{-5}	1.44×10^{-5}	5.05×10^{-3}	7.3×10^{-16}
1700	6.82×10^{-5}	1.98×10^{-5}	7.69×10^{-6}	1.47×10^{-4}	4.44×10^{-5}	7.58×10^{-3}	1.73×10^{-14}
1800	2.04×10^{-4}	5.30×10^{-5}	2.39×10^{-5}	3.56×10^{-4}	1.21×10^{-4}	0.01077	2.88×10^{-13}
1900	7.69×10^{-4}	1.81×10^{-4}	6.58×10^{-5}	7.90×10^{-4}	2.96×10^{-4}	0.0148	3.57×10^{-12}
2000	1.31×10^{-3}	2.84×10^{-4}	1.64×10^{-4}	1.62×10^{-3}	6.64×10^{-4}	0.0197	3.43×10^{-11}
2100	2.91×10^{-3}	5.83×10^{-4}	3.74×10^{-4}	3.10×10^{-3}	1.38×10^{-3}	0.0256	2.66×10^{-10}
2200	5.97×10^{-3}	1.12×10^{-3}	7.90×10^{-4}	5.61×10^{-3}	2.69×10^{-3}	0.0324	1.70×10^{-9}
2300	0.01155	2.05×10^{-3}	1.57×10^{-3}	9.63×10^{-3}	4.93×10^{-3}	0.0400	9.30×10^{-9}
2400	0.0211	3.53×10^{-3}	2.94×10^{-3}	0.0158	8.61×10^{-3}	0.0488	4.40×10^{-8}
2500	0.0366	5.85×10^{-3}	5.23×10^{-3}	0.0250	0.0144	0.0586	1.84×10^{-7}
2600	0.0610	9.33×10^{-3}	8.90×10^{-3}	0.0382	0.0231	0.0690	6.86×10^{-7}
2700	0.0974	0.01434	0.01467	0.0565	0.0359	0.0806	2.32×10^{-6}
2800	0.151	0.0214	0.0230	0.0814	0.0539	0.0930	7.19×10^{-6}
2900	0.226	0.0311	0.0353	0.114	0.0788	0.1061	2.06×10^{-5}
3000	0.330	0.0441	0.0525	0.157	0.1125	0.1203	5.51×10^{-5}
3100	0.469	0.0610	0.0721	0.223	0.1567	0.1350	1.38×10^{-4}
3200	0.653	0.0829	0.1078	0.279	0.2138	0.1515	3.26×10^{-4}
3400	1.191	0.1447	0.203	0.465	0.377	0.183	1.56×10^{-3}
3600	2.025	0.237	0.356	0.732	0.625	0.218	6.26×10^{-3}
3800	3.25	0.369	0.588	1.099	0.984	0.255	0.0217
4000	4.97	0.549	0.924	1.585	1.477	0.293	0.0662
4500	12.13	1.273	2.39	3.435	3.49	0.393	0.693
5000	24.6	2.49	5.09	6.39	6.95	0.496	4.5

These equilibrium constants have been calculated by us from other equilibrium constants (e.g. $p_H^2 \cdot p_O / p_{H_2O}$ and $p_N \cdot p_O / p_{NO}$) tabulated by Gurvich et al. (1962). They are based on $L(C) = 169.585$, $D(CO) = 255.70$, $D(O_2) = 117.973$ and $D(N_2) = 225.072$ kcal/mole

FLAME TEMPERATURE: III. CALCULATED VALUES

TABLE 12.3

Some heats of formation for atoms and compounds (in the gaseous state) from their elements in their standard states at 25°C (298.15 K) are given below. These use $D(O_2) = 117.97$ and $D(OH) = 101.5$ kcal/mole at 0 K. Some values are from Gurvich et al. (1962) and others from Rossini et al. (1947) or Gaydon (1968). Values in kcal/mole; 1 kcal = 4.18 J

O	59.56	NH_3	− 10.97	propane	− 24.82
O_3	34.5	N_2H_4	22.8	n-hexane	− 39.96
H	52.12	C	170.9	ethylene	+ 12.58
OH	9.1	C_2	199 ± 4	acetylene	+ 54.25
H_2O	−57.75	CO	− 26.39	benzene	+ 19.82
N	113.0	CO_2	− 94.01	formaldehyde	− 27.2
NO	21.64	CH	142.1	cyanogen	73.9
N_2O	19.57	methane	− 17.80	CN	106 ± 5
NO_2	8.05	ethane	− 20.24		

TABLE 12.4

Heat content of gases in kcal/mole

T(K)	H	H_2	O	O_2	OH	H_2O	N_2	NO	CO	CO_2
0	−1.48	−2.02	−1.61	−2.07	−2.11	−2.36	−2.07	−2.19	−2.07	−2.24
298	0.00	0.00	0.00	0.00	0.00	0.00	0.00	0.00	0.00	0.00
400	0.51	0.71	0.53	0.72	0.72	0.82	0.71	0.73	0.71	0.94
600	1.50	2.10	1.54	2.21	2.14	2.51	2.13	2.19	2.14	3.09
800	2.49	3.51	2.55	3.78	3.55	4.30	3.60	3.72	3.63	5.45
1000	3.49	4.94	3.55	5.43	5.00	6.21	5.13	5.31	5.18	7.98
1200	4.48	6.40	4.55	7.11	6.49	8.24	6.72	6.96	6.79	10.63
1400	5.47	7.91	5.55	8.83	8.02	10.38	8.35	8.64	8.45	13.36
1600	6.47	9.45	6.54	10.58	9.60	12.63	10.01	10.36	10.13	16.15
1800	7.46	11.04	7.54	12.35	11.21	14.96	11.71	12.09	11.84	18.99
2000	8.45	12.66	8.54	14.15	12.85	17.37	13.42	13.84	13.56	21.86
2200	9.45	14.31	9.53	15.97	14.53	19.84	15.14	15.60	15.30	24.75
2400	10.44	16.00	10.53	17.80	16.23	22.37	16.88	17.37	17.05	27.67
2600	11.44	17.71	11.52	19.66	17.95	24.94	18.63	19.14	18.81	30.61
2800	12.43	19.45	12.52	21.54	19.70	27.55	20.39	20.93	20.58	33.56
3000	13.42	21.21	13.52	23.45	21.46	30.20	22.16	22.72	22.36	36.53
3200	14.42	22.99	14.52	25.36	23.23	32.87	23.93	24.52	24.14	39.51
3400	15.41	24.79	15.53	27.30	25.03	35.57	25.71	26.33	25.93	42.50
3600	16.40	26.62	16.54	29.25	26.83	38.30	27.50	28.14	27.72	45.51
3800	17.40	28.46	17.55	31.22	28.65	41.04	29.29	29.95	29.51	48.52
4000	18.39	30.32	18.57	33.20	30.49	43.80	31.08	31.77	31.32	51.54
4500	20.87	35.04	21.13	38.20	35.13	50.78	35.59	36.34	35.84	59.12
5000	23.36	39.87	23.72	43.26	39.85	57.83	40.12	40.92	40.37	66.75
5500	25.84	44.79	26.34	48.35	44.63	64.45	44.67	45.54	44.93	74.43
6000	28.33	49.80	28.98	53.48	49.46	72.13	49.24	50.19	49.51	82.17

These values are based on those by Gordon (1957) with a few alterations following a personal communication from Prof. J. S. Gordon. They agree with data by Gurvich *et al.* (1962) except that the latter's values are a little lower for OH and NO at the highest temperatures (by 0.3 kcal at 6000 K), and appreciably higher for H_2O at high temperature (by 0.35 kcal at 3000° and by 2.9 kcal at 6000 K).

Discussion of the effect of varying mixture strength

It is worth having a closer look at the variation of composition and temperature with mixture strength, as illustrated for ethylene/oxygen in Fig. 12.3 and for ammonia/oxygen in Fig. 12.5. For ethylene the temperature is highest slightly on the rich side of stoichiometric. The reason for this is that

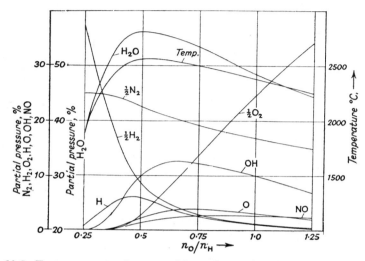

Fig. 12.5. Temperature and composition of NH_3/O_2 flame against mixture strength.

ethylene is an endothermic compound and releases heat on decomposition. An oxy-hydrogen flame, on the other hand, would reach its maximum temperature near the stoichiometric point. In the C_2H_4/O_2 flame p_H becomes very great on the rich side, up to 10% of atomic hydrogen being present; this is because p_{H_2} is big and the temperature is still high. p_O has its maximum on the lean side, where p_{O_2} is steadily increasing, and the fall in p_O for very lean mixtures is due to the fall in temperature. p_{OH} reaches its maximum somewhere between those for p_H and p_O. For the ammonia flame, the temperature reaches its maximum slightly on the lean side, because NH_3 is an exothermic compound. Here the temperature drops very rapidly on the rich side, so p_H has its maximum nearly at the stoichiometric point. NO reaches a partial pressure of about 3%; this value may not be very accurate in practice because NO is formed by a slow process, but there is no doubt that some NO is formed, as it can be detected by its absorption spectrum in these flames. O_2 ceases to be present in appreciable concentration in all flames just beyond the stoichiometric point.

We have made similar calculations on acetylene/oxygen flames at 1 atm (Fig. 12.6); in this flame the temperature maximum comes well to the rich side, due to acetylene being a very endothermic compound.

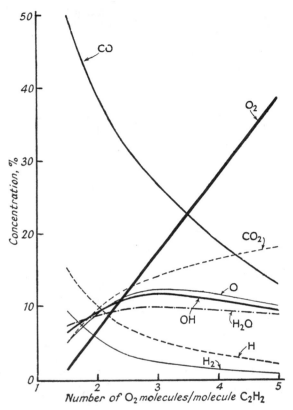

Fig. 12.6. Calculated composition of a C_2H_2/O_2 flame at 1 atm. For richer mixtures see Table 8.2, page 218.

The effects of pressure

The temperature of a flame depends on pressure because if the mole number changes, as is usual, the amount of dissociation will vary with pressure. Any increase in p_O, p_H or p_{OH} affects the heat balance. Thus we find that a stoichiometric oxy-acetylene flame at 1/100 atm pressure is 610° cooler than the same flame at 1 atm. Flames with air are much less sensitive to change of pressure, as the degree of dissociation in these flames is relatively small. However, for the rather hot acetylene/air and cyanogen/air flames the pressure still has an appreciable effect on the temperature.

Table 12.5, from Brown, Everest, Lewis & Williams (1968), shows the effect of pressure on the adiabatic temperature for a selection of flames.

TABLE 12.5

Effect of pressure on adiabatic flame temperature, in K. Values for flames with air are for stoichiometric mixtures

	0·1	1	10	100 atm
C_2H_2/air	2420	2545	2657	2748
C_2N_2/air	2475	2601	2717	2814
H_2/air	2301	2388	2450	2489
CH_4/air	2170	2232	2276	2304
$C_2H_2 + 2\frac{1}{2}O_2$	3006	3341	3737	4199
$C_2H_2 + 1\frac{1}{2}O_2$	3076	3431	3849	4324
$CH_4 + 2O_2$	2782	3053	3356	3688
$N_2H_4 + O_2$	2763	3026	3312	3606
$H_2 + \frac{1}{2}O_2$	2795	3077	3395	3740
$H_2 + N_2O$	2722	2963	3216	3463
$C_2H_2 + 5N_2O$	2884	3152	3440	3738
$N_2H_4 + 2F_2$	3542	3934	4393	4913
$N_2H_4 + ClF_3$	3137	3444	3786	4142
$C_4N_2 + 1\frac{1}{3}O_3$	5190	5516	5850	6170

Temperatures and compositions of some typical flames

In addition to the data for flames of ethylene, ammonia and acetylene with oxygen, shown graphically in figures 12.3, 12.5 and 12.6, we have calculated the adiabatic flame temperatures and compositions for some other flames, and list these, together with a few others from the literature, in Table 12.6. These values are sometimes based on old, slightly inaccurate, thermochemical data, but serve to indicate the composition and temperature fairly well.

Other useful flame temperatures (maximum values for flames with air) are: ammonia 2600, ethane 2244, propane 2250, *n*-butane 2256, benzene 2365 and ethylene oxide 2411 K.

TABLE 12.6

Values of calculated temperature and composition of burnt gases for some typical flames at atmospheric pressure. Values are for stoichiometric mixtures, except for the cyanogen/oxygen flame, which is equimolecular

Fuel oxidiser	H_2 +air	H_2 +$\frac{1}{2}O_2$	H_2 +NO	CH_4* +air	CH_4† +$2O_2$	C_2H_2 +air	C_8H_{18} +$12\frac{1}{2}O_2$	CO +$\frac{1}{2}O_2$	CO +N_2O	H_2‡ +Cl_2	C_2N_2§ +O_2
Temp., K	2380	3083	3113	2222	3010	2523	3082	2973	2898	2503	4850
H_2O	0·32	0·57	0·35	0·18	0·37	0·07	0·26	—	—	—	—
CO_2	—	—	—	0·085	0·12	0·12	0·14	0·46	0·25	—	0·00
CO	—	—	—	0·009	0·15	0·04	0·22	0·35	0·20	—	0·66
O_2	0·004	0·05	0·04	0·004	0·07	0·02	0·09	0·15	0·08	—	0·00
H_2	0·017	0·16	0·12	0·004	0·07	0·00	0·05	—	—	0·04	—
OH	0·010	0·10	0·08	0·003	0·14	0·01	0·14	—	—	—	—
H	0·002	0·08	0·08	0·0004	0·05	0·00	0·05	—	—	0·01	—
O	0·0005	0·04	0·04	0·0002	0·03	0·00	0·05	0·04	0·02	—	0·008
NO	0·00	—	0·02	0·002	—	0·01	—	—	0·02	—	3×10^{-4}
N_2	0·65	—	0·27	0·709	—	0·73	—	—	0·43	—	0·32

* From Friedman & Levy (1963).
† From Sachse & Bartholomé (1949); the values of OH appear a little high.
‡ This flame also contains 0·87 HCl, 0·08 Cl and 0·0006 Cl_2.
§ This flame also contains about 0·005 C, 0·002 CN, 0·012 N and 10^{-5} C_2.

Chapter XIII

Ionisation in Flames

At the time of the first edition of this book, ionisation in flames was a scientific curiosity rather than a subject of any practical importance. Interest in the subject has since greatly increased. The possibility that electrically conducting flames might be used for power generation (by MHD) and the fact that rocket exhaust plumes were highly ionised and interfered with radio guidance signals led to a lot of research in the 1960–70 decade and lead to a complete revision of this chapter for the third edition. In recent years the emphasis on flame ionisation research has been more on flame control – stability, heat transfer and soot formation. These investigations into practical problems have led to a lot of fundamental work on laboratory flames to explain the processes of ion formation and decay. It has been found that kinetic considerations are of major importance and that equilibrium ionisation is seldom obtained. These subjects are discussed here, but it has not been found necessary to make major alterations in the chapter; the book by Lawton & Weinberg (1969) treats the subject in much greater detail, although with rather different emphasis.

Flame ionisation is really a subdivision of plasma physics, yet it has developed independently because generally flames are at a much lower temperature than plasmas; also the composition of flame gases leads to a greater variety of possible reactions than is generally the case in plasmas. This chapter will be restricted to ionisation phenomena in reacting systems which can be described as flames. Diagnostic methods will be discussed first, followed briefly by a discussion of ionisation equilibrium. Recent work on the formation and removal mechanisms of ions in flames is then discussed, as well as some unexplained ionisation phenomena. Finally a brief survey of ionisation in rocket exhausts is presented.

Measurement of ionisation

Measurement by microwave absorption and phase shift. The most reliable method of determining the electron concentration in flames or hot gases is by measurement of microwave absorption and phase shift, as the microwave beam does not disturb the flame plasma. The normal arrangement uses two focusing dielectric lenses, so that the flame is traversed by a parallel beam or, alternatively, the beam may be focused into the flame and, in such cases, the focus of the beam may be as narrow as two to three times the wavelength for the chosen frequency.

Physically the electrons in the flame are accelerated by the electric field of the microwave beam and lose energy through collisions with neighbouring heavier atoms and molecules. The heavier ions, i.e. the positive ions mainly, have a much smaller amplitude and so collide less with other atoms due to their excursions in the electric field, and can generally be neglected in the calculation of microwave absorption. Thus attenuation for a given microwave frequency depends on the number density of electrons which, in turn, determines the plasma frequency ω_p (the natural oscillation frequency) in the flame, and the electron collision frequency ν.

The plasma frequency is,

$$\omega_p = \left(\frac{n_e e^2}{m\varepsilon_0}\right)^{\frac{1}{2}} \qquad (13.1)$$

where $\varepsilon_0 = 10^{-9}/36\pi$ the permittivity of space, n_e is the number density of electrons/cm^3, e is the electronic charge, and m is the mass of the electron. Inserting all atomic constants one finds (Huddlestone & Leonard, 1965),

$$f_p = \frac{\omega_p}{2\pi} = 8980\,(n_e)^{\frac{1}{2}} \qquad (13.2)$$

with f_p in Hz and n_e in electrons/cm^3.

The collision frequency depends on the mean thermal speed of the electrons (i.e. their temperature, which is usually assumed to be the equilibrium flame temperature) and the mole fraction of the flame species. Explicit formulae for the collision frequency are misleading as the collision cross-sections of the molecules depend on velocity and this dependence varies for different atoms and molecules. An approximate formula is

$$\nu \approx P\,\sigma^2 \left(\frac{\pi}{2mkT}\right)^{\frac{1}{2}} \qquad (13.3)$$

with P being the pressure, T the temperature, σ the mean diameter of the molecule, which is velocity dependent, and k the Boltzmann constant ($1 \cdot 38 \times 10^{-23}$ joules/K). For $P = 10^6$ dyne cm^{-2} and $T = 2000$ K we have $\nu \approx 7 \times 10^{10}$. (For an experimental determination of ν see below.)

Maxwell's equations permit the calculation of the index of refraction and the extinction coefficient for a microwave beam traversing a plasma. These values depend on the plasma frequency and the collision frequency, as well as on the microwave frequency. In a microwave experiment one measures the absorption in db/metre and a phase shift in radians/metre and these values are related to the index of refraction and the extinction coefficient.

Balwanz (1959) has prepared very useful graphs that relate the electron density, microwave frequency, collision frequency and absorption or phase shift. In these graphs, of which figures 13.1 and 13.2 are examples, the dimensions are, unfortunately, in part not the same as used in texts on plasma physics. They are: electron density in electrons m^{-3}, collision frequency in sec^{-1}, microwave frequency in Hz (not angular frequency as stated in the original publication; private communication from Mr. Balwanz), absorption in db/metre, where a db is defined as ten times the logarithm of the ratio of initial to transmitted power, and phase shift in radians/metre.

As a first illustrative example, assume a collision frequency of 7×10^{10} sec^{-1} and a wavelength of 1 cm (30 gigahertz). Then from the graph one reads for three different electron densities

electrons cm^{-3}	electrons m^{-3}	db m^{-1}
10^7	10^{13}	$1 \cdot 05 \times 10^{-4}$
10^{10}	10^{16}	$0 \cdot 9$
10^{13}	10^{19}	$1 \cdot 5 \times 10^3$

Thus for 10^7 electrons cm^{-3} the absorption is negligible, whereas for 10^{13} cm^{-3} the gas is opaque.

As a second example, assume the same collision frequency of 7×10^{10} but a wavelength of 10 cm (3 GHz); we then have,

electrons cm^{-3}	electrons m^{-3}	db m^{-1}
10^7	10^{13}	6×10^{-3}
10^{10}	10^{16}	6
10^{13}	10^{19}	$2 \cdot 1 \times 10^3$

Thus we see that for a given microwave frequency and length of absorbing gas there is a range of approximately two orders of magnitude of electron density that can be measured by this method. A change in frequency enables a different range of electron density to be covered, so that under favourable conditions a range of between 10^9 and 10^{13} electrons cm^{-3} may be measured. If we are dealing with small flames, then we are limited to high frequencies because the focused microwave beam must be small compared to the dimensions of the flame.

Fig. 13.2 (from Balwanz, 1969) is a normalised plot that allows the phase shift, in radians/metre, to be deduced. A detailed discussion of microwave absorption and reflection can be found in Heald & Wharton (1965).

As electron density can be measured at two or more microwave frequencies, one can consider the collision frequency to be an adjustable para-

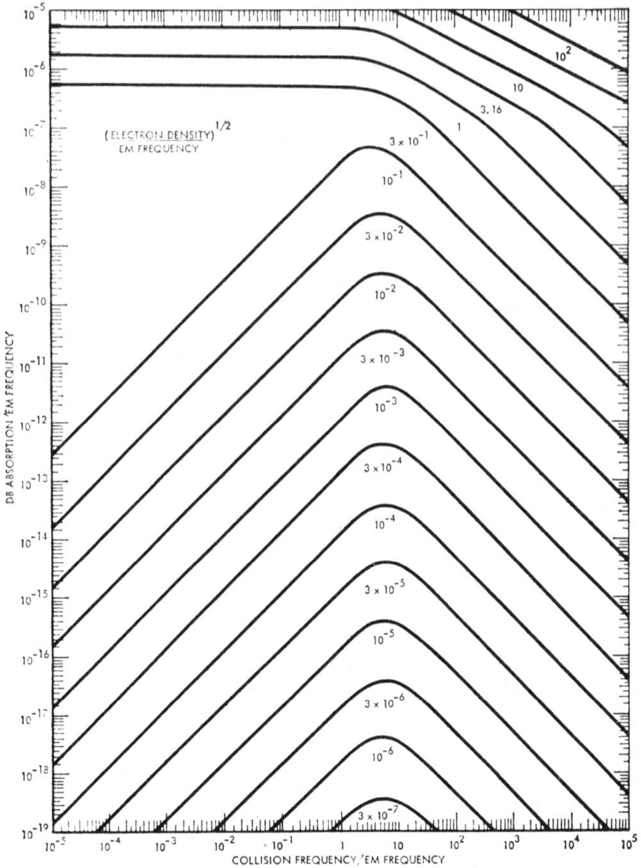

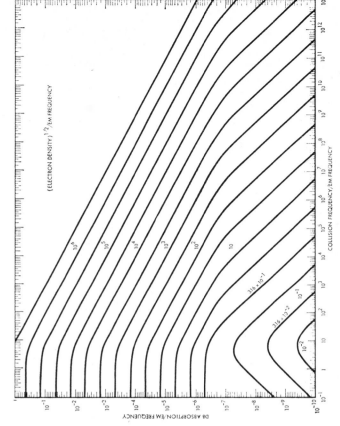

Fig. 13.1a. Cross plot of absorption in decibels, electron density, collision frequency and EM frequency for a microwave beam traversing a plasma. [Redrawn from Balwanz (1969)]

Fig. 13.1b. Continuation of Fig. 13.1a.

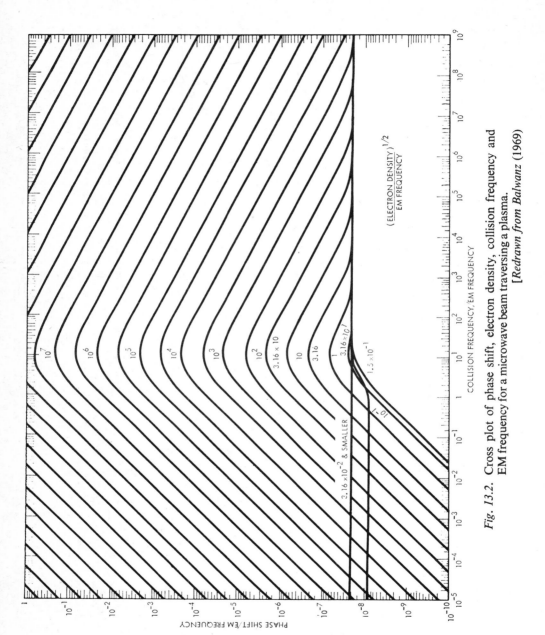

Fig. 13.2. Cross plot of phase shift, electron density, collision frequency and EM frequency for a microwave beam traversing a plasma.
[Redrawn from Balwanz (1969)]

meter and determine its value so that the electron density is the same at all frequencies.

Langmuir probes. Microwave-absorption techniques measure an average electron density across a flame or stream of hot gas. Often one wants to make point-by-point measurements or, in the case of a premixed flame, one wants to study ionisation in the thin reaction zone where non-equilibrium ionisation often occurs. In these cases microwave techniques cannot be applied, and Langmuir probes have been extensively used. Such a probe is experimentally simple, consisting of a bare wire inserted into the flame; a negative or positive voltage is applied to the wire, relative to a second much larger electrode in the flame. Often the walls of a vessel or the tip of a burner may serve as the second electrode, provided good electrical contact with the flame gases is present.

A typical curve of voltage versus probe current can be seen in Fig. 13.3. This curve is called the probe characteristics curve.

At negative potentials all electrons are repelled and positive ions are attracted to the wire. The ion current is then a measure of the positive ion concentration in the plasma or flame. If the probe dimensions are small compared to the mean free paths of electrons and ions, then the following simple expression is valid:

$$J_+ = \frac{I_+}{A_p} = \frac{n_+ e v}{4} = \frac{n_+ e}{4}\sqrt{\frac{2k T}{m_+}} \quad (13.4)$$

where J_+ is the random ion current density (amp m^{-2})

I_+ is the random ion current (amp), as indicated in Fig. 13.3 (taken at the inflection of the curves)

A_p is the area of the probe (m^2)

m_+ is the mass of the positive ion (kg)

n_+ is the ion density (m^{-3})

e is the electronic charge, 1.6×10^{-19} coulomb

v is the mean velocity of the ions (m/sec)

and k is the Boltzmann constant, 1.38×10^{-23} joules/K.

At low pressures, where the mean free path is larger than the probe dimensions, the electron temperature, ion temperature and kinetic gas temperature will generally not be the same and equation (13.4) should be modified (Wharton, 1963) to

$$J_+ \approx 0.4\, n_+ e \sqrt{\frac{2k T_e}{m_+}} \text{ (amp m}^{-2}\text{)} \quad (13.5)$$

where T_e is the electron temperature.

If the probe potential is sufficiently positive, then all electrons within the plasma sheath will be collected at the surface of the probe. The electron saturation current J_e is given by:

$$J_e = \frac{I_e}{A_p} \approx \frac{n_e e}{4} \sqrt{\frac{2k T_e}{m_e}}. \tag{13.6}$$

The ratio of electron to positive ion current, i.e. the ratio of equations (13.6) to (13.5), is:

$$\frac{J_e}{J_+} \approx \sqrt{\frac{m_+}{2m_e}}. \tag{13.7}$$

The region of the curve corresponding to the retarding field (see Fig. 13.3) allows the determination of the electron temperature, T_e. At the

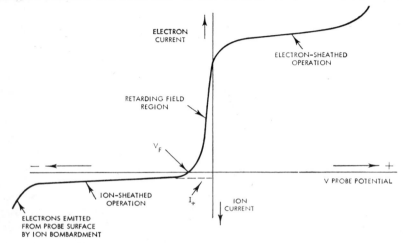

Fig. 13.3. Langmuir probe characteristics. [Redrawn from De Leeuw (1963)]

negative potential V_F some electrons reach the probe, cancelling the positive ion current. For a thermal plasma, i.e. $T_e = T_{ion}$, $V_F \approx \frac{1}{2}kT_e$ (with T_e in eV). In the steeply rising part of the curve the relation between electron current and probe potential is given by:

$$\frac{d \ln J_e}{d V_p} = -\frac{e}{kT_e} \tag{13.8}$$

with V_p being the probe potential.

J_e is the difference between the measured current density and the ion current density. The latter can be determined by an extrapolation of the ion current, measured at large negative potential. This extrapolation leads to only small errors, as J_+ is usually small compared to J_e. Thus equation (13.8) enables the electron temperature T_e to be derived.

Often it is more reliable to use double probes instead of single Langmuir probes. The characteristic curve (Fig. 13.4) is then symmetrical as, depending on the probe potential, one or the other probe reaches the ion saturation current. The saturation current is again given by equation (13.5) which enables the determination of n_+.

It must be pointed out that all the relations so far derived apply only for probes whose dimensions (i.e. diameters) are small compared with the mean free path of the positive ions or electrons. Thus only low-pressure flames can satisfactorily be studied with Langmuir probes.

If these conditions are not fulfilled, in observations at higher pressures, the occurrence of non-equilibrium or of flow velocities approaching the kinetic velocities, then probe theories become very complex indeed

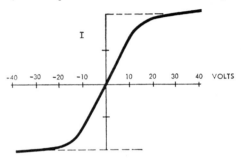

Fig. 13.4. Double-probe characteristics.

and, in part, obscure. The reader is referred to Loeb (1960) and Lawton & Weinberg (1969). In flames, probes have often been used to measure the decay of ionisation above the reaction zone. A correction for work at high pressure has been proposed by King & Calcote (1955). This is

$$n_+ \text{ (corrected)} = (r/\lambda) \, n_+ \text{ (experimental)}$$

where r is the radius of the probe and λ the mean free path.

Applying these corrections, reasonable ion densities have been measured, in general agreement with those derived by microwave methods. Caution is necessary in applying such simple corrections to probe measurements and the derived ion densities should be considered as more qualitative than quantitative. The use of electrical probes has also been investigated by Travers & Williams (1965) who applied the theory to higher pressures and spherical probes.

Other electrical methods. A great variety of methods has been used to measure electron concentrations in flames or in wakes and boundary layers of hypervelocity projectiles. This subject goes far beyond the purpose of this book and the reader is referred to texts on plasma physics e.g. Huddlestone & Leonard (1965), Anderson et al. (1963). It must suffice to mention the general principles of such methods. Often the plasma can be contained in a microwave resonant cavity where the plasma detunes the resonant frequency and the change in 'Q' and resonant frequency are a measure of the electron density. A variation of this method has been described by Williams (1959) who wrapped a self-resonant coil around a flame which was loosely coupled to a generator and detector coil. As in the resonant cavity, the detuning of the coil is a measure for the plasma properties. In general such methods are more sensitive than single-path microwave absorption methods, allowing the determination of lower electron concentrations. As the resonance coils could be made rather small, Williams was able to measure the decay of the electron concentration downstream of the reaction zone of a low-pressure flame. Fuhs (1963) describes a conductivity technique which allows the electron concentration to be determined in the boundary layer of a hypervelocity vehicle. A sensing coil is situated between the primary coils. The magnetic field of the primary coils penetrates into the plasma flow of the boundary layer. A current is then induced in the sensing coil and this depends on the velocity of the plasma and its conductivity, thus permitting an estimate of electron concentration. For details see the original paper.

Measurement of ionisation by Stark broadening. High temperature flames seeded with metal vapours of low ionisation potential may attain electron densities up to 10^{15} or even 10^{16} electrons cm^{-3}, and only optical methods can be used to determine the degree of ionisation. One of these methods is the measurement of Stark broadening and shift of spectrum lines.

The theory of line broadening and line shift caused by the interaction of atoms with ions and electrons is complex and the reader is referred to Griem (1964) for details. A practical formula is given by Wiese (1965). The half-width of a neutral atomic line is:

$$\Delta \lambda_{\frac{1}{2}} \approx 2[1 + 1\cdot75 \times 10^{-4} N_e^{1/4} \alpha (1 - 0\cdot068 N_e^{1/6} T^{-\frac{1}{2}})] 10^{-16} W N_e$$

and the shift of the line centre is:

$$\Delta \lambda_s \approx [(d/W) + 2\cdot0 \times 10^{-4} N_e^{1/4} \alpha (1 - 0\cdot068 N_e^{1/6} T^{-\frac{1}{2}})] 10^{-16} W N_e.$$

N_e is the electron density per cm^3, the width and shift are given in angstroms, α, d/W and W are atomic constants which can be taken from Griem (1964) and the values in Table 13.1 allow the calculation of the effects for some sodium and caesium lines. Griem's book gives values for all flame additives of interest.

One can see that temperature has little influence on line width and shift. If it is possible to observe lines involving higher terms, then the effects become large and easily measurable. Even the sodium line 5683 Å, which occurs quite strongly in high-temperature flames has a half-width of 4·2 Å

TABLE 13.1

Sodium

Line (Å)	2500 K			5000 K		
	W	d/W	α	W	d/W	α
5890/96	9.13×10^{-3}	1.772	0.039	1.11×10^{-2}	1.738	0.033
8183/95	1.83×10^{-1}	1.364	0.069	2.27×10^{-1}	1.114	0.059
5683/88	1.76×1	0.640	0.230	1.68×1	0.529	0.238
4979/83	4.75×1	0.604	0.371	4.47×1	0.483	0.388
4665/69	1.03×10	0.588	0.526	9.64×1	0.463	0.553

Caesium

Line (Å)	2500 K			5000 K		
	W	d/W	α	W	d/W	α
8943 ($n = 6$)	6.86×10^{-2}	1.672	0.053	8.73×10^{-2}	1.450	0.044
4593 ($n = 7$)	1.61×10^{-1}	1.450	0.102	2.08×10^{-1}	1.129	0.084
3899 ($n = 8$)	4.80×10^{-1}	1.292	0.160	6.04×10^{-1}	0.985	0.135
3617 ($n = 9$)	1.20×1	1.181	0.226	1.47×1	0.896	0.195
3480 ($n = 10$)	2.62×1	1.099	0.301	3.13×1	0.831	0.264
3238 ($n = 18$)	1.09×10^2	0.748	1.178	1.17×10^2	0.550	1.121

for an electron density of 10^{16} cm^{-3}. Thus for this line, a density of 10^{15} cm^{-3} should be measurable, and for caesium even lower densities.

This method is restricted to flames with high ionisation; however, when applicable, the measurement is free from absolute errors and only limited by the accuracy of the constants listed in Table 13.1.

Ionisation in local thermodynamic equilibrium (LTE)

The Saha equation. The ionisation of an atom or molecule may be treated like the dissociation of a molecule. Saha & Saha (1934) derived the original formula by considering the equilibrium: atom $\rightleftharpoons$ ion + electron. A more satisfactory derivation is given by Unsöld (1955). Practical formulae which we shall use here can be found in Allen (1963). We only consider single ionisation, as higher ions are not usually present in flames.

$$\log \frac{N_i}{N_n} N_e = - U \frac{5040}{T} - \frac{3}{2} \log \frac{5040}{T} + 20.937 + \log 2 \frac{U_i}{U_n} \quad (13.9)$$

where
- N_i is the positive ion density (cm^{-3})
- N_n is the neutral atom density (cm^{-3})
- N_e is the electron density (cm^{-3})
- U is the ionisation potential (eV)
- T is the absolute temperature (K)
- U_i and U_n are the partition functions of the ion and neutral atom.

The partition functions can be replaced by the statistical weight at normal flame temperatures; in high-temperature plasma this is not permissible and the partition functions must be used (Allen, 1963). The factor 2 is the statistical weight of the electron.

If a flame is seeded with a metal vapour, so that all ions are, for example, Cs^+, then the number density of Cs^+ and electrons is the same, and equation (13.9) enables $N_e = N_i$ to be calculated. For appreciable ionisation N_n is not, of course, the original number density of seed atoms, but has to be corrected for depletion due to the ionisation.

The presence of two or more ionisable atoms in hot gases leads to two or more independent equations. Thus for atoms A and B we can calculate.

$$\frac{N_{iA} \times N_e}{N_{nA}} \text{ and } \frac{N_{iB} \times N_e}{N_{nB}}$$

with $N_e = N_{iA} + N_{iB}$. Thus the ion density of each element and the electron density can be calculated.

Statistical weights and ionisation potentials are given by Allen (1963) and for the convenience of the reader some are abstracted in Table 13.2

TABLE 13.2

Element	Statistical weight neutral atom	Statistical weight, ion	Ionisation potential, eV
Li	2	1	5·39
Na	2	1	5·138
K	2	1	4·339
Rb	2	1	4·176
Cs	2	1	3·893
Be	1	2	9·320
Mg	1	2	7·644
Ca	1	2	6·111
Sr	1	2	5·692
Ba	1	2	5·21
Al	6	1	5·98
Sc	10	15	6·56
Ti	21	28	6·83
V	28	25	6·74
Cr	7	6	6·764
Ga	6	1	6·00
In	6	1	5·785
La	10	21	5·61

Normal combustion products are not appreciably ionised in equilibrium because H_2O, CO_2, CO, O_2, H_2 etc. all have ionisation potentials above 12 eV. The only exception is NO at 9·25 eV. In the complete absence of impurities nitric oxide will be slightly ionised at normal flame temperatures. In equilibrium, the amount of NO in flames does not depend very much on temperature and is generally around one to two mole per cent. An NO concentration of one per cent leads to 1.5×10^7 ions cm^{-3} at 2000 K and to about 9×10^{10} at 3000 K. In general, in equilibrium, ionisation is due to metallic impurities, even if these are only present in parts per million. Table 13.3 presents number densities of electrons for K, Na and Ca in flames.

TABLE 13.3
Number of free electrons per cm^3

Metal	K			Na			Ca	
P, atm	10^{-2}	10^{-4}	10^{-6}	10^{-2}	10^{-4}	10^{-6}	10^{-4}	10^{-6}
T, K								
1000	22×10^5	22×10^4	22×10^3	27×10^4	27×10^3	27×10^2	1800	180
1500	13×10^{10}	13×10^9	13×10^8	59×10^8	59×10^7	59×10^6	27×10^6	27×10^5
2000	9×10^{12}	9×10^{11}	9×10^{10}	10×10^{11}	10×10^{10}	10×10^9	11×10^9	11×10^8
2500	14×10^{13}	14×10^{12}	12×10^{11}	20×10^{12}	20×10^{11}	20×10^{19}	41×10^{10}	41×10^9
3000	68×10^{13}	66×10^{12}	23×10^{11}	15×10^{13}	15×10^{12}	13×10^{11}	46×10^{11}	46×10^{10}
3500	23×10^{14}	16×10^{13}	21×10^{11}	65×10^{13}	63×10^{12}	20×10^{11}	25×10^{12}	16×10^{11}

Equation (13.9) can only be expected to apply at high pressures. Although the forward reaction $A + M \rightarrow A^+ + e + M$ is bi-molecular, the reverse reaction is ter-molecular and therefore slow. Kinetic considerations will be discussed in a later section.

Negative ions. Some atoms and molecules form stable negative ions and O^- and O_2^- are examples. Usually the attachment energies are small (1.47 and 0.44 eV in the above cases), so that at temperatures above 2000 K these ions will dissociate (e.g. $O^- \rightarrow O + e$) and the depletion of electrons, formed by the ionisation of metallic additives, will be minor, except in the presence of halogens (see below). To calculate the concentration of a negative ion, such as O^-, one can again use Saha's equation to calculate

$$\frac{N_O}{N_{O^-}} N_e$$

using $U = 1.47$ eV, the statistical weight term being log 3.

Some negative ions have larger attachment energies and Table 13.4 is based on data given by Gurvich *et al.* (1974). There is still some uncertainty about some of the values.

In flames containing fluorine or chlorine, negative ions are very important, as the attachment energy for these elements is so high. The most important role of the negative ions is, however, not their influence on the equilibrium electron density, but on the decay of ionisation in a slowly cooling flame. As the gases cool by conduction and radiation, the temperature falls, electrons begin to get attached and mutual neutralisation by the process $A^+ + B^- \rightarrow A + B$ occurs; the effective cross-section is large and recombination by this route is fast.

It should be mentioned here that many reactions in flames involve

TABLE 13.4

Electron affinities, in eV

Species	Attachment energy	Species	Attachment energy
H	0·754	O_2	0·44
C	1·27	OH	1·83
O	1·47	CH	2·6
F	3·45	C_2	3·4
Cl	3·61	C_2H	3·0
Br	3·37	C_3	1·8
I	3·0	CHO	2·0

charge exchange or negative ion formation. In these cases equilibrium considerations are not applicable. Examples of such reactions are:

$$e^- + H_2O \rightarrow OH^- + H$$
$$e^- + HCl \rightarrow Cl^- + H$$
$$e^- + C_2H \rightarrow C_2^- + H$$
$$CH^* + O_2 \rightarrow CHO^+ + O^-.$$

Often the heats of formation and rate constants are only approximately known. Calcote & Jensen (1966) have discussed a number of these reactions.

The role of chemical reactions in the ionisation of metal additives

So far it has been tacitly assumed that any metallic vapour introduced into a flame would be available for ionisation. This is often not so. It is known from spectroscopic studies that the introduction of alkaline-earth metals into a flame leads to the emission of bands due to the metal oxide and hydroxide; thus the concentration of free metal is depleted. The only safe way to ascertain the concentration of neutral metal atoms seems to be by quantitative absorption spectroscopy. We shall discuss first the alkali metals and then the alkaline-earth metals.

The alkali metals. Padley & Sugden (1962) have investigated the formation of metal compounds in flames in great detail. The experiments have been made in hydrogen–oxygen–nitrogen flames at 1 atm and around 2400 K. The flame is seeded with an ACl solution (A being an alkali metal). Electron concentrations are measured by a microwave cavity method. In this, one has, in effect, a multiple-path microwave beam through the flame. However, rather than measure absorption, one determines the detuning

IONISATION IN FLAMES 353

of the cavity, which is related to the electron density. The ionisation of the metal vapour is often slow, so measurements have to be made a considerable distance above the reaction zone to reach equilibrium. In these flames $t_{\frac{1}{2}}$ (the time to reach 50% of the ionisation level) is of the order 0·1 to 1·0 msec at 1 atm. Alternatively, 1% of acetylene may be introduced so that the chemi-ions in the reaction zone can charge exchange with the metal atoms and speed up their ionisation.

Although alkali hydroxides may be formed they have no known spectra, so indirect methods have to be used for their measurement. One possible way is indicated in Fig. 13.5. The intensity of the atomic resonance line (in an optically thin layer) is plotted against $1/T$ (Jensen & Padley, 1966).

The slope of the straight line corresponds to the excitation energy of the first resonance doublet. At the higher temperatures the intensity of the atomic line is reduced by depletion due to hydroxide formation, and in addition by ion formation. The ion formation can be suppressed by adding

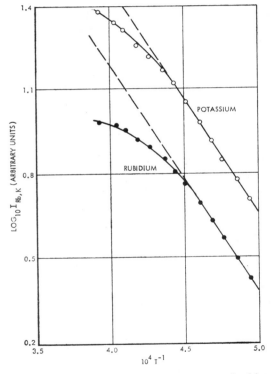

Fig. 13.5. Plots of the intensity of the first resonance doublets of Rb and K against temperature.   *Redrawn from Jensen & Padley* (1966)

an excess of a further metal with low ionisation potential. The rise in N_e due to this second metal will shift the equilibrium of the first, as the product $N_e N_A^+ / N_A$ has to stay constant. In effect, therefore the deviation in Fig. 13.5 can be taken as a measure of AOH formation. Measurements of this kind led to the conclusion that for lithium, in the flame studied, the hydroxide (LiOH) had eight times the number density of the metal atom. For sodium, hydroxide formation was negligible.

The equilibrium constant governing the hydroxide formation is

$$K = [AOH][H]/[A][H_2O] = \Phi [H][H_2O].$$

[H] and [H$_2$O] are the equilibrium values of the concentrations in the flame and $\Phi = [AOH]/[A]$. Values for K derived by the above-mentioned flame method can be seen in Fig. 13.6.

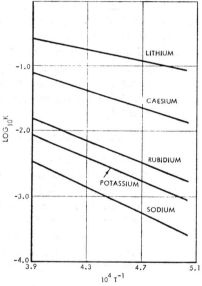

Fig. 13.6. Plot of the equilibrium constant governing hydroxide formation against temperature. [*Redrawn from Jensen & Padley* (1966)

Dissociation energies derived in this way for the process $AOH_g \rightarrow A_g + OH_g$ are:

LiOH	101 ± 3 kcal/mole
NaOH	77 ± 4 kcal/mole
KOH	81 ± 2 kcal/mole
RbOH	83 ± 2 kcal/mole
CsOH	91 ± 3 kcal/mole

The rate of ionisation of alkali metals is the same in hydrogen as in carbon monoxide flames (Hollander et al., 1963). The activation energies are the ionisation energies, within experimental error. Thus ionisation is by the reaction:

$$A + M \rightarrow A^+ + e^- + M$$

and reactions such as

$$A + OH \rightarrow A^+ + OH^-$$
$$A + e^- \rightarrow A^+ + 2e^-$$

and $A + H_2O \rightarrow A \cdot H_2O^+ + e^-$ can be rejected.

Alkaline-earth metals. Reactions that lead to ionisation are more complicated for alkaline-earth metals than for alkali metals. It was found for example, that barium and strontium gave more ionisation in a hydrogen flame than sodium, despite their higher ionisation potentials (Jensen, 1968).

Alkaline-earth metals form very stable oxides with high boiling-points, so that condensation to solid oxide particles, which may have low work functions, has to be considered. In the presence of hydrogen, very stable hydroxides are formed and both the oxide and hydroxide usually emit strong bands in the spectrum. A further complication is the fact that the formation of the gaseous oxide* by

$$M_g + O_2 \rightarrow MO_g + O \text{ (M being an alkaline-earth metal)}$$

is endothermic so that equilibrium may be achieved only very slowly, or not at all, in low-temperature flames in the absence of hydrogen.

Jensen (1968) and Schofield & Sugden (1965) investigated ionisation from alkaline-earth metals in detail. They found the MOH^+ (and also the $M(H_2O)^+$ ion) as well as the M^+ ion in flames. The primary process of ionisation is $M + OH \rightleftharpoons MOH^+ + e^-$ or $MO + H \rightleftharpoons MOH^+ + e^-$. One cannot distinguish between these two reactions as $M + OH \rightleftharpoons MO + H$ is balanced. This is accompanied by the neutral reactions:

$$M + H_2O \rightleftharpoons MOH + H$$
$$MOH + H_2O \rightleftharpoons M(OH)_2 + H$$
$$OH + H_2 \rightleftharpoons H_2O + H$$
$$MOH + H \rightleftharpoons MO + H_2.$$

The metal atom may be ionised by secondary ionisation, i.e. $MOH^+ + H \rightarrow M^+ + H_2O$.

The most abundant alkaline-earth metal compound is not really evident, but Jensen (1968) believes this to be $M(OH)_2$. In any case, two facts emerge

* Barium is an exception.

for alkaline-earth metals: the metal is mostly present as compounds rather than as free metal atoms (neutral or ionised), and the rate of ionisation is fast because it is due to a molecular rearrangement rather than to direct collision with an inert molecule such as H_2O, N_2 or O_2, as is the case for alkali metals.

By plotting the observed ionisation against temperature and considering rather involved kinetic arguments (for details see Jensen (1968)), one can get approximate heats of formation of the hydroxide ions. They are:

$Ca + OH \rightarrow CaOH^+ + e^-$ $\quad \Delta H_0^0 = +35 \pm 10$ kcal/mole
$\quad K = 1.8 \times 10^{-3} \exp(-35\,000/RT)$

$Sr + OH \rightarrow SrOH^+ + e^-$ $\quad \Delta H_0^0 = +25 + 8$ kcal/mole
$\quad K = 2.0 \times 10^{-3} \exp(-25\,000/RT)$

$Ba + OH \rightarrow BaOH^+ + e^-$ $\quad \Delta H_0^0 = 6 \pm 10$ kcal/mole
$\quad K = 1.7 \times 10^{-3} \exp(-6\,000/RT)$.

It should be pointed out that the present data and conclusions are only tentative; also it is not clear how far equilibrium is attained in these flames or whether negative ions play an important role. More recent studies (Hayhurst & Kittelson, 1974) have shown that reactions of the type

$$Ca^+ + H_2O = CaOH^+ + H$$

are rapid enough to be balanced everywhere in the flame and that the ionisation potentials of the monohydroxides are slightly lower than of the corresponding metals themselves, e.g. $I\,(CaOH) = 5.7 \pm 0.3$ and $I\,(SrOH) = 5.4 \pm 0.3$ eV.

In the absence of hydrogen (for example in carbon monoxide flames) the rate of ionisation of alkaline-earth metals would be slow and due to the process

$$M + CO_2 \text{ (or } CO, N_2, O_2) \rightarrow M^+ + e^- + CO_2\,(CO, N_2, O_2)$$

as is the case with alkali metals where hydroxide ions do not occur.

Ionisation by inert collisions. The rate of this type of ionisation can be investigated in two ways. One can measure the increase in ionisation with time, i.e. in the direction of flow of the gases, or one can investigate the reverse process, that is the recombination, and relate the rates through the equilibrium constant calculated by Saha's equation. The first experiment would involve the measurement of N_e along the flame axis with a microwave cavity, or else measurement of the atomic resonance line emission as a function of flame temperature to derive the depletion of neutral atoms by ion formation (see above); such experiments are easiest with sodium for which hydroxide formation may be neglected.

The second, reverse, process can be investigated if a hydrocarbon is added to hydrogen flames, so that the metal is ionised in the reaction zone beyond the equilibrium amount by the process $M + S^+ \rightarrow M^+ + S$, where S^+ is an ion such as CHO^+ formed in the reaction zone by chemi-ionisation. Then, as excess ionisation of the metal in the reaction zone decays with time, the rate of recombination can again be measured by microwave techniques. Jensen & Padley (1966; 1967) measured the rate constants both ways. They found that the ionisation process is given by $M + (M) \rightarrow M^- + e^- + (M)$ where M represents metal vapour atoms and (M) diatomic or polyatomic flame-gas molecules. The activation energy is found, within experimental error, to be the ionisation potential of the metal atom. The cross-section calculated on the basis of bi-molecular collisions is several orders of magnitude larger than the gas-kinetic cross-section. The reason for this large cross-section is unknown at this time. However, it should be remarked that, although the rate of ionisation is faster than expected, it is slower than for alkaline-earth metals in hydrogen flames where ionisation occurs by a chemical step, i.e. hydroxide-ion formation. For example, in flames at 1 atm and between 2000 and 2500 K the rise time to half the final ionisation is between 0·1 and 1·0 msec.

Jensen & Padley also found that the rate for ionisation by inert collisions was first order with respect to the metal atoms as well as to the flame molecules. Therefore ionisation reactions such as $M + OH \rightarrow M^+ + OH^-$ and $M + e^- \rightarrow M^+ + 2e^-$ are excluded. These findings agree with those from a study of ionisation in carbon monoxide flames (Hollander, 1968).

As can be seen from the above discussion, ionisation reactions are complex. If the possibility of chemical ionisation exists this will be the preferred route. In the absence of chemical ionisation, slow processes prevail in which only particles of sufficient kinetic energy are able to remove an electron. Of particular interest is the observation that the rate of ionisation is independent of electron concentration, in contrast to the fact that electrons are more efficient than atoms and molecules in exciting electronic states.

Ionisation in hydrocarbon flames

Positive ions. The products of combustion in a hydrocarbon flame all have high ionisation potentials, so that the levels of ionisation should be quite small in equilibrium. The ionisation potentials of some common flame products are given in Table 13.5.

TABLE 13.5
Ionisation potentials of flame species, in eV, from Gurvich et al. (1974)

O_2	12·08	NO	9·26	CH_3	9·8
H_2	15·43	NO_2	9·78	CH_4	12·9
OH	13·18	NH	13·1	C_2H	12·2
H_2O	12·61	C_2	11·9 ± 0·6	C_2H_2	11·41
CO	14·01	C_3	11·9 ± 0·6	CHO	9·8 ± 0·2
CO_2	13·79	CH	10·6 or 11·1	CH_2O	10·87
N_2	15·58	CH_2	10·4		

NO has the lowest ionisation potential and may produce some ionisation (*see* page 350) but even this is relatively unimportant and also the formation of NO itself from nitrogen and oxygen is a slow process.

It has been known for a long time that premixed organic flames have ionisation levels much higher than could be accounted for on equilibrium considerations. This was shown by flame-deflection studies in an electric field, as well as by measurements with Langmuir probes where it was found that the ionisation reached a maximum level in the reaction zone (Calcote, 1949; Calcote & King, 1955; King, 1957; Kinbara & Nakamura, 1955).

The effect of non-equilibrium ionisation resembles in many ways the chemiluminescence of lines and bands in the reaction zones of flames discussed in previous chapters and was thus named chemi-ionisation. This phenomenon of chemi-ionisation is the basis of modern gas chromatographic systems in which very small quantities of organic vapours are detected by the increased ionisation which is produced when they are introduced into a flame of pure hydrogen.

It is beyond the scope of this book to follow all the experimental work that has been done in the last few years, but we present a summary of what appear at this time to be the agreed processes. Excellent summaries have been published by Miller (1968; 1973), from whom some data are extracted, and Lawton & Weinberg (1969). The experimental tools for these investigations are Langmuir probes, microwave absorption and ion-mass spectrometers.

As equilibrium is insufficient to account for the ionisation level, one has to search for reactions that may produce positive ions and electrons. It is also known that hydrogen and carbon-monoxide flames have no significant chemi-ionisation so reactions such as

$$CO(^1\Pi, \text{excited state}) + O \rightarrow CO_2^+ + e^- \quad \Delta H = +6 \text{ kcal/mole}$$
and
$$H + H + OH \rightarrow H_3O^+ + e^- \quad \Delta H = +27 \text{ kcal/mole}$$

may be eliminated. This leaves a long list of possible reactions involving molecules having C_2 or CH bonds. They are listed in Table 13.6.

TABLE 13.6

	Reaction	ΔH, kcal/mole
(1)	C_2 ($A^3\Pi_g$, excited state) $+ O_2 \to CO + CO^+ + e^-$	$+16$
(2)	$C_2 + HO_2 \to C_2O_2H^+ + e^-$	?
(3)	CH ($x\,^2\Pi$, ground state) $+ O \to CHO^+ + e^-$	$+20$*
(4)	CH^* ($A^2\Delta$, excited state) $+ O \to CHO^+ + e^-$	-46
(5)	$CH^* + O_2 \to CHO_2^+ + e^-$	-45
(6)	$CH^* + HO_2 \to CHO^+ + OH^-$	-23
(7)	$CH^* + HO_2 \to H + CHO_2^+ + e^-$	$+2$
(8)	$C_2H + O_2$ ($B^3\Sigma_u^-$, excited) $\to CO_2 + CH^+ + e^-$	$+47$
(9)	$C_2H + O_2$ ($b^1\Sigma_g^+$, excited) $\to C_2O_2H^+ + e^-$	?
(10)	$CH^* + C_2H_2 \to C_3H_3^+ + e^-$	-2
(11)	$C_2H + O_2$ ($B^3\Sigma_u^-$, excited) $\to CO + CHO^+ + e^-$	-60

* Later photo-ionisation studies (Matthews & Warneck, 1969) suggest $\Delta H = 4 \pm 5$ kcal/mole; reactions involving CH^* would also then be 16 kcal/mole more exothermic.

All reactions involving more than one molecule in an excited electronic state have been omitted. It can also be seen that all ionisation reactions are strongly endothermic unless at least one molecule is in an excited electronic state. The $O_2(B^3\Sigma_u^-)$ is the upper state of the Schumann-Runge system. The bands of this system are emitted in hot flames supported by oxygen; in cooler flames with air they have not been seen and the number density of O_2 ($B^3\Sigma_u$) is presumably small. There is a similarity in the occurrence of CH emission in flames and chemi-ionisation, so that reactions (4) to (7) may be important and also, of course, reaction (3).

Fig. 13.7 shows the relative positive ion density in a low-pressure flame whose reaction zone is relatively thick so that details of formation and decay can be seen (Calcote et al., 1965).

The reaction zone, extending to 3 cm above the burner, contains a great variety of different positive ions. Most of them recombine or lose their charge by charge-exchange at the end of the reaction zone. Only H_3O^+, its hydrate $H_5O_2^+$ and, to a smaller extent, CHO^+ extend into the burnt gas. Present evidence,* summarised by Miller (1973), indicates that the main reaction is (3), $CH + O = CHO^+ + e^-$, followed by proton transfer to water to form H_3O^+. Other papers on the $CH + O$ process are by Peeters (1973) and Ay, Ong & Sichel (1975). Fig. 13.7 shows $C_3H_3^+$ occurring early and strongly, but there are a number of arguments (see Miller) for rejecting this as a main species leading to chemi-ionisation in hydrocarbon

* A still more recent note by Tse, Michaud & Delfau (1978) confirms that in oxy-acetylene flames $C_3H_3^+$ precedes CHO^+ and H_3O^+ and it coincides with C_2 and CH emission. The isotope C^{13} has been used to sort out the identity of the positive ions. The reaction, suggested by Bowser and Weinberg,
$$C_2^* + CH_3 = C_3H_3^+ + e^-$$
is strongly favoured.

oxidation, although it may be a subsidiary reaction. Ionisation has been found to occur (Bowser & Weinberg, 1976) during pyrolysis of ethylene, free from oxygen except perhaps as a trace impurity; even here there is some difficulty in using the $CH^* + C_2H_2 = C_3H_3^+ + e^-$ reaction as production of CH^* is believed usually to involve oxygen in the $C_2 + OH = CH^* + CO$ process. There is some evidence that in acetylene/oxygen flames reaction (11), $C_2H + O_2^* = CHO^* + e^-$ and also another reaction $C_2 + O = C_2O^+ + e^-$ may be partly responsible for the high chemi-ionisation in these flames.

The rate constant for ion formation, $K[CH][O]$, is only approximately

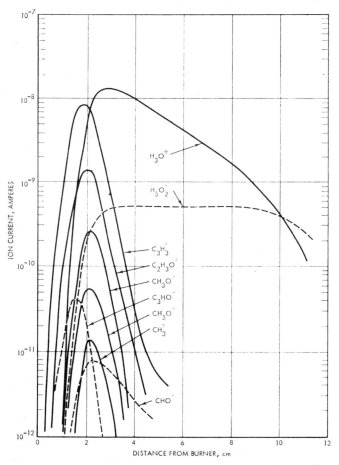

Fig. 13.7. Positive ion profile in acetylene-oxygen flame at 2 torr.

Redrawn from Calcote et al. (1965)

known because of difficulties of determining [CH] but seems to be of the order 10^{-13} cm^3 sec^{-1}; Peeters (1973) gives 1.4×10^{-13} at 2200 K. It can be seen from Fig. 13.7 that all the ions containing carbon decay at about the same rate. It is strange that all these ions should recombine or exchange charge uniformly; thus it has been postulated that the ions are in partial equilibrium among themselves and only one ion such as CHO$^+$ decays to form H$_3$O$^+$ by the reaction CHO$^+$ + H$_2$O → H$_3$O$^+$ + CO.

Negative ions. Langmuir probes are unreliable for the determination of electron concentrations, so the possibility exists that most negative charges are present as negative ions (*see* page 352 for electron attachment potentials). Feugier & Van Tiggelen (1965) were able, by reversing the electrical and magnetic fields in their mass spectrograph, to identify a great number of negative ions. Their abundances change greatly with the type of fuel, but a representative number of negative ions from an acetylene flame is given below in order of increasing abundance:

$$C_2H^-, CO_2H_2^-, COH_3^-, C_2OH_2^-, CO_3H_2^-, O_2^-, CO_3H_3^-, OH^-,$$
$$C_2OH^-, C_2O_2H_4^-, C_2H_2^-, C_2^-.$$

It should be noted that the chemical identification is only tentative and other species having the same mass number could be the true ions.

The ratio of positive to negative ions is subject to some uncertainty. Knewstubb (1965) found 50% of all negative charges to be negative ions in the reaction zone of a flame at 1 atm, with a much smaller percentage in the burnt gases. Calcote *et al.* (1965), using an acetylene flame at 1 torr found a distribution as shown in Fig. 13.8. Here N_- is approximately $0.01\ N_+$. Some ions, such as O$^-$, OH$^-$ and O$_2^-$ are persistent, whereas C$_2^-$ has a strong peak in the reaction zone. Again, as in the case of positive ions, charge-exchange will play a role, but direct attachment is also possible. For example:

$$e^- + OH + M \rightarrow OH^- + M$$
$$O + e^- \rightarrow O^- + h\nu$$

or
$$H_2O + e^- \rightarrow OH^- + H.$$

The reactions involved are in no way decided, and the reader is referred for details to Calcote *et al.* (1965) or Miller (1968).

Ion removal. Low-pressure flames, mostly used in this type of ionisation experiment, are not suitable for the study of ion-removal mechanisms as electrons and ions diffuse very fast out of the flame by ambipolar diffusion. The recombination rate has usually been found to be independent of pressure (at least for pressures above 0·1 atm) so that three-body recombination is ruled out. Bi-molecular processes can include radiative

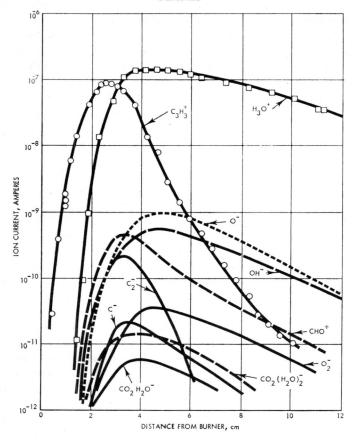

Fig. 13.8. Negative ion profiles in acetylene-oxygen flame at 1·0 torr.
[*Redrawn from Calcote* et al. (1965)

recombination, which generally has a small cross-section, and dissociative recombination such as

(1) $H_3O^+ + e^- \rightarrow H + H_2O$
(2) $H_3O^+ + e^- \rightarrow H + H + OH$
(3) $H_3O^+ + OH^- \rightarrow 2\ H_2O$
(4) $H_3O^+ + OH^- \rightarrow H + OH + H_2O$.

It should be remarked that as the flame cools, more negative ions will be formed, as those with low attachment potentials are dissociated at high temperature but have higher concentrations in the cooler gas. Thus in cooler flames recombination will be mainly by processes (3) and (4) whereas at high temperature recombination by reactions (1) and (2) will dominate.

Further investigations of ion reactions in flames

After discussing the principal processes of ion formation in flames, we will now mention some further investigations where ions play an important and interesting role. Hurle, Sugden & Nutt (1969) investigated carbon monoxide and hydrogen flames to which 0·5 mole per cent of acetylene and varying small amounts of nitric oxide were added. They probed the flames with an ion mass spectrometer and followed the formation and removal of ions. The greatest surprise is the presence of appreciable amounts of $O_2{}^+$ in CO, but not in H_2 flames. Hydrogen flames with added C_2H_2 generally follow the course of reactions discussed in the previous section on hydrocarbon flames, i.e. CHO^+ is formed and decays by charge-transfer to H_3O^+ which persists. On adding NO a sharp peak of NO^+ is observed in the reaction zone; NO^+ subsequently decays and then increases again. The first peak is presumably due to charge transfer from CHO^+ whereas the second slow increase in NO^+ is the thermal ionisation of NO.

In carbon monoxide flames the reactions are more complicated. NO^+ increases with addition of NO, at the expense of $O_2{}^+$ as well as H_3O^+ and HCO^+, and two NO^+ maxima are observed. For details the reader is referred to the original paper.

There are similarities between chemi-ionisation and chemi-excitation (discussed in Chapter IX), as noted by Gaydon & Wolfhard (1951). Hydrocarbon-air flames exhibit chemi-excitation strongly, whereas excitation in hydrocarbon-nitric oxide flames is mainly thermal. Cummings & Hutton (1967) measured the ionisation in mixed hydrocarbon–air–NO flames to test for such similarities. They find indeed a reduction of the positive ion current in the reaction zone as air is in part replaced by NO; however, as the hydrocarbon–NO flame is much hotter than the air flame (3000°C instead of 2000°C) and thermal ionisation is possible in the pure NO flame, the test is not decisive for large replacements of air by NO. For small replacements by NO the strength of C_2, CH and OH bands decreases, as does the ion saturation current, and these facts do not exclude the mechanism previously postulated that the chemi-ions are formed by $CH + O = CHO^+ + e^-$. However, the relationship between CH (ground state) and the chemi-excitation, as exhibited by reversal temperatures exceeding the adiabatic flame temperature, is tenuous. As chemi-excitation and ionisation are presumably due to specific reactions, they need not have a common origin in all cases and similarities between the two processes may be more superficial than real.

Let us consider a chemiluminescent and chemi-ionisation process in more detail:

(a) $A + B \rightarrow AB^* \rightarrow AB + h\nu$ (chemi-excitation)
(b) $A + B \rightarrow AB^+ + e^-$ (chemi-ionisation).

The molecule AB* is generally a short-lived collision complex (about 10^{-13} sec). As the average time for an electronic transition to occur is around 10^{-8} sec or longer, it can be seen that chemi-excitation by a direct association of this type is a rare event for any one specific collision. (This would, not, however apply to a more complex reaction such as $CH + O_2 = CO + OH^*$.) In contrast, in chemi-ionisation, potential energy curves have to cross, with a rearrangement of the electronic structure. This electronic rearrangement takes place, if it takes place at all, in a time which is short compared with the relative movement of the atomic nuclei; thus chemi-ionisation can be a fast process, despite the fact that higher energies are involved than for excitation processes.

Chemi-ionisation can probably be studied best in cold atomic flames as thermal ionisation processes are then excluded. Fontijn & Vree (1967) and Fontijn (1969) studied ion formation in atomic oxygen and nitrogen flames with carbon-containing compounds. In collisions with O-atoms the, by now, well-known reaction $CH + O = CHO^+ + e^-$ is operative. Nitrogen atoms alone, together with the hydrocarbon, lead to no ionisation or only an insignificant amount, whereas a mixture of N and O atoms leads to much greater amounts of ionisation than for pure O atoms. The mechanism of the process is not yet known.

In Chapter XIV we will be discussing some flames of exceptionally high temperature, due to burning metals. If further metals of low ionisation potential are introduced, then very high concentrations of ions can be achieved (up to 10^{16} ions cm^{-3}). An example of such a system would be a mixture of aluminium powder and caesium nitrate ($CsNO_3$); this mixture burns like a solid propellant with the products Al_2O_3 and caesium vapour which becomes nearly totally ionised. A similar high-temperature flame can be achieved by burning cyanogen, or its derivatives, to carbon monoxide. Friedman & Macek (1965) used a solid mixture of tetracyanoethylene, hexanitroethane and caesium azide which burnt to CO, N_2 and Cs; the products were all gaseous, in contrast to the aluminium mixture, and this had advantages in applying diagnostic methods. The flame temperature was close to 4000 K. The ion density was determined by the Stark broadening of the emitted caesium lines and concentrations up to 10^{15} ions cm^{-3} were measured. It is, of course, possible to increase the temperatures of flames by means other than using high energy propellants. The burnt flame gases can be electrically boosted by applying electrical discharges, and ionisation can be driven to very high amounts. This subject is important for MHD generation but goes beyond the scope of this book. *See also Reactions under Plasma Conditions* by Venugopalan (1971).

Ions may play an important role in carbon formation, and this subject has already been discussed in Chapter VIII.

The effect of electric fields on flames

Electric fields are known to have some marked effects on flames and flame propagation. With a stationary flame in a transverse direct-current field, the flame is distorted and bent towards the negative electrode. With round burners the flame usually tends to lift on one side, and the limits of stability between blow-off and flash-back are narrowed. With spherical explosion flames, the flame speed varies with the direction of travel, the front being accelerated in the direction in which positive ions are drawn and retarded in the other direction. With flames propagating along tubes a variety of effects have been observed. Sometimes, especially for rich mixtures of hydrocarbons, the flame is held close to one side of the tube and is slowed down or even quenched. In other cases, especially for high frequency alternating fields, the speed of the flame front may be increased. Transverse fields may prevent the development of detonations.

This subject has been fully treated by Lawton & Weinberg (1969) and a further full coverage will not be attempted here. Most of the effects of applied fields can be explained by the well known Chattock electric-wind effect. For an applied field E, the force F exerted on the gas due to ions being accelerated and colliding with gas molecules is

$$F = Ee(N_+ - N_-)$$

where e is the charge on the electron, N_+ is the number concentration of positive ions and N_- that of the negative species (usually electrons). For a neutral gas, for which $N_+ = N_-$, this does not lead to any net force on the gas molecules. However, because of the high mobility of electrons, around 4000 cm^2 sec^{-1} V^{-1}, compared with the relatively low mobility of about 1 cm^2 sec^{-1} V^{-1} for the positive ions in flame gases, there is charge separation, both by thermal diffusion and especially by the action of an applied field. Normally positive space charges occur due to removal of electrons by the electrodes or by diffusion to the burner walls and so the flame is deflected towards the negative electrode. Calcote (1949) studied butane/air flames on a rectangular burner with a transverse field and followed the gas flow lines by track photography of aluminium particles; the particles were assumed to follow the gas flow, although if the particles were appreciably charged this might not be strictly true. Fig. 13.9 shows diagrammatically the form of the flame, bending towards the negative electrode on the right, and the flow lines.

These electric wind effects can change the flame area and also any air entrainment and so alter the flame stability. Heat transfer to surrounding surfaces may be increased if the flame is deflected towards them. In the combustion of oil sprays, the diffusion flame round each droplet may be similarly distorted. Strong direct-current fields tend to extinguish droplet

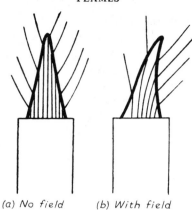

(a) No field   (b) With field

Fig. 13.9. Distortion of flame front and flow lines by an electric field.

flames, but alternating fields usually increase heat transfer to the droplets thus increasing their evaporation and burning rate.

For propagation in tubes, distortion of the flame shape may increase the flame area, leading to a higher rate of consumption of the mixture and higher speed of flame travel. There have been a number of suggestions, especially in the older literature, but also some more recently (e.g. Jaggers & von Engel, 1971) that the applied field actually leads to a small increase in burning velocity. However, other measurements (Blair & Shen, 1969; Bowser & Weinberg, 1972; Fox & Mirschandani, 1974) on the effect of fields on burning velocities made with a cooled porous-plate burner do not support this. It is very difficult to make measurements of burning velocity which are unaffected by the change in flame shape when an electric field is applied, and although it seems very unlikely on present evidence that ions and electrons contribute significantly to flame propagation or that fields affect the burning velocity, it is difficult to say categorically that there is no effect at all. Active gases, which include free atoms as well as ions, from a plasma jet, do assist in maintaining flame stability (Harrison & Weinberg, 1971) and there is some evidence that microwaves may assist the performance of internal combustion engines. The influence of applied fields and also of easily ionised metallic additives on soot formation in hydrocarbon flames has already been discussed in Chapter VIII.

Ionisation and electron temperatures

As an example of the experimental work mentioned previously and of the theoretical knowledge available, we will now briefly discuss the ionisation in rocket exhausts. Fig. 13.10 shows the structure of a rocket exhaust flame when the nozzle exit pressure is equal to the ambient pressure. The gas goes through a series of shock compressions and expansions. When the gas enters the shock diamonds, its pressure and temperature are raised and it becomes luminous. A mixing region surrounds the shock structure. Rocket propellants are fuel-rich so as to decrease the molecular weight in the exhaust, in order to increase the specific impulse. In a hydrocarbon–oxygen rocket up to 50% of the exhaust gases are hydrogen, carbon monoxide and unburnt hydrocarbon. In the mixing zone these products mix with air, and secondary reactions take place producing luminous flames which raise the temperature of the mixing zone.

The absorption (in decibels) of microwave beams is measured at various frequencies. In Fig. 13.11 for example, X, K and S bands, corresponding to approximately 9, 24 and 3 GHz are used to study a 1500-pound thrust alcohol–oxygen rocket (Balwanx, 1965). As the microwave beams cannot be focused precisely because of the limitations discussed in a previous section the step increases in ionisation as one goes through the shock structures cannot be completely resolved, but the overall effect of the shocks can clearly be seen. It is, of course, questionable whether the stepwise increase in temperature and pressure at the shock fronts leads to a stepwise increase in ionisation. In reality the increase may be quite smooth due to the finite ionisation relaxation time. There does not seem to be sufficient time for recombination to occur in the expansion following the shocks, so the overall effect is a repeated increase of ionisation.

The ionisation seems to be mainly due to impurities. For Fig. 13.12 sodium was intentionally added (Balwanz, 1965) and it is clearly seen that overall ionisation and microwave attenuation are coupled to the impurity content, at least at the high amounts of impurity used here. Often the structure of ionisation due

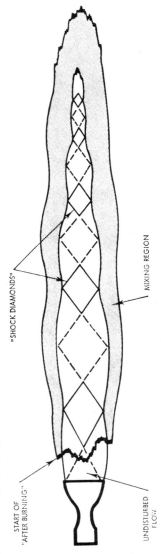

Fig. 13.10. Structure of a rocket exhaust flame when the nozzle exit pressure is equal to the ambient pressure.

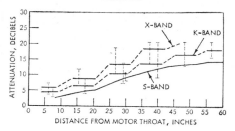

Fig. 13.11. Attenuation of X, K and S band microwave beams in a rocket exhaust. [*Redrawn from Balwanz* (1965)]

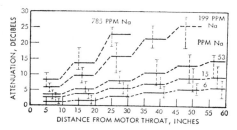

Fig. 13.12. Microwave attenuation at K band in a rocket exhaust containing sodium impurity. [*Redrawn from Balwanz* (1965)]

to the shock diamonds cannot be seen. Afterburning of the excess fuel in the mixing zone raises the flame temperature above the nozzle-exit temperature and ionisation (due to chemi-ionisation and impurities) increases rapidly beyond the nozzle, reaches a maximum further downstream and then decays as exhaust gases mix with more air and are cooled thereby.

Pergament & Calcote (1967) have made extensive calculations using rate constants as described above for ionisation and charge transfer. Their calculation applies to the first part of the exhaust flow, before internal shocks greatly modify the flow, so that the exhaust structure they consider is limited to flow from the nozzle and mixing with the ambient air (Fig. 13.13).

The exhaust gases flow from the nozzle exit into ambient air of known velocity and temperature. In the mixing region unburnt CO and H_2 burn with air to release heat. Unburnt hydrocarbons, represented by CH_4, form the relatively stable H_3O^+ and e^- by chemi-ionisation; impurities, represented by potassium, ionise in equilibrium to K^+ and e^-.

The flow can be calculated; however, a major difficulty is the accuracy of the eddy diffusivity. For details the reader is referred to Pergament & Calcote (1967). Chemi-ionisation is calculated by assuming that one in 10^6 hydrocarbon molecules leads to a chemi-ion, generally H_3O^+ and e^-. In the actual examples quoted below it is further assumed that the mole fraction of CH_4 is 10^{-3}.

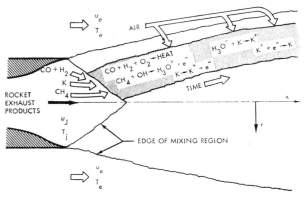

Fig. 13.13. Model of chemical reactions in afterburning rocket exhaust.
[*Redrawn from Pergament & Calcote* (1967)]

Potassium (1 part per million) ionises by

$$K + M \to K^+ + e^- + M$$

with two possible rate constants,

(1) $K = 5 \times 10^{-7} \exp(-100\,000/RT)$ (fast reaction)

or (2) $K = 5 \times 10^{-9} \exp(-100\,000/RT)$ (slow reaction)

and charge transfer can occur

$$H_3O^+ + K \to K^+ + H_2O + H.$$

Negative ions are neglected in the region of the flow for which calculations were made. Recombination of chemi-ions is considered, i.e.

$$H_3O^+ + e^- \to H_2O + H$$

as it is bi-molecular.

Fig. 13.14 shows calculations of radial temperature profiles at 50 000 feet altitude (i.e. 0·1 atm ambient pressure) for a rocket exhaust with afterburning. The increased temperature in the mixing zone as well as the overall increase of temperature with distance, x, along the centre line are evident. The mole fraction of positive ions and electrons along the centre line is given in Fig. 13.15. One sees the slow ionisation of potassium up to 20 exit diameters (this calculation uses the fast-rate constant (1)); thus chemi-ionisation predominates close to the nozzle, but then decreases in importance as the hydrocarbons are consumed.

Overall results of these numerical calculations show that chemi-ionisation is important if hydrocarbon fragments are present in the exhaust.

370   FLAMES

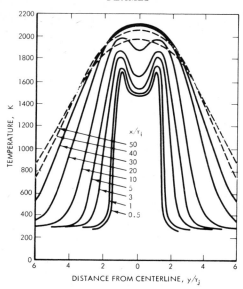

Fig. 13.14. Radial temperature profiles in afterburning rocket exhaust. r_j is the nozzle-exit diameter.

[*Redrawn from Pergament & Calcote* (1967)]

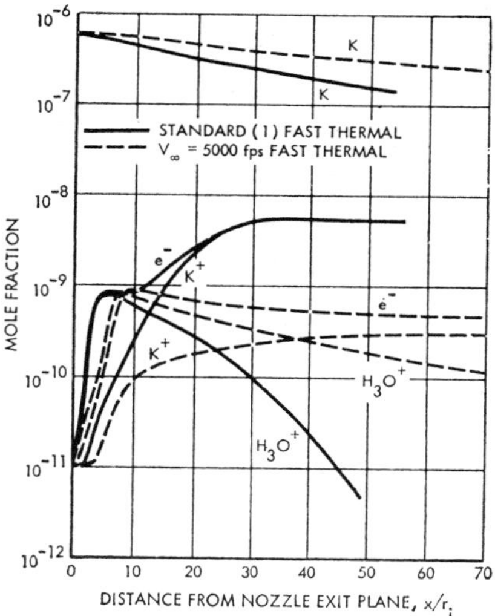

Fig. 13.15. Concentration of neutral and charged species along centre line of afterburning rocket exhaust.

[*Redrawn from Pergament & Calcote* (1967)]

Also, equilibrium calculations are inadequate and would overestimate ionisation close to the nozzle. Even small amounts of metallic impurities are important because the charge on the chemi-ion transfers to the metal atom which, in turn, has a smaller rate of recombination.

Chapter XIV

Combustion Processes of High-energy and Rocket-type Fuels

This chapter is not designed to deal with the question of rocket propulsion as a whole, but to show the characteristics of the combustion processes occurring when high energy or other unusual fuels, or oxidisers other than oxygen, are used. These flames reveal many new features not observed in hydrocarbon/oxygen combustion and it is hoped that their study will prove useful not only for the understanding of jet propulsion, but for combustion as a whole.

Flames supported by oxides of nitrogen

Reactions of oxides of nitrogen with fuels are of great importance, not only because many oxidisers used in jet propulsion belong to this group, but because many commercial explosives have a structure which incorporates NO_2 or NO_3 groups. Although these explosives will not be considered here, it is clear that their decomposition and explosion reactions will be rather similar to those described below. As the decomposition reactions of nitric oxide are of prime importance for all flames using oxides of nitrogen, NO will be considered first.

Decomposition flame of NO. Nitric oxide is a compound which liberates 21·6 kcal/mole upon decomposition into nitrogen and oxygen. It owes its stability to its large activation energy of decomposition of about 60 kcal. The heat of decomposition would be enough to heat the products up to about 2600 K. Henkel, Hummel & Spaulding (1949) predicted that a pure NO decomposition flame could be obtained if the unburned gas was preheated so that the temperature of the product gases exceeded 3000 K. Parker & Wolfhard (1953) succeeded in burning this precarious flame. To do this, NO was heated rapidly to 800–1000°C in a vertical furnace of 3 cm diameter. Ignition was a problem as homogeneous reactions could only be initiated above 3000 K. This was achieved by adding hydrogen to the NO in the furnace and letting a H_2/NO flame establish itself at the furnace exit. As hydrogen was then withdrawn, the flow had to be carefully adjusted because of the reduced burning velocity of pure NO. Unfortunately the flame was so precarious and local overheating took place so fast where the flame anchored, that no burning velocity measurements were possible. The spectrum of the flame shows only Schumann-Runge bands of O_2 with some OH due to the moisture in the surrounding air (*see* Plate 18*a*).

The mechanism of NO decomposition is somewhat doubtful. In making their burning velocity calculations, Henkel *et al.* (1949) assumed a direct bimolecular decomposition without radical chains; Kaufman & Kelso (1955) found the decomposition to be bimolecular ($2NO \rightarrow N_2 + O_2$) between 1400 and 1600 K with an activation energy of 63·8 kcal. However, beyond 1800 K, they assumed that radical reactions such as $O + NO \rightarrow N + O_2$ became important and finally rate-controlling. Fenimore & Jones (1957) measured the decomposition of NO downstream of a nitrous oxide/fuel flame but their evidence is conflicting and the presence or absence of N atoms could not be determined.

The hydrogen/nitric oxide flame. A hydrogen/nitric oxide flame is again not easy to ignite. A match held into the gaseous mixture does not initiate a flame. Similarly, if hydrogen burns as a diffusion flame in the surrounding air and nitric oxide is introduced into the hydrogen, the flame normally goes out. However, if the burner diameter is of the order of 3 cm or if the pressure is well above one atmosphere, the premixed H_2/NO flame is initiated by the preceding hydrogen flame. A whitish reaction area forms within the hydrogen/air diffusion flame upon introduction of NO. This area subsequently contracts and moves down to sit on the burner where it forms a conventional reaction zone. The extremely large quenching diameter makes it possible to maintain the flame up to very high pressures. Cummings (1958), measuring the burning velocity with a burner flame over a wide pressure range, found it to be independent of pressure. His values are 56 cm/sec $\pm$ 7 between 1 and 40 atm. Strauss & Edse (1959) used a soap bubble method and found the burning velocity to increase from 55 cm/sec at 1 atm to 80 cm/sec at 90 atm. The maximum burning velocity occurs close to stoichiometric. The quenching distance has been determined by Litchfield & Blanc (1958) with the Bureau of Mines spark-gap apparatus; it is about 0·7 cm, i.e. the quenching diameter is about 1·0 cm. The minimum ignition energy also is anomalously high at 8·7 mJ. Interestingly enough, the mixture strength for which the measured ignition energy and distance were smallest corresponded to 23% hydrogen, i.e. it was far to the NO-rich side. This suggests that hydrogen atoms do not take part in the NO decomposition. This conclusion is also supported by the nature of the spectrum of the flame obtained by Wolfhard & Parker (1955), which shows only OH and O_2 bands and probably weak NO bands in emission. The NH bands are spurious and disappear on preheating the H_2/NO mixture suggesting that the very weak NH emission is due to some N_2O impurity. This is in contrast to the very strong NH bands of hydrocarbon/nitric oxide flames. It is difficult to believe that NH radicals would not be formed if N atoms were produced in the reaction zone by a chain reaction during the NO decomposition. Thus the spectral evidence supports a straight bimolecular NO decomposition similar to the NO decomposition flame, leading directly to N_2 and O_2 with the latter reacting with hydrogen. This receives additional support from absorption spectroscopy, which shows that nitric oxide disappears just above the reaction zone for all mixture strengths, including NO-rich flames. Even moderate dilution of hydrogen/ nitric oxide flames soon leads to extinction; it seems that virtually the full adiabatic flame temperature of 3100 K has to be available to overcome the high activation energy of the NO decomposition.

Flames of other fuels with nitric oxide. Moist carbon monoxide/nitric oxide flames can be maintained on very wide burners. The same reaction scheme as for H_2/NO flames seems to be indicated. The spectrum shows the CO flame bands, O_2 bands and weak OH bands; no NH bands are present. The flame is slower than the H_2/NO flame as the adiabatic flame temperature is also reduced. Formaldehyde/nitric oxide flames also resemble hydrogen flames closely. No CN bands are emitted (Hall, Mc-Coubrey & Wolfhard, 1957).

The most striking contrast to the hydrogen/nitric oxide flame is the ammonia/nitric oxide flame. If the same mechanism were operative, one would expect that this flame could not burn at all because of the reduced flame temperature. However, the ammonia flame is easy to ignite and burns on quite small burners, i.e. the quenching diameter is small. Adams, Parker & Wolfhard (1955) postulated that ammonia or its decomposition products, such as NH_2 and NH react with NO, the thermal NO decomposition not occurring at the low temperature. This can be demonstrated spectroscopically; NH absorption is weaker in a NH_3/NO flame than in the NH_3/O_2 flame, the radical NH being removed by reaction; NH is, however, still strong in emission. In absorption it can be seen that NO is not totally decomposed in flames rich in NO; presumably not enough radicals are available for complete decomposition. The plot of burning velocities against flame temperatures (Fig. 5.8) shows that the NH_3/NO flame is far outside the range of other NO flames. Details of the reaction mechanism in NH_3/NO flames are not yet clear as ammonia itself has a high activation energy for thermal decomposition. However, it is known that ammonia begins to decompose rapidly beyond 2000°C, as proved by the appearance of NH absorption bands in a furnace (Frank & Reichardt, 1936).

The hydrazine/nitric oxide flame has such a small quenching diameter that it can be maintained down to 13 torr on a 2-cm burner (Hall & Wolfhard, 1956). This contrasts with the 2-cm burner required by H_2/NO at 1 atm. The flame emits NH and NH_2 bands much more weakly than from a NH_3/O_2 flame, as expected for fast removal of NH_2 and NH by NO.

Hydrocarbon/nitric oxide flames seem to involve both the thermal and radical decomposition mechansim of NO. This shows up sometimes as a double reaction zone. For example, in lean hydrocarbon/nitric oxide flames a first reaction zone emits strong C_2, CH, CN, NH and NO γ bands. Above this luminous zone a bluish second zone is visible and at this point OH bands start to be emitted strongly (Plate 18*f*) and also some Schumann-Runge bands of O_2. In this second flame zone the thermal decomposition mechanism is operative as the temperature is close to that of H_2/NO flames. It is not clear which of the radicals present in the first flame zone is most effective in decomposing NO. NH is quite strong in emission, but cannot be detected in the absorption spectra, presumably again due to its rapid removal. It is very doubtful, however, whether enough NH or NH_2 radicals are formed to have a large effect on NO decomposition. CN does not seem to be particularly effective as it accumulates in some flames to such an extent that it becomes readily detectable in absorption (Plate 18*b*). It is interesting to note that methyl bromide burns more readily with NO than does methane, although the adiabatic flame temperature is lower. Methyl bromide also emits C_2 and CH bands much more strongly than does methane. CS_2/NO flames burn readily and there is strong radiation due to emission of S_2 bands. In addition, there is strong continuous radiation, the cause of which is not yet clear. The reactions of this flame are not readily explained as the flame temperature cannot be very high. Diborane nitric oxide flames have also been obtained. The decomposition of NO is no problem here as the flame temperature is very high. Details of this flame will be discussed under diborane flames.

Hydrocarbon/nitric oxide flames are also interesting in so far as they lack the excess electronic excitation so prominent in flames with oxygen and nitrous oxide (Plate 18*d* and *e*). This is due to the small reaction rates in these flames, as evidenced by the low burning velocity, coupled with the high adiabatic flame temperature (*compare* Chapter V). The rotational temperature of OH is also normal in NO flames (Plate 18*f* and *g*).

Spontaneous ignition temperatures correlate closely with the ideas expressed above (Wolfhard & Strasser, 1958). Carbon monoxide/nitric oxide mixtures are the most difficult to ignite and only mixtures rich in NO could be ignited at 1450°C. If only the thermal decomposition of NO is rate-determining, one would expect H_2/NO mixtures to have exactly the same ignition temperature; however, the value was 1320°C, indicating a small but definite deviation from a pure thermal mechanism. Hydrocarbons had lower values, in the order of the availability of radicals in the reaction zone. Thus methane had a value of 1110°C and ethylene of 1045°C. Ammonia/nitric oxide mixtures had a value of 800°C, which is surprisingly low considering the stability of both ammonia and nitric oxide. Carbon disulphide/nitric oxide mixtures also have a low ignition temperature of 775°C (Roth & Rautenberg, 1956). Reaction rates measured just below this temperature indicated the reaction to be of second order with respect to NO and first order with respect to CS_2. The mode of NO decomposition is not clear.

The hydrogen/nitrogen dioxide flame. The hydrogen/nitrogen dioxide flame can be maintained on burners as small as 1 mm. The burning velocity is nearly 200 cm/sec at 1 atm, in contrast to the burning velocities of NO flames which are all below 100 cm/sec. The flame has a single yellow reaction zone. The stoichiometric flame emits only continuous radiation in the visible, and OH bands are absent. In absorption, it can be seen that NO is formed in the reaction zone but escapes further reaction. The measured flame temperature is 1550°C, in contrast to the adiabatic temperature of 2660°C. Nitrogen dioxide seems to decompose in the flame front into NO and oxygen. Only oxygen sustains combustion, nitric oxide behaving as an inert diluent. The reason for the inability to form a second flame zone for the reaction of nitric oxide with part of the hydrogen seems to be mainly the great speed of the first zone. so that a second zone would continuously blow-off. The H_2/NO_2 flame, if plotted into Fig. 5.8, falls on to the line representing hydrogen/oxygen/nitrogen systems.

Carbon monoxide will burn with NO_2 and the light emission seems to be similar to that of the H_2/NO_2 flame. The reaction proceeds only to nitric oxide,

$$CO + NO_2 \rightarrow CO_2 + NO.$$

The burning velocity is very small and is affected by moisture. Formaldehyde behaves in all respects as a mixture of H_2 and CO; flame speeds are large (Pollard & Wyatt, 1950). Ammonia cannot be burned easily as a premixed flame with nitrogen dioxide because a white deposit forms at once at the point where the gases meet. Carbon disulphide is miscible with nitrogen dioxide and is a powerful explosive when ignited.

Hydrocarbon/nitrogen dioxide flames. Methane and ethane burn more readily with nitrogen dioxide than with nitric oxide, although the burning velocities are practically the same (about 20 cm/sec). For fuel-rich and stoichiometric mixtures, the flame clearly has two reaction zones. A picture of such a flame can be seen in Plate 17a. The first zone emits a characteristic yellow-orange light which is identical with the light emitted by the H_2/NO_2 flame. In this zone part of the hydrocarbon reacts with oxygen from the decomposed NO_2. Strong NO bands can be seen in absorption between the first and second zone (Plate 19c). This second zone resembles in every detail a hydrocarbon/nitric oxide flame, with CH, CN, NH, OH and $NO\gamma$ emission. The formation of the second zone does not depend on the presence of secondary air. In fuel-lean flames the first (yellow-orange) and second (purple) zones close up and tend to merge; however, a third reaction zone appears above the first two where excess NO decomposes, as was the case in nitric oxide flames. These NO_2 flames have been little explored. It is possible to extinguish the second zone by momentarily introducing a wire gauze into the flame; the first zone then continues to burn but with reduced velocity, obviously leading to partial combustion only. Burning velocities of about 20 cm/sec for butane/NO_2 mixtures have been reported by Miller & Setzer (1957), and stability regions for a burner flame have been mapped. Wharton et al. (1957) have measured emission spectra across a butane/NO_2 reaction zone.

Myerson et al. (1957) investigated the spontaneous ignition of propane/nitrogen dioxide mixtures. They find that at 400°C and with less than 74% NO_2 (propane-rich), only a pale orange cool flame can be initiated. This of course corresponds to the first reaction zone of the propagating flame. At higher NO_2 concentrations, a whitish blue flame follows the orange flame after a short time interval. The similarity to the propagating flame seems therefore to be complete. The intensity of the first orange flame depends on mixture strength up to about 74% NO_2, when the second stage sets in; for higher concentrations, the intensity of the first flame remains constant. Analyses have been made of the product gases for a variety of NO_2 concentrations, i.e. of one- or two-stage flames of various relative intensities. The cool flame involves considerable reaction, methane, ethylene, HCN, NO and CO being found when the second stage is fully developed but not when it begins to form. Spontaneous ignition in alcohols also reveals a two-stage character (Gray & Yoffe, 1950). Pentane/nitrogen dioxide mixtures at 332°C show only blue luminescence at a pressure of 12.5 cmHg. At 14.2 cmHg, a white flash and explosion are observed (Yoffe, 1953). Yoffe also reports that all hydrocarbons from ethane to pentane show two stages in ignition, the first stage being a blue luminescence. This contradicts the results of Myerson, who found that the first stage for butane emitted orange radiation. This point needs clearing up, as blue luminescence would imply emission of formaldehyde bands, whereas the first stage of a propagating flame emits continuous radiation. In general, it seems that the spontaneous ignition temperatures of fuel/nitrogen dioxide mixtures are lower than for the corresponding oxygen mixtures.

The properties of NO and NO_2 flames have been discussed rather fully here. This is necessary as they are fundamental to the reaction of organic nitrogen compounds. Gray & Yoffe (1955) present useful thermodynamic data for nitrogen dioxide.

Nitric acid flames

Nitric acid flames differ in many ways from nitrogen dioxide flames. Propane-rich flames at 1 atm, when surrounded by nitrogen, show only one reaction zone of magenta colour. In air, outer diffusion flames form; these are themselves divided into a doublet, the first flame being yellow, the second blue (Boyer & Friebertshauser, 1957). For acid-rich flames, the primary reaction zone splits up into two regions, the first being orange, the second blue. The relative intensity of these regions depends on mixture strength. In none of these flames is combustion complete; at best 22% of the nitrogen is converted to N_2, the rest remaining as undecomposed NO. The reason for the failure of NO to decompose in acid flames, in contrast to the more complete reduction in NO_2 flames, is the reduced adiabatic flame temperature of the acid flame. One would therefore expect a more complete NO reduction in acid flames under high pressure conditions, such as pertain in rocket motors or upon preheating the unburned mixture. Flame speeds are higher in acid flames than in NO_2 flames; e.g. propane/nitric acid mixtures have a burning velocity of about 100 cm/sec (Mertens & Potter, 1958). These authors found also that C_2, CH, CN and NH bands are emitted from acid-rich propane flames. Interestingly enough, OH bands are emitted only when secondary air from a jacket assists combustion.

Wayman & Potter (1957) and Potter & Wayman (1958) found that hydrocarbon/nitric acid mixtures are able to detonate. At the high pressures of detonation a more complete reduction of NO is indicated.

Nitrate and nitrite flames

The complicated nature of nitrate combustion is best illustrated by the behaviour of ethyl nitrate, $C_2H_5ONO_2$. Pure ethyl nitrate can be made to burn as a decomposition flame. At low pressure the reaction zone has only one stage of grey-blue colour and is only visible in a dark room (Hall & Wolfhard, 1957). Upon premixing air with this flame, no change in structure or burning velocity (7 cm/sec) occurs and the flame extinguishes before a stoichiometric mixture is reached. However, if a rich ethyl nitrate/air mixture is sparked, a flame with two reaction zones and increased burning velocity can be obtained. The first zone is blue and emits CH_2O bands, whereas the second zone is yellow-orange and emits continuous radiation similar to the first stage of NO_2 flames. If further air is added to this two-stage flame, a sudden increase in burning velocity occurs and a third purple-blue reaction zone forms; this emits C_2, CH, NH and OH bands.

The pure decomposition flame at low pressure leads to incomplete combustion with most of the nitrogen being present as NO. The flame temperature is only about 500°C (Wolfhard, 1955). At 1 atm, the decomposition flame consists of a thin red first reaction zone, followed after several millimetres by a diffuse orange glow which slowly goes over into a faint green afterglow (Needham & Powling, 1955). Thermocouple measurements reveal that the temperature increases from the red to the orange zone, where it reaches about 800°C. Further temperature increases occur within the orange-glow region. Major products of the first flame zone are CO, NO, H_2O and CH_2O, all present in roughly equimolar proportions. Subsequently in the orange glow, the main chemical reactions are a decrease of CH_2O, with a corresponding increase in H_2 and CO, and a small decrease in NO concentration, coupled with the formation of some CO_2 and N_2. The mode of decomposition of this small amount of NO in the flame remains unexplained, but it is vital for the existence of the flame, as without NO reduction no heat would be liberated by decomposition of $C_2H_5ONO_2$.

A similar problem arises with methyl nitrite, CH_3ONO. The existence of the decomposition flame of this compound was discovered by Gray, Hall & Wolfhard (1955a). The single orange-reaction zone is just visible; the burning velocity is about 3 cm/sec at 1 atm. The flame products were analysed by Gray & Pratt (1957) and Arden & Powling (1957). Within the reaction zone, CH_3ONO and CH_2O disappear. About 50% NO is reduced and approximately 1 mole of CO and H_2 and 0·5 mole of H_2O are formed, besides 0·25 mole N_2. The mode of NO reduction is again the main difficulty. Arden & Powling suggest that H atoms reduce NO according to: $H + NO \rightarrow HNO$; $2HNO \rightarrow N_2O + H_2O$, etc. Gray, however, believes that the methoxy radical CH_3O is responsible for the NO reduction.

Methyl nitrate, CH_3ONO_2, at low pressures, forms a decomposition flame which has two zones (Gray, Hall & Wolfhard, 1955b). The first is bright blue due to formaldehyde emission, the second gives continuous radiation and has an orange colour. Apart from formaldehyde, no bands are emitted, not even OH. This indicates again that only partial combustion occurs and that NO is not reduced. The heat evolved would be due to

$$CH_3ONO_2 \rightarrow NO + H_2O + \tfrac{1}{2} H_2 + CO + 33 \text{ kcal/mole}$$

whereas complete combustion would yield

$$CH_3ONO_2 \rightarrow \tfrac{1}{2} N_2 + H_2O + \tfrac{1}{2} H_2 + CO_2 + 122 \text{ kcal/mole}.$$

As the first two zones have a high-burning velocity, a third zone seems unable to establish itself as in the case of the H_2/NO_2 flame (*see above*). Higher flame temperatures are attained in explosion flames in closed vessels due to constant volume combustion. This speeds up the thermal decomposition of NO. Under such conditions, it was found by Zeldovich & Shaulov (1946) and Adams & Scrivener (1955) that a second whitish-blue flame follows a low-intensity flame and complete reduction of NO is achieved. The spectrum of this explosion flame (Hall & Wolfhard, 1957) reveals OH bands and a strong continuum. In contrast to flames of organic fuels with NO, no CH, CN or NH bands are emitted. This favours the view that the first two stages of CH_3ONO_2 decomposition do indeed lead to NO, H_2O $\dot{H}_2$ and CO or possibly CH_2O, but not to complicated organic fragments as in flames of CH_3ONO or $C_2H_5ONO_2$/oxygen. Thus, the second whitish-blue flame corresponds to a $H_2/CO/NO$ reaction which also explains its reluctance to form.

Spectra of nitrate and nitrite/oxygen mixtures also reveal important differences. The three-zone structure of the ethyl nitrate/air flame has already been mentioned. The last stage emits CN and NH bands, indicating reaction of organic fuel fragments with NO. Methyl nitrite is very similar to ethyl nitrate in this respect. However nitro-methane, CH_3NO_2, which has practically the same heat of decomposition as CH_3ONO, is very different. It has no decomposition flame at 1 atm and in oxygen mixtures it does not show the first stage, which emits formaldehyde bands. When the last stage is properly initiated, methyl nitrate/oxygen mixtures emit only OH bands, indicating early consumption of organic fragments and reaction of NO with H_2 and CO only. Plate 19 shows some characteristic spectra.

Dinitrates again show very specific behaviour (Powling & Smith, 1958). The burning velocity of the gaseous flame burning on a liquid butane-2-3-idiol dinitrate ($CH_3 - \overset{ONO_2}{CH} - \overset{}{\underset{ONO_2}{CH}} - CH_3$) surface is lower than the burning velocity of ethyl nitrate, despite a higher heat of explosion. Products formed just before the visible flame front are NO_2 and acetaldehyde. The very first flame reaction thus resembles the slow decomposition at low temperature. Further combustion in the visible flame front and behind is therefore mainly due to fuel/NO_2 reactions, in contrast to the reactions of the mononitrates. Steinberger & Schaaf (1958), analysing the temperature profile of an ethylene glycol dinitrate ($C_2H_4(ONO_2)_2$) flame, find heat-releasing reactions before and after the visible flame front, but little heat release within it.

Propellant flames

The burning rates of a great number of liquid nitrates, fuel-nitric acid mixtures, or fuel-tetranitromethane mixtures have been determined by measuring the rate of recession of a liquid meniscus in a capillary. Such measurements are often subject to quenching or turbulent burning and are therefore difficult to evaluate, especially for effects of pressure on burning velocity. They will not be discussed here.* Similarly, the development and propagation of explosions in solids is not discussed here, but the reader may be referred to the book by Bowden & Yoffe (1958).

Double-base solid propellants are usually mixtures of nitroglycerin and nitrocellulose and minor components. For experimental purposes they are burned in so-called strands, i.e. rods of 5 to 10 mm diameter. The flame consumes only the end surface as in a cigarette. The surface decomposes on burning (foam zone) and gas escapes forming a dark space above the surface. At high pressures a luminous second flame forms, its distance from the surface depending on pressure. Heller & Gordon (1955) measured temperature profiles and found a steep temperature gradient close to the surface. In the absence of a secondary flame, the temperature in the dark space is constant, but increases at increasing pressures, until the secondary flame forms. The gases in the dark space consist mainly of CO and NO, with lesser amounts of CO_2 and H_2. Little nitrogen is present as N_2. The spontaneous ignition of the secondary flame is difficult to understand as the available temperatures are only about 1100°C. The propellant flame has been investigated spectroscopically by Rekers & Villars (1956). The spectrum is determined mainly by impurities such as CaOH bands. C_2 bands are claimed to be present. The second flame spectrum is mainly a continuum. No CH or OH bands could be found. NO_2 could not be detected by absorption in the dark space.

Composite solid propellants often have ammonium perchlorate as oxidiser and this type of oxidiser has been investigated as to its behaviour in flames. The active oxidiser is a chlorine-oxygen molecule. ClO_2 is an endothermic compound that decomposes into chlorine and oxygen liberating 25 kcal/mole; thus a decomposition flame can be observed, and burning velocities have been measured by Laffitte et al. (1967). The theoretical flame temperature is about 1670 K and the burning velocity is 315 cm/sec. Comparing these data with those plotted in Fig. 5.7, one can see that the ClO_2 decomposition flame is the most reactive flame so far encountered, with an overall activation energy of only about 8 kcal/mole. The decomposition flame of perchloric acid is described as pale fawn (Combourieu et al., 1969) and it has a burning velocity of only 12 cm/sec, referred to the unburnt gas at 210°C; the activation energy is 25 kcal/mole.

In an effort to understand reactions of chlorine dioxide and of perchloric acid ($HClO_4$), flames of hydrocarbons and other fuels supported by these oxidants have been studied (Cummings & Hall, 1965; Heath &

* For details, see Behrens (1951c; 1952b) and Whittaker et al. (1956; 1957).

Pearson, 1967; Combourieu & Moreau, 1969; Hall & Pearson, 1969; Combourieu *et al.*, 1969). Premixed flames have mostly been studied at reduced pressure on special burners; for flames with perchloric acid at ambient pressure it is necessary to heat the whole burner assembly. For ClO_2/CH_4 the burning velocity has the high value of 620 cm/sec and the activation energy is only 10 kcal/mole. The flame is complex, under some conditions having three zones, a blue inner cone with a 'white-pinkish' disc above, surrounded by a bright green plume. The spectrum shows strong cool flame bands (CH_2O) in the blue cone; C_2 and CH are strongest in the disc zone and OH extends throughout the flame; the green zone has a mainly continuous spectrum. Flames of fuels burning with perchloric acid again have a high burning velocity, up to 4 m/sec for methane, and again flames tend to be complex with multiple zones, especially for rich mixtures or when diluted with an inert gas. For methane the cool-flame bands are strong in the base or lowest zone, with OH and some rather weak C_2 and CH higher up. Methyl alcohol–perchloric acid shows strong cool-flame bands, but ethane emits the hydrocarbon-flame bands (HCO). For perchloric acid–methane flames, mass-spectrometric sampling showed a relatively high rate of CO_2 formation in the very early stages and the consumption of methane was controlled by reactions of ClO radicals. In general, it seems that reactions of ClO radicals and O atoms are important in the subsequent reaction processes of both ClO_2 and perchloric acid. It was not possible to study premixed flames of perchloric acid with ammonia because solid ammonium perchlorate was formed as soon as the gases mixed.

Ammonium perchlorate is a solid; it is endothermic and will therefore form a decomposition flame. When burning strands of NH_4ClO_4 it was found that the decomposition flame had a lower as well as an upper pressure limit (Levy & Friedman, 1962). The former is 22 atm, whereas the upper limit depends greatly on the experimental conditions and may be due to convective heat losses. The low pressure limit of 22 atm is very surprising considering the ease with which fuels burn with $HClO_4$. One has, however, to remember that NH_4ClO_4 is a solid. A decomposition flame may have the following structure: the solid is vaporised, on heating, to ammonia and $HClO_4$. The two gases then react to form a flame as discussed in the previous paragraph. This flame above the solid strand will conduct heat to the solid to ensure the continued vaporisation of the solid. The distance between flame and solid is therefore critical as it determines the heat flux to the solid. Thus the low pressure limit is presumably due to a failure to supply sufficient gases by vaporisation of the solid, rather than a decrease in reaction rate in the gaseous flame itself. This point is also strengthened by the fact that the low pressure limit can be extended to one atmosphere by heating the burning surface of the strand with a focused light beam.

Composite solid propellants are a mixture of ammonium perchlorate and a fuel binder. The burning properties of these propellants depend greatly on the fuel and oxidiser ratio as well as the grain size of the ingredients and the additives. An extensive literature exists on the burning behaviour of composite propellants and the reader is referred to the papers that have appeared in the Symposia on Combustion during the last few years.

The burning rates, measured as regression of the solid propellant strand (usually around 0·5 to 3·0 cm/sec) are generally larger for the composite propellant than for NH_4ClO_4 alone. This is easily understood as flame temperatures are higher and therefore the heat flux to the solid surface is increased. This also has the effect that the low pressure limit for propellant burning is reduced. At high pressure the burning rate may increase with pressure, may be constant, or may even have a negative pressure exponent depending on the fuel binder being used (Adams et al., 1962) and no theory so far has been able to account for all of these facts.

Halogens as oxidisers

Fluorine and chlorine trifluoride are very reactive oxidisers, leading generally to much higher flame temperatures than oxygen. Other halogens are less interesting from a propulsion standpoint; however, as some propellant combinations contain chlorine, the effect of chlorine on flames will be discussed briefly.

Fluorine. At room temperature hydrogen and fluorine usually explode upon premixing. This seems to be due to surface reactions and to the presence of small amounts of HF and O_2 in fluorine. Burning velocity measurements of stoichiometric mixtures of fluorine with hydrogen or fuel are not available therefore. Grosse & Kirshenbaum (1955) however, were able to run premixed burner flames up to 33% fluorine by volume, if the gases were precooled to 90 K, and in this range the burning velocity was ten times greater than that of a hydrogen/oxygen flame. However, their inference that the maximum burning velocity would therefore be 35 m/sec precooled, and about 100 m/sec at room temperature, is misleading as the H_2/F_2 and H_2/O_2 burning velocities will attain their maxima at different mixture strengths. The reaction zone is a pale pink colour.

Diffusion flames of a wide variety of fuels burning with fluorine were first investigated by Durie (1952a). Skirrow & Wolfhard (1955) used a flat diffusion-flame technique to study flame structures in detail; for reasons explained below they used mainly chlorine trifluoride. The system hydrogen, carbon, nitrogen, fluorine and chlorine is of interest as radicals such as C_2 and CN are quite stable at high temperatures. In the presence of even small amounts of oxygen, the equilibrium would shift towards a higher concentration of CO, and light emission would be expected to change radically.

Despite the very high temperature of the burned gases (about 4000 K), the hydrogen/chlorine trifluoride diffusion flame emits only a weak blue continuous radiation, due to the lack of suitable emitters. Small amounts of hydrocarbons mixed with the hydrogen gives rise to a dazzlingly bright flame which clearly has two zones. On the hydrogen plus hydrocarbon side, there is continuous radiation from carbon particles, as is to be expected by analogy with hydrocarbon/oxygen diffusion flames (*see* Chapter VI). On the oxidiser side, the chlorine trifluoride flame emits very strong C_2 and CN bands (nitrogen from impurities), but no CH in the 'main reaction zone', in contrast to oxygen flames which show blue continuous radiation in the 'main reaction zone', with very weak C_2 and CH bands between the carbon and 'main reaction zone'. If the hydrogen in the above experiment is gradually replaced by a hydrocarbon, such as ethane, the C_2 and CN bands first increase in intensity but then become weaker; their position closes up and partially overlaps with the carbon zone. Pure ethane/chlorine trifluoride flames emit only continuous carbon radiation; the flame is sluggish and carbon growth clogs the burner within seconds. Three spectra (Plate 20*a*, *b*, *c*) show the changes occurring during such a replacement of hydrogen by a hydrocarbon. The colour of the carbon radiation suggests temperatures of the order of 2000 K rather than of 4000 K. We do not yet understand all the reasons for this exceptional behaviour; however, the high density of carbon particles in such flames may well introduce important changes as carbon particles are slow to react with fluorine. Carbon formation in a methane/chlorine trifluoride diffusion flame is much more copious than, for example, in acetylene or benzene/air flames.

Traces of O_2, even in very small amounts, have a marked influence on fluorine flames. The reason for this has been discussed. If we burn a mixed hydrogen/methane fuel with chlorine trifluoride we have, as already explained, very strong C_2 and CN radiation separated from the carbon zone. Introduction of a small amount of oxygen into ClF_3 can reduce the light intensity of the flame by as much as a hundredfold. The hottest zone (previously emitting C_2 and CN) now emits blue continuous radiation and OH bands. Between this zone and the carbon zone, C_2 and CN bands remain but are very reduced in intensity. Oxygen or OH radicals obviously consume C_2 and CN, and only those close to the carbon zone escape reaction. It is interesting to note that CH radicals are now present, as well as C_2 and CN, this supporting the view that C_2 reacts with OH to produce CH and CO. The sensitivity of fluorine flames to oxygen is of course the reason why chlorine trifluoride is preferable to fluorine in experimental investigations, as all commercially available fluorine contains traces of oxygen. Plate 20*d* and *e* shows a hydrogen (plus hydrocarbon)/chlorine trifluoride flame, with and without a small amount of oxygen in the oxidiser; note that the exposure time is the same in both spectra.

Carbon monoxide alone, or mixed with hydrogen, burns with a blue-grey flame. No C_2 bands are emitted, which seems to indicate that CO is not dissociated prior to reaction with fluorine; COF_2 may be an intermediate. Water vapour reacts vigorously with fluorine and the flame emits selectively the Schumann-Runge bands of oxygen. The explosive reaction of chlorine trifluoride with hydrocarbons has been investigated by Baddiel & Cullis (1962).

Chlorine. The temperature of flames with chlorine is quite low as compared with those with fluorine. Nevertheless, all hydrocarbons burn readily with chlorine as premixed and diffusion flames. Both types give abundant carbon formation, which seems to be quantitative under most conditions, i.e. all carbon appears as soot. Hydrogen is the most reactive fuel. Bartholomé (1949) found that H_2/Cl_2 mixtures have a maximum burning velocity on the fuel-rich side of 400 cm/sec. It is interesting to note that chlorine, in contrast to bromine, does not inhibit hydrogen/air flames (Simmons & Wolfhard, 1957*a*). In fact the limits of flammability of the H_2/Cl_2 air system follow Le Chatelier's rule. Carbon monoxide does not burn with chlorine, although this would be feasible, as the formation of phosgene is very exothermic; this is probably due to the strong inhibition of the carbon monoxide/air flame by chlorine.

Premixed hydrocarbon/chlorine flames are exceptional for many reasons. Despite very small burning velocities (ethane/Cl_2 burns at about 5 cm/sec), the flames can be held on small burners and are extremely stable. Plate 17*e* shows an ethane/chlorine flame. Soot formation has already been mentioned. Strong carbon radiation can be used to make detailed temperature measurements in the flame front. Heat losses by radiation are of overriding importance. Thus, for example, a stoichiometric ethylene/chlorine flame reaches a maximum temperature of 1000°C, whereas the adiabatic flame temperature is close to 2000°C. Flame radiation leads not only to heat losses, but also to irradiation of the premixed gases. For large flames this irradiation could trigger ignition away from the front, leading to uncontrolled flame propagation. This is suggested by a simple experiment in which a flashlight is switched on close to a burning ethane/chlorine flame, causing it to flash-back momentarily. Acetylene can be burned with chlorine under proper precautions, especially exclusion of air. Acetylene and chlorine are self-igniting at room temperature in the presence of oxygen. This can be demonstrated with an acetylene/air mixture flowing from a burner; a small jet of chlorine ignites the mixture. The spectroscopy of chlorine flames will not be discussed here; details are given by Simmons & Wolfhard (1957b).

Ozone

Ozone is an endothermic compound ($\Delta H = 34 \cdot 0$ kcal) and, when used as an oxidiser, it leads to higher flame temperatures than does oxygen. This has to be balanced against the instability which leads to great hazards. It has been found that gaseous O_3/O_2 mixtures can detonate for ozone concentrations down to 10% by volume. Liquid ozone has a detonation velocity of over 5000 m/sec (Streng et al., 1958). Important details on the handling of ozone can be found in Thorp (1955). The decomposition flames of pure and oxygen-diluted ozone were investigated by Streng & Grosse (1957); the flames are pale blue and only readily visible for concentrations greater than 50%. Burning velocities are 475 cm/sec for pure gaseous ozone, dropping to 9 cm/sec for 17 mole per cent O_3 in O_2. Diffusion flames of hydrogen or carbon monoxide with ozone/oxygen mixtures resemble the corresponding oxygen flame.

Metal alkyl flames

These flames are of interest as they allow a study of metal combustion under premixed gaseous conditions. Trimethylaluminium can be premixed with oxygen at low pressure and a flame of multiple reaction zone structure is observed (Vanpee & Seamans, 1967; also Vanpee, et al. 1965). A flame at 1·6 torr pressure has a first zone of yellow-green colour followed by a dark zone, after which the reaction zone gets bluish (second zone) and this turns finally into red. The first zone emits OH bands and a continuum with a maximum at around 4750 Å, a second continuum seems to peak at 1 to 2 μ in the infra-red. No AlO, CH or C_2 bands are emitted in this zone. The second luminous zone emits Al lines, AlO, C_2 and CH bands as well as the afore-mentioned continua. The reactions are therefore extremely complex. The continuum in the blue may be the same as observed in night-time trimethylaluminium releases from rockets into the upper atmosphere, whereas the continuum in the red and infra-red is due to the radiation from solid Al_2O_3 particles. The emission of OH in the first zone is largely unexplained.

The spontaneous ignition of metal alkyls in air at somewhat higher pressures is of importance in air breathing engines and supersonic combustion (*see* page 399); the ignition behaviour has been studied by Marsel & Kramer (1959).

High-energy fuels

Hydrazine. Hydrazine (N_2H_4) burns readily with oxygen, with nitric oxide, or even as a pure decomposition flame. The decomposition flame is interesting in so far as it is an example of a flagrant violation of equilibrium. The heat of decomposition into H_2 and N_2 is $\Delta H = 26$ kcal and this leads to an adiabatic flame temperature of about 1250°C. Murray & Hall (1951) found that the actual flame temperature was higher and that the products of combustion were not $N_2 + 2H_2$, but $\frac{1}{2} N_2 + \frac{1}{2} H_2 + NH_3$. In equilibrium, ammonia should decompose; its formation and persistence therefore account for extra heat release according to

$$N_2H_4 \rightarrow \tfrac{1}{2} H_2 + \tfrac{1}{2} N_2 + NH_3 \qquad \Delta H = 36 \text{ kcal.}$$

The hydrazine decomposition flame has been extensively investigated as it promised to be a simple model for flame propagation theories, assuming a first-order law. However, this assumption has not been experimentally verified as the burning velocity was found to be independent of pressure (Hall & Wolfhard, 1956 and Gray, Lee et al., 1957). We must therefore assume that the fission of N_2H_4 into $2NH_2$ is not rate-controlling. The burning velocity of 110 cm/sec is very high even for an adiabatic flame temperature of about 1650°C. Plotting this point in Fig. 5.8 brings it nearly on to the $H_2/O_2/N_2$ line.

Alone and mixed with oxygen, the hydrazine flame has an interesting structure. The pure decomposition flame at low pressure has a yellow-brown reaction zone with an afterglow. NH_2 and NH bands are emitted in both zones. When oxygen is added to the hydrazine, the afterglow increases in intensity and thickness, indicating that, at least for small oxygen additions, oxidation is slower than decomposition, giving the impression of a two-stage mechanism. For larger oxygen additions, OH and NO radiation begins at the first visible reaction zone, with some overlapping of oxidation and self-decomposition. Gray & Lee (1959) measured burning velocities for N_2H_4/O_2, N_2H_4/N_2O and N_2H_4/NO mixtures in a constant-volume apparatus. The oxygen flame has a maximum burning velocity of 280 cm/sec and an adiabatic flame temperature of 2700 K. There is a considerable rise in pressure, as an increase in mole volume occurs. It should be noted that the temperature and especially the burning velocity are much lower than for a hydrogen/oxygen flame.

The flame with nitrous oxide is remarkable as its maximum burning velocity is only about 160 cm/sec, which is not much higher than that of the pure N_2H_4 decomposition flame (at 110 cm/sec), although the latter has a flame temperature of ~2650 K compared with 1900 K for the decomposition flame. The ease with which hydrazine decomposes NO (*see under* nitric oxide flames) has already been discussed. It also shows up in the burning velocity measurements. The maximum burning velocity of a $N_2H_4/$NO flame occurs on the fuel-rich side and is 245 cm/sec, which is close to that of an oxygen flame; the flame temperature is 2740 K.

Alkyl hydrazines have been extensively studied as possible fuels. Harshman (1957) has summarised their physical and chemical properties. They retain some of the desirable properties of hydrazine, such as spontaneous ignition with some oxidisers.

Boron hydrides. Due to the high heat of combustion of boron, the combustion of boranes leads to very high flame temperatures. The heat of formation of B_2O_3 has been determined as 302 kcal (Eckstein & van Artsdalen, 1958); diborane is endothermic by 7·5 kcal.

The spontaneous ignition temperatures of diborane (B_2H_6) and pentaborane (B_5H_9) have been investigated by Whatley & Pease (1954), Roth & Bauer (1955), and Baden, Bauer & Wiberley (1958). Both compounds exhibit upper and lower pressure limits. Ignition temperatures are very low; at low pressures the peninsula occurs below 200°C for diborane, and below room temperature for pentaborane. At 1 atm only diborane can be premixed with air without spontaneous ignition.

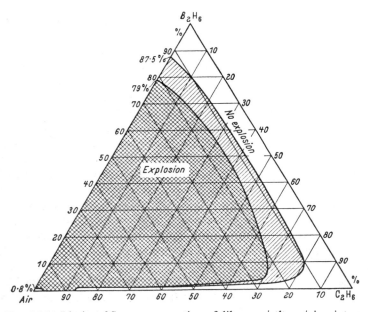

Fig. 14.1. Limits of flame propagation of diborane/ethane/air mixtures.

Fig. 14.1 shows the limits of flame propagation in the B_2H_6/C_2H_6/air system (Parker & Wolfhard, 1956). The flame temperature is only about 500°C at the lean and rich limits for diborane/air combustion; this is far below that of hydrocarbons, and even below hydrogen. The rich limit does not follow Le Chatelier's rule; a mixture consisting of 80% C_2H_6, 10% B_2H_6 and 10% air is flammable. This can only be explained if diborane reacts preferentially with air, with ethane acting largely as diluent. Otherwise no heat would be evolved. The lightly shaded area of Fig. 14.1 is, in fact, the region of a hardly-visible flame behind which white oxide smoke can be seen. The burning velocity of diborane/air flames was measured on 0·4 mm burners. It reaches a maximum of 5 m/sec for near stoichiometric mixtures. This value is high for an air flame, but is not anomalous for a flame temperature of about 2600°C (*compare* Fig. 5.8).

The reaction zone of the diborane/air flame emits the α bands of BO, the BH band at 4332 Å and the doublet of atomic boron near 2500 Å. The burned gases emit the very intense fluctuation bands now attributed to BO_2. No OH bands could be found. Berl *et al.* (1956) burned diffusion flames of pentaborane/air so highly diluted with nitrogen that the maximum flame temperature was below 500 K; only BO and weak BO_2 bands were observed under these conditions.

Berl *et al.*, (1957) succeeded in measuring burning velocities of pentaborane/air flames. To prevent self-ignition upon premixing the gases, pentaborane was first diluted with nitrogen and then oxygen was rapidly admitted. Spherical flames, produced by spark ignition, were allowed to propagate in a free premixed gas stream. The burning velocity could thus be deduced. As for diborane, the maximum burning velocity was again close to 5 m/sec. An increase of oxygen from 21% to 26% doubled the burning velocity. Other fuels are not as sensitive to the amount of oxygen. For example, hydrogen increases its burning velocity in the above range by only 20%.

Interesting flame structures can be seen in premixed hydrocarbon/ diborane/air flames. When diborane is added to a burning hydrocarbon/ air flame, a visible reaction zone forms below the existing one; its strength depends on the amount of diborane added (Berl & Dembrow, 1952). The second stage, i.e. the hydrocarbon reaction zone, can be extinguished and partial combustion occurs with the first zone burning alone. These flames have been investigated in detail by Breisacher, Dembrow & Berl (1959). To get better separation, the flames were run at low pressure (110 torr) on a sintered disc. Fig. 14.2 shows a stability diagram for a diborane/methane/ air flame. In region (A), only one reaction zone is present, whereas (B) supports two. In (C), one or two reactions can be observed depending on the ignition source. (LBO denotes lean blow-off.) Hydrocarbons are partially decomposed in the first (diborane) reaction zone, but the diborane is not completely consumed, as hydrogen reacts much more slowly than boron. In flames of mixed hydrocarbon and diborane the formation of a second zone depends on whether the products of the first zone are hot

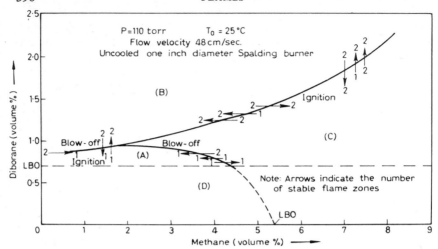

Fig. 14.2. Stability diagram of diborane/methane/air system.
[*After Breisacher, Dembrow & Berl* (1959)]

enough to ignite the second zone. Similar interdependence of reaction zones was observed by Wolfhard (1955) for ethyl nitrate.

Diborane/nitric oxide mixtures burn very violently (Roth, 1958). Flame temperatures are so high that it is immaterial whether NO decomposes thermally or by radical reactions. It may be remarked here that high-energy fuels are not only desirable for their high flame temperature and heat release, but also for their wide limits of flammability and small quenching diameters. Boranes are unsurpassed in this respect.

Other high-energy fuels. Cyanogen (C_2N_2) is not a high-energy fuel in the proper sense; however, the products of its combustion, according to the equation $C_2N_2 + O_2 = 2CO + N_2$, are so stable that little dissociation occurs and flame temperatures are very high. Using the vibrational levels of CN, Thomas, Gaydon & Brewer, (1952) found a flame temperature of about 4800 K. Conway *et al.* (1955) were able to maintain this flame at 100 atm, producing temperatures above 5000 K. Carbon subnitride (C_4N_2) is even more extreme than cyanogen, and flame temperatures up to 5260 K are possible at 1 atm.

Acetylene flames differ very little from other hydrocarbon flames, except that they are hotter due to the endothermic nature of C_2H_2. Acetylene is, however, of interest because of its ability to form a decomposition flame. The explosive decomposition of gaseous acetylene has only been observed at pressures slightly above atmospheric (Jones *et al.*, 1944). This pressure limit is mysterious, as for all other flames it has been found that lower burning pressures can be achieved by using larger burners and flame dimensions. For acetylene there seems to exist an absolute pressure limit. It is even more surprising that pure acetylene can be detonated well below 1 atm (Duff, Knight & Wright, 1954). The luminous reaction zone follows 9 cm behind the shock front. This seems to be the first example where detonation limits are wider than flame propagation limits. We venture to offer an explanation of this anomalous behaviour: the flame of acetylene, ignited with a fusing wire in a long closed tube of about 5 cm diameter, propagates very slowly, the burning velocity being 10 to 20 cm/sec. The reaction zone is several millimetres thick, so that carbon particles radiate strongly and rapidly cool down, leaving only dark black carbon visible. This is exactly the situation as in hydrocarbon/chlorine flames (Simmons & Wolfhard, 1957a) where the carbon appears quantitatively as carbon particles. It has been shown for hydrocarbon/chlorine flames that the radiation is black-body (i.e. the emissivity is unity), and that the radiation losses are so great that the temperature of the flame is 1000°C instead of 2000°C. The temperature of an acetylene decomposition flame has not been measured, but a visual estimate based on brightness and colour would put it at less than 1500°C. This again is far removed from the adiabatic flame temperature, which is close to 2800°C for the reaction $C_2H_{2\,gas} \rightarrow 2C_{solid} + H_2$. The flame zone of the acetylene flame is therefore excessively cooled by radiation. For a given volume of flame zone, the relative heat loss is larger at lower pressure, because less chemical heat is produced. The low-pressure limit is therefore the pressure at which radiation losses cool the flame front to such an extent that propagation ceases. It can easily be seen that enlarging the flame size does not help to reduce radiation losses. However, detonation is still possible, as chemical heat release is speeded up. It is interesting to note that the adiabatic flame temperature of pure acetylene is much higher than that of an acetylene/air flame.

Combustion of solid fuels

The combustion of solid fuel particles or metal foil was the subject of some early investigations because the high light yield associated with this type of combustion made it useful as a light source for photography. We are more concerned here with the fact that the heat of formation of many oxides from solid materials such as aluminium, beryllium, carbon, boron, magnesium, etc., is very large, i.e. combustion is vigorous and leads to high flame temperatures. It is not usually recognised that even the combustion of solid carbon with oxygen to form carbon monoxide leads to a flame temperature of about 3200°C. In many types of propulsion unit the specific impulse is important and the lighter fuels, with a higher heat release per gram, may be favoured.

Solid carbon burns easily in air due to its low thermal conductivity. Bulk aluminium conducts heat away at a rate such that the burning surface is usually cooled below the ignition temperature and flame cannot be sustained. However, small aluminium particles burn rapidly. The temperature of metal-powder flames burning in air usually coincides with the boiling-point of the metal oxide. This was observed by Westermann (1930) for lead, tin, zinc, cadmium, bismuth and probably antimony and copper. It is also true for aluminium and magnesium (Wolfhard & Parker, 1949a), but not for boron. The heat of vaporisation of most metal oxides is so large that the boiling point puts a practical limit on flame temperature. In the case of aluminium and magnesium, the burned product therefore consists of partially vaporised oxides together with condensed oxides present as tiny liquid spheres. Upon cooling, these become small solid oxide particles. The size of these particles is below the wavelength of light, which accounts for the blue tint of the smoke from such flames.

The combustion of solids often depends on the rate of inward diffusion of oxygen to the hot surface and the outward diffusion of products; the boiling points of the solid and its oxide are important. For carbon, which is non-volatile but has a gaseous oxide, the diffusion of oxygen is rate-determining at high temperature (*see* page 233). For more volatile solids like magnesium the metal may vaporise and diffuse outwards to meet the incoming oxygen so that the situation, e.g. for magnesium dust, is rather similar to that of burning oil droplets, each particle supporting a surrounding diffusion flame. In the burning of bulk material, however, accumulation of a non-volatile oxide may impede further burning. Formation of solid protective films of oxide, formed by slow oxidation processes, may also impede combustion; these films may break up if either the solid itself or the oxide melts.

In discussing the flame behaviour of various solids it is useful to have some important characteristics tabulated. Table 14.1 shows ignition temperatures and energies (in millijoules) of some solid materials in air or oxygen, both for the bulk and the powder form. Table 14.2 gives the heat of formation of the oxide as well as its melting and boiling points.

Ignition and flammability of solids

Solid fuels seem to ignite most readily when in dust layers, as any heat generated by slow oxidation is not dissipated. Dust suspended in air and flowing through a furnace of predetermined temperature has a well-defined ignition temperature. Zirconium clouds may ignite at room temperature; however there is a suspicion that ignition occurs by electrostatic discharge. The actual ignition temperature of a dust depends of course on parameters such as dust concentration, oxygen index of the atmospere, particle size

TABLE 14.1

Element	Melting point, °C	Boiling point, °C	Spontaneous ignition of dust in air, °C (a)	Ignition of bulk material in O₂, °C (b)	Minimum ignition energy of dust clouds, mJ (a)	Remarks
Aluminium, Al	660	2467	645	>1000	20	
Magnesium, Mg	650	1107	520	625	20	Can be ignited by spark in CO_2
Lithium, Li	179	1317	—	190	—	
Beryllium, Be	1280	2970	—	—	—	
Boron, B	2300	2530	730 (c)	—	—	No ignition by spark
Silicon, Si	1414	2355	775	—	—	
Carbon, C	>3540	4827	610	—	40	Coal
Titanium, Ti	1730	3260	330	—	25	Ignite in N_2 and CO_2 under favourable circumstances
Zirconium, Zr	1845	3578	Room temp.	—	5	

(a) Hartmann, Nagy & Brown (1943) and Hartmann, Nagy & Jacobson (1951).
(b) Grosse & Conway (1958).
(c) Personal communication by Dr. Hartmann.

TABLE 14.2

Oxide	Heat of formation $\Delta H°_{f-298}$, kcal/mole	Melting point, K	Boiling point, K
Li_2O	−143	1843	~2836
BeO	−143	2820	~4060
MgO	−144	3100	3466
B_2O_3	−304	723	2316
Al_2O_3	−398	2323	~3800
SiO_2	−218	1996	~2500
TiO_2	−226	2143	~3000
ZrO_2	−262	2950	4548

These data are mainly extracted from JANAF Thermodynamic Tables.

and extent of surface oxide-layer and shape. For this reason the smallest measured values are quoted in Table 14.1.

There is a similar dependence on conditions for ignition by condensed sparks. The minimum ignition energy may change by a factor of more than ten if the particle diameter changes from 0·5 to 0·05 mm. But even for particles having the same size, i.e. passing through the same sieve, large differences occur between atomised, milled or stamped powders. Hartmann (1948) reports that atomised magnesium in dust layers has a minimum ignition energy of 0·05 J, whereas stamped dust layers have one of only 0·000 24 J. It is clear that dust can behave in a great variety of ways depending on its condition. In this connection it is interesting to note that zirconium, magnesium and titanium are easily ignited in pure CO_2. In the absence of an oxide layer, many powdered materials ignite in air at room temperature. For this reason, aluminium flakes are milled under the protection of inert gas.

Cassel & Liebman (1958) measured ignition temperatures of single spherical particles and mono-disperse dust clouds of magnesium in air. Single particles have an increasing ignition temperature varying from 645°C at 55 μm diameter to 780°C at 15 μm. Below 10 μm, single particles no longer ignite. Ignition temperatures of dust clouds are lower for clouds of fine particles, when the dust concentration is high. For low dust concentrations (less than 60 milligrams per litre air) the situation is reversed. This can be understood from a consideration of the average distance between particles and the co-operative effects of the particles on ignition. It is interesting to note that ignition temperatures for single particles are the same in air and oxygen.

The limits of flammability are expressed as minimum concentrations in gram per litre air. Representative values for aluminium, magnesium, carbon (coal), titanium and zirconium are 0·035, 0·020, 0·035, 0·045 and 0·040, respectively (Hartmann, 1948). If the dust is uniformly dispersed in air and if oxide and air are heated to a uniform temperature during combustion, then the limit temperature for magnesium is below 400°C. The mechanism of flame propagation under limit conditions is therefore obscure as the limit temperature is far below the ignition temperature. It is clear, however, that metal powders have very wide limits. This shows up also in the 'safe' concentration of oxygen–nitrogen mixtures which no longer support combustion. For aluminium this is reported to be 9% oxygen but for magnesium it is only 2%; in this latter case nitrogen seems no longer to be inert.

Flame propagation in dust clouds appears to be a discontinuous process. If we assume that the particles are arranged in a regular pattern, then if one row of particles ignites and burns, heat is conducted to the next layer which ignites and in turn passes on the ignition reaction. Dust flames can be burned on Bunsen burners. Plate 17 shows a variety of such flames. These flames sometimes have a well-defined reaction zone so that their burning velocities can be deduced as in the case of premixed flames (Cassel, Das Gupta & Guruswamy, 1949). The burning velocity seems to be relatively independent of dust concentration. It is about 20 cm/sec for atomised aluminium (particle size 0–40 μm). However, smaller average particle sizes lead to burning velocities close to 100 cm/sec.

Explosion pressures of dust–air mixtures in closed vessels are just as great or even greater than those of gaseous fuel/air mixtures; however, the rate of pressure rise is smaller. Gliwitzky (1936) found pressure ratios up to 12·6 for aluminium-flake/air explosions. Hartmann (1954) and Palmer (1973) describe in detail many aspects of dust explosions.

The burning characteristics of solids

The combustion of individual particles of beryllium has been observed by Macék et al. (1964). They imbedded a small number of 30 to 45 μm particles in a solid propellant. The temperature of the solid propellant flame can be varied by changing the ratio of the fuel and oxidiser. This unfortunately also changes the oxygen partial pressure and due to experimental limitations (from the toxicity of Be) the total pressure of the system. At low oxygen partial pressure, ignition of the particles occurs at around 2800 K and the luminous particle tracks are straight and uninterrupted, a characteristic of surface combustion. At high oxygen pressure ignition starts at around 2400 K and the particle tracks are diffuse and often intermittent and this is interpreted as gaseous combustion.

The surface reaction of bulk beryllium (size about 1 cm) with oxygen and water vapour was investigated by Blumenthal & Santy (1967). With oxygen in a closed system the reaction becomes fast above 1300 K. At about 1600 K half the sample oxidises in 15 min. Reaction with water vapour is much faster and starts at a lower temperature (around 1050 K) and this suggests that H_2O plays an important role in the combustion of beryllium. Samples of Be immersed in a fuel-rich hydrogen–oxygen flame show very vigorous surface combustion; oxygen-rich flames give a very hot vapour-phase flame with the sample itself reaching temperatures above 2500 K and BeO smoke is formed. The reactions leading to the gas-phase formation of BeO are not clear, but water vapour seems to play an important role.

The combustion of aluminium has been extensively studied. Dusts suspended in air or oxygen burn vigorously and a stationary flame can be achieved by loading air continuously with dust. Wolfhard & Parker (1949a) studied such a premixed aluminium-flake/air flame of about 20 cm cross-section. Heat radiation from this flame is so high that it is impossible to stand near; the spectrum of the flame is continuous, with weak AlO bands. The colour temperature of the continuum is about 3900 K. This coincides roughly with the boiling-point of Al_2O_3 which Brewer & Searcy (1951) reported as 3800 $\pm$ 200 K. This boiling-point is an upper limit for the adiabatic flame temperature, as the heat of vaporisation of Al_2O_3 is so large that only partial vaporisation occurs with the available heat of combustion. Although the colour temperature of the flame coincides with the adiabatic flame temperature, it is unrealistic to assume that the colour temperature is a reliable guide for temperature measurements. The source of the continuous light obviously is the very small oxide particles that lose heat mainly by radiation until they are cooled down to a temperature at which radiation becomes insignificant. Thus, radiation measurements should give an average temperature, as the light beam focused on a spectrograph represents radiation of oxide particles in many stages of their cooling curve. Caldin (1946) has in fact considered temperature corrections due to the cooling down of solid or liquid particles. Proof that the colour temperature does not represent the true temperature can be obtained by measuring the absorption through the burning cloud. This shows that absorption is much heavier in the ultra-violet part of the continuous spectrum than in the visible, i.e. radiation is not grey or black and the colour temperature is far higher than the true temperature. This type of radiation is to be expected when the radiating particles are smaller than the wavelength of light (Mie, 1908). Temperature measurements should therefore be made by comparing the absolute radiation intensity at a given wavelength with a black body (the crater of a carbon arc is a reasonable approximation) and taking the emissivity into consideration.

Brzustowski & Glassman (1964) studied the combustion of aluminium foil in a flash bulb. After ignition the AlO bands appear strongly on a continuum. As the combustion progresses the bands become weak on a much increased continuum, resembling the one described above. The cooled combustion products are mostly alumina particles of 5 μm size with some larger hollow particles up to 100 μm. One can assume that in the confined space of a flash bulb liquid Al_2O_3 particles coalesce on the wall and that originally particles may have been much smaller. In any case the evidence points to a gas-phase reaction in which aluminium is vaporised prior to oxidation, and this situation seems to be typical of free-burning aluminium flames.

This mode of oxidation is however not the only possible one and heterogeneous oxidation as discussed under magnesium might also be possible. If bulk material is heated in a furnace in the presence of oxygen, only slow oxidation occurs below the ignition temperature. Above the ignition temperature, aluminium begins to form pools of burning molten metal that float on top of the molten oxide. These burning pools emit light of great brilliance. Combustion can be propagated indefinitely by feeding bulk aluminium in to the furnace (Grosse & Conway, 1958). Oxygen seems to be essential for the combustion of aluminium in bulk as combustion dies out when air is admitted. Ignition temperatures determined in this way are included in Table 14.1.

Aluminium is often added to solid propellants to raise the flame temperature and thus the specific impulse. In order to model this type of combustion Drew *et al.* (1964) introduced aluminium particles into a glass-blowing torch. The flame was directed downwards so that the alumina particles could be collected and observed. As the particles left the flame they entered a helium atmosphere to interrupt burning and to quench the particles. Without quenching by helium, aluminium particles of 70 μm size formed hollow thin-shelled spheres of alumina together with smaller particles, some of which had not oxidised. High-speed motion pictures reveal that some aluminium particles after ignition expel a stream of smaller particles of approximately 5 μm size. 70 μm size particles, if quenched early in the flame, form a luminous track of sub-micron size particles. The ignition and fragmentation of aluminium particles has also been observed by Friedman and Macek (1964). Wilson & Williams (1971) found that 50 μm particles burnt like droplets, each particle being surrounded by a detached flame envelope. It thus appears that aluminium particles oxidising in the flame of a parent propellant can undergo a great variety of physical and chemical steps and it is no surprise that similar complexity exists for the combustion of aluminium particles in a solid-propellant engine. Particles collected from within such an engine are generally smaller than 1 μm size, whereas particles collected outside are about ten times larger and this indicates that coagulation occurs in the nozzle (Brown & McArty, 1962).

Magnesium is most conveniently burnt from easily available ribbons. On ignition with a match a very bright flame forms, surrounding the ribbon; thus Mg vapour forms a diffusion flame with the surrounding air. The emitted light is mainly a continuum from glowing oxide particles, but weak bands of MgO and some Mg lines are also detectable, especially if the ambient pressure of the air is reduced. The flame gives, therefore, the impression of a gas-phase reaction in close analogy to a burning liquid droplet. This view is, however, not fully justified, as the ribbon leaves a solid ash after the flame has passed, indicating that some surface combustion must have been involved. The continuous light emitted by the luminous flame has a colour temperature of 3900 K (Wolfhard & Parker, 1949a) and this temperature exceeds the boiling-point of the oxide. As the heat of combustion of magnesium is not sufficient to vaporise all the oxide, it has to be assumed that the high colour temperature is due to the small original size of the condensed oxide particles, which causes the emissivity to be wavelength-dependent in thin optical layers for particles with $\frac{2\pi r}{\lambda} < 1$ (where r is the radius of the particle and λ the wavelength of light).

The heterogeneous combustion of magnesium has been investigated by Markstein (1963). Magnesium is vaporised at a few torr and the vapour is mixed with oxygen. A flame of bluish colour can be seen emitting continuous radiation and the flame temperature is estimated to be about 100 K. This flame is very different in nature from the gas-phase reactions discussed above and it is assumed that the oxidation of the Mg vapour occurs on the surface of nuclei, which grow in the process. In the case of magnesium (and possibly also aluminium) we deal therefore with three oxidation mechanisms. The first is a gas-phase reaction between Mg vapour and oxygen to form gaseous oxide that subsequently condenses. In the presence of nitrogen (air as oxidiser) not all oxidation can go through this route as the heat of reaction is insufficient to heat all the oxide to beyond its boiling-point and the details of the physical steps are not clear. The second route of oxidation is a surface reaction whereby oxygen reacts with solid or liquid magnesium on the surface of the bulk material. The third possibility is the oxidation of gaseous Mg and oxygen on the surface of an oxide particle and the cooling of the particle by radiation. It is fair to comment that, in most cases of magnesium (and possibly aluminium) combustion, all three steps may be operative and, depending on the conditions, one or the other mechanism may be overriding.

When calcium wire is burned (Gouldin & Glassman, 1971) there is a distributed reaction zone within the visible flame envelope, and a brightness-emissivity measurement of temperature yields a value of 2550 K, well below the adiabatic 3200 K. It is interesting to note that the CaO smoke particles form chains, very like those in soot.

The burning of single boron particles, ignited by a laser beam, again shows diffusion-type combustion (Macek & Semple, 1969; 1971) but the burning rate exceeds the expected rate by a factor of 2 to $2\frac{1}{2}$; this is attributed to some gas-phase combustion and to movement of the particles through the gas, under gravity, assisting the diffusion process.

The combustion of titanium and zirconium can also be studied by burning wires or ribbons (Harrison, 1959). 1-mm diameter titanium wire could not be burned in 'air' with an oxygen index of less than 0·5; however, thinner wires required a lesser O_2 concentration. Propagation of flame along the wire is fast, and in pure oxygen it may attain 7 cm/sec. This propagation is not continuous; molten globules form at the end of the wire and drop off periodically as the surface tension can no longer support the increasing weight of the droplet. Zirconium wire burns in much the same way as titanium except that at low oxygen indices (below 0·5) no globules drop off. The wire acquires a solid oxide layer during combustion which keeps the core of the wire from burning. As the heat of combustion can no longer melt the oxide layer, combustion is incomplete.

The differences between these two metals show up also when they burn in a furnace in pure oxygen. A bundle of titanium wires burns completely, leaving brittle oxides in the furnace, whereas combustion of zirconium is interrupted in an early stage by a coating of oxides. The temperature close to a burning titanium bundle was estimated by placing gold or platinum wire close by and observing melting or vaporisation. Thus it was established that the temperature near the bundles was above 2950° (boiling-point of gold) but below 3500°C (the probable boiling-point of platinum).

Very little smoke is formed during the burning of titanium and zirconium. This suggests that the chemical reaction is restricted to the metal surface and that the oxides that form do not vaporise.

Some experimental results are also available for the surface combustion of graphite, molybdenum and tungsten. These data, together with further information on metal combustion, are contained in the selection of technical papers issued as *Progress in Astronautics and Aeronautics* (1960), Vol. 1, and (1964), Vol. 15. Kanury (1975) also discusses the combustion of solids.

Supersonic combustion

Combustion in a rocket chamber is subsonic, and supersonic flow is only achieved in the nozzle where the gases expand isentropically and are cooled in the process. Turbo-engines also use subsonic combustion whether the vehicle flies sub- or supersonic, as the gases are compressed by the compressor and the air velocity is thus reduced before combustion takes place. Such subsonic combustion is generally maintained by recirculating part of the burnt hot gases which then, in turn, ignite the unburnt fuel-air mixture.

For true supersonic combustion no recirculation of the burnt gases can be tolerated as this would cause excessive pressure losses. The air temperature is also close to ambient (without compression) so that one has to use fuels that ignite spontaneously with air (hypergolic fuels). Such fuels are, for example, trimethylaluminium, triethylaluminium, pentaborane or aluminium borohydride (AlB_8H_{12}). A typical arrangement is depicted in Fig. 14.3. A wedge of 10° angle is exposed to supersonic air of Mach 5. A bow shock forms and is reflected by the plate opposite the wedge. After traversing both shocks the air has a Mach number of approximately 3.2 and is slightly heated. Without fuel injection isentropic expansion occurs behind the knee of the wedge and the Mach number increases again. Fuel is injected and penetrates the boundary layer and combustion proceeds in the expansion section, raising the pressure and therefore providing thrust.

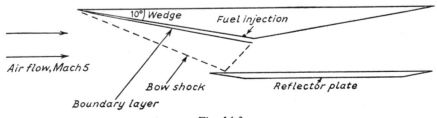

Fig. 14.3.

Supersonic combustion can be used in at least two ways; either, as depicted above, to provide forward thrust, or alternatively one could use it to provide lateral forces to increase lift under a supersonic wing. We do not intend to treat this subject in any more detail, especially as we found it equally impossible (due to space limitations) to deal with the combustion processes in a rocket chamber. For references on supersonic combustion see Billig (1965) and Fletcher (1967).

Chapter XV
Flame Problems

The study of flames is grouped along two lines of interest. The first is the concern of scientists, mainly at universities, who are able to tackle problems of inherent scientific interest and in the process provide suitable material for post-graduate studies. The second line of interest should be classified as applied research and it concerns problems of industrial processes or defence needs. This second kind is often more challenging and difficult, as a specific problem has to be solved independently of its complexity. In practice both problem areas are often intermixed.

Let us dwell on applied combustion problems first. Typical of such areas are (with combustion problems indicated):

The internal-combustion engine (spark ignition, spontaneous thermal ignition, knocking, turbulent flame propagation and, for diesel engines, droplet combustion, solid carbon formation).

Industrial burners (droplet combustion, heat release rate in turbulent flames, infra-red radiation, heat transfer, propagation through dust clouds, ignition of coal particles, fluidised beds, monitoring of flame blow-out).

Incinerators (surface smouldering versus gaseous flames, inhibition of combustion by inorganic compounds).

Air pollution by combustion (nitric oxide formation, carbon formation, ejection of unburnt fuel and toxic inorganic compounds, role of combustion products in photochemical reactions leading to ozone smogs, oxidation of SO_2 to SO_3, chilling out of intermediate combustion products, non-equilibrium effects).

Combustion in jet engines (flame piloting, droplet combustion, heat release in turbulent flow, minimum ignition energies in turbulent flow at altitude, carbon formation, flame noise).

Combustion in rocket engines (droplet breakup by shear forces, rates of reaction of high energy fuels, burning rates of solid propellants, combustion instability, mixing and heat release rates in turbulent flames at high pressure).

Rocket exhaust plumes (afterburning, ionisation, relaxation phenomena in expanding flows).

Hypersonic flight (upper atmosphere chemistry, ozone/nitric oxide reactions, ignition and combustion behind shock fronts).

Combustion of low-grade fuels (effects of preheating on flammability limits and burning velocity, catalytic combustion, electrical augmentation, combustion in fluidised beds).

MHD and EHD gas generators (ionisation in combustion products, vaporisation of seed particles, corrosion by flame gases).

Explosion hazards (minimum ignition energy, flame propagation through confined spaces, minimum ignition temperature, shock formation, detonation, flame traps, explosion relief methods).

Explosives (sensitivity, ignition, propagation).

Fire hazards (spontaneous ignition, flammability, flame spread, ignition and combustion of plastic materials, flame detection using radiation or ionisation).

Fire fighting (flame propagation through solid fuels, chemical extinguishants including halogenated compounds, mechanical extinguishants, foam extinguishants).

Gas generators for pneumatics (low-temperature combustion).

Carbon black (formation of carbon particles in flames, pyrolysis).

It is fair to comment that most of the applied combustion problems have been more or less solved by trial and error methods without a clear understanding of the underlying physical and chemical processes. Basic research has, however, not been without merit. For example, in explosion hazards the first level of brute force solution has involved modification of explosives used in mines, ventilation of mines to exclude methane, and prevention of sparks. However, as the principles of minimum ignition energy, quenching distance and hot-gas ignition become known, safety testing (which is, after all, trial and error) could commence at a higher level of sophistication. Thus combustion research has not solved hard applied combustion problems, but has rather been the guiding hand for the trial and error methods by which a specific applied problem was solved, or at least was controlled. This statement is probably true for all applied combustion problems listed above.

On the more fundamental side there are a number of problems which are still outstanding. Most of them have already been referred to in the text, and many of these have an eye to practical applications. The problem of soot formation and control is still with us; we need more quantitative data on the rates of the various processes which have been postulated; electrical effects, especially those involving additives of low ionisation potential, when fully understood, could lead to better control. The intriguing problem of the formation of nitric oxide in the reaction zone of organic flames (prompt NO_x) needs further work. The effects of turbulence on flame propagation have been discussed. Is the wrinkled-flame concept a sufficient approximation, or, as some recent work suggests, is there a real contribution from micro-turbulence within the reaction zone? Are large coherent structures important? Similarly we know that most electrical effects on flame propagation are due to Chattock-type electric winds, but do applied high-frequency electric fields really increase the burning velocity, and if so how? Are reports of very high electron temperatures in some flames genuine?

A combustion-driven laser, using the departures from equilibrium which are known to occur in some flame gases, has for some years been a pipe dream. So far (Bronfin, 1975) some low-power infra-red chemical lasers, mostly of pulsed type, have been developed, but the ideal high-power continuous laser, preferably at shorter wavelengths, releasing an appreciable fraction of the combustion energy in a laser beam, has not been produced. It does not seem impossible.

Other problems, discussed here but still not fully solved, are the nature of chemi-ionisation and chemiluminescence processes, and some of the reactions leading to abnormal rotational temperatures of excited radicals such as C_2 and OH. By making measurements on a variety of flames, one can define under what condition this non-equilibrium radiation occurs and to what degree, but it is very doubtful whether research on flames will permit a complete unravelling of the physics and chemistry involved. Fortunately, energy transfer phenomena are not unique to combustion. They are important in upper atmospheric phenomena, lasers and hypersonic flow. Equally, experimental tools are available, other than flames, that allow energy transfer phenomena to be studied. These are, among others, shock tubes, flash photolysis, arc-jets and molecular beam methods.

What is suggested here is, that flames may not always be the best experimental medium for the understanding of flames, and many individual chemical and physical steps that in total make a flame should be studied individually. This is not to deny that often a flame is a very suitable medium for detailed research. It is easy to maintain over a wide range of pressure and temperature, and atoms and molecules can be added to the flow and observed in emission and absorption, by microwave absorption or by mass-spectrometric sampling. Thus we tend to view flames as intriguing subjects of research, yet this research is of phenomenological or only semi-quantitative character. Often research closely connected with combustion may be quantitative science such as heat transfer, flow, thermodynamics and spectroscopy. Often attempts were made in the past to quantify, for example, laminar flame propagation. Due to our ignorance of the details of the propagation mechanism, one has to use global reaction rates. By proper adjustment of parameters in the differential equations, one can come up with solutions for the burning velocity that may be correct over a wide pressure range. However, the question arises as to what physical understanding one has derived from such a procedure. The same argument can be applied to the concept of 'excess enthalpy' that allows the derivation of minimum ignition energies. This elegant theory did indeed allow physical insight into combustion phenomena. In a quantitative sense, however, it was bound to fail as it ignored diffusion; thus an important interplay of physical forces was neglected.

Our contention, therefore, is that combustion phenomena, with the exception of possibly the simplest systems such as $H_2 - O_2$, $H_2 - Cl_2$ etc., are too complex for a detailed and basic study using flames alone as a medium. Studies should involve all methods of physics and chemistry to learn more about the rates of reaction and energy transfer in collisions. Global phenomena such as ignition energies, burning velocities, carbon formation, ionisation, etc., are too complex and too much subject to extraneous circumstances to be usefully measured to better than, say, 10% accuracy, and such numbers are only useful for comparative purposes.

As we see combustion research in the future, we think that much further work is needed in the applied areas outlined above, but this work is generally of a global, phenomenological and semi-quantitative nature. Basic research should recognise that flames are only part of the infinite variety of gas kinetic reactions and physical interplay of forces, and should be studied as such.

The tools for the study of applied as well as basic research have become much more refined in the last decade. The most remarkable development has been that of the laser which has revolutionised interferometry and holography – both valuable tools for flame studies – and has additional applications in laser-Doppler anemometry and laser-Raman spectroscopy for temperature and concentration measurements without disturbing the flame. Lasers are also being used for ignition studies and for excitation of fluorescence of flame species. Tunable dye lasers are a good prospect for the future. High-intensity nano-second flash tubes are likely to prove valuable tools for combustion research, extending the already useful tool of flash photolysis. Electronic techniques, such as those used in developing fast-response thermocouples and in monitoring devices to study flame blow-out using correlation of flame flicker, are likely to have further applications. Various types of equipment used for space research is proving useful in many other fields. Zero-gravity studies on flames in satellites could further research on buoyancy effects in flames. Developments in computer techniques, as in recent work for handling many reactions simultaneously in methane flames and in quantitative study of carbon nucleation rates, are likely to prove still more valuable in the future (provided accurate and complete data are fed into the computer!). We can therefore expect further refinement and sophistication in combustion research.

This book has aimed to give those scientists who study applied research an overall understanding, in a phenomenological sense, of the great variety of phenomena that can be observed in flames.

References

References are given in the usual form, usually based on the *World List of Scientific Periodicals*, except for the following *special abbreviations*.

C. & F.	Combustion and Flame
J.C.P.	J. Chem. Phys.
J.O.S.A.	J. Opt. Soc. Amer.
P.R.S.	Proc. Roy. Soc. Lond. A.
T.F.S.	Trans. Faraday Soc.
Z.p.C.	Z. phys. Chem.
3rd Symposium to 16th Symposium	These refer to the biennial series of *Symposia* (*International*) *on Combustion*. The 3rd, 4th and 8th symposia were published by Williams, Wilkins and Co, Baltimore; the 5th and 6th by Reinhold, New York; the 7th by Butterworths, London; the 9th by Academic Press, New York; and the 10th and subsequent ones by The Combustion Institute, Pittsburgh.

ADAMS, G. K., NEWMAN, B. H., & ROBINS, A. B., 1962, *8th Symposium*, p. 693.
—— PARKER, W. G., & WOLFHARD, H. G., 1955, *T.F.S.*, **14**, 97.
—— & SCRIVENER, J., 1955, *5th Symposium*, p. 656.
ADDECOTT, K. S. B., & NUTT, C. W., 1969, *Amer. Chem. Soc. Div. Petrochem.*, **14**, A69.
ADOMEIT, G., 1963, *Proc. 1963 Heat Transfer and Fluid Mechanics Inst.* (Ed. A. Roshko *et al.*) Stanford Univ. Press, p. 160.
—— 1965, *10th Symposium*, p. 237.
AGNEW, W. G., 1968, *P.R.S.*, **307**, 153.
—— & AGNEW, J. T., 1965, *10th Symposium*, p. 123.
AGOSTON, G. A., 1962, *C. & F.*, **6**, 212.
AKITA, K., 1973, *14th Symposium*, p. 1075.
—— & YUMOTO, T., 1965, *10th Symposium*, p. 943.
ALLEN, C. W., 1963, *Astrophysical Quantities*, The Athlone Press.
ANACKER, F., & MANNKOPFF, R., 1959, *Z. Phys.*, **155**, 1.
ANAGNOSTOU, E., & POTTER, A. E., 1963, *9th Symposium*, p. 1.
ANDERSEN, J. W., & FEIN, R. S., 1949, *J.C.P.*, **17**, 1268.
—— —— 1950, *J.C.P.*, **18**, 441.
ANDERSON, T. P., SPRINGER, R. W., & WARDER, R. C., 1963, Editors, *Physico-Chemical Diagnostics of Plasmas*, Northwestern University Press, Evanston.
ANDREWS, G. E., & BRADLEY, D., 1972, *C. & F.*, **18**, 133.
—— —— & LWAKABAMBA, S. B., 1975, *C. & F.*, **24**, 285.
ARDEN, E. A., & POWLING, J., 1957, *6th Symposium*, p. 177.

REFERENCES

ARTHUR, J. R., 1950, *Nature*, **165**, 557.
──── COMMINS, B. T., GILBERT, J. A. S., LINDSEY, A. J., & NAPIER, D. H., 1958, *C. & F.*, **2**, 267.
ASHFORTH, G. K., 1950, *Fuel*, **29**, 285.
ATALLAH, S., 1965, *C. & F.*, **9**, 203.
ATEN, C. F., & GREENE, E. F., 1956, *Disc. Faraday Soc.*, **22**, 162.
AY, J., ONG, R. S. B., & SICHEL, M., 1975, *Combustion Sci. Techn.*, **11**, 19.
──── & SICHEL, M., 1976, *C. & F.*, **26**, 1.

BABCOCK, W. R., BAKER, K. L., & CATTANEO, A. G., 1967, *Nature*, **216**, 676.
BADAMI, G. N., & EGERTON, A. C., 1955, *P.R.S.*, **228**, 297.
BADDIEL, C. B., & CULLIS, C. F., 1962, *8th Symposium*, p. 1089.
BADEN, H. C., BAUER, W. H., & WIBERLEY, S. E., 1958, *J. phys. Chem.*, **62**, 331.
BAKER, R. J., BOURKE, P. J., & WHITELAW, J. H., 1973, *14th Symposium*, p. 699.
BALLAL, D. R., 1979, *P.R.S.*, in press.
──── & LEFEBVRE, A. H., 1975, *P.R.S.*, **344**, 217.
──── ──── 1977a, *16th Symposium*, in press.
──── ──── 1977b, *P.R.S.*, **357**, 163.
BALLINGER, P. R., & RYANSON, P. R., 1971, *13th Symposium*, p. 271.
BALWANZ, W. W., 1959, U.S. Naval Research Laboratory, Washington, D.C., NRL Report 5388.
──── 1965, *10th Symposium*, p. 685.
BARNES, M. H., & FLETCHER, E. A., 1974, *C. & F.*, **23**, 399.
BARR, J., & MULLINS, B. P., 1949, *Fuel*, **28**, 181, 200, 205 & 225.
──── 1954, *Fuel*, **33**, 51.
BARTHOLOMÉ, E., 1949a, *Z. Elektrochem.*, **53**, 191.
──── 1949b, *Naturwissenschaften*, **36**, 171 & 206.
──── *et al.*, 1950, *Z. Elektrochem.*, **54**, 165, 169 & 246.
──── & SACHSE, H., 1949, *Z. Elektrochem.*, **53**, 326.
BAUER, E., 1969, *J. Quant. Spectr. Rad. Transfer*, **9**, 499.
BEAMS, J. W., 1955, *Physical Measurements in Gas Dynamics and Combustion*, Oxford University Press, p. 26.
BEATTIE, I. R., *et al.*, 1978, *C. & F.* in press.
BECHERT, K., 1949, *Ann. Phys.*, Leipzig, **4**, 191.
──── 1950, *Ann. Phys.*, Leipzig, **5**, 349; **7**, 113.
BEHRENS, H., 1950a, *Z.p.C.*, **195**, 24.
──── 1950b, *Z.p.C.*, **196**, 78.
──── 1951a, *Naturwissenschaften*, **38**, 187.
──── 1951b, *Z.p.C.*, **197**, 6.
BELL, J. C., & BRADLEY, D., 1970, *C. & F.*, **14**, 225.
BENEDICT, W. S., 1951, *U.S. Nat. Bur. Stand.*, Report 1123, p. 15.
BENNETT, R. G., & DALBY, F. W., 1960, *J.C.P.*, **32**, 1716.
──── ──── 1962, *J.C.P.*, **36**, 399.
BERL, E., & WERNER, G., 1927, *Z. angew. Chem.*, **40**, 243.
BERL, W. G., & DEMBROW, D., 1952, *Nature*, **170**, 367.
──── GAYHART, E. L., MAIER, E., OLSEN, H. L., & RENICH, W. T., 1957, *C. & F.*, **1**, 420.
──── ──── OLSEN, H. L., BROIDA, H. P., & SHULER, K. E., 1956, *J.C.P.*, **25**, 797.

BERLAD, A. L., 1954, *J. phys. Chem.*, **58**, 1023.
BLAIR, D. W., & SHEN, F. C. T., 1969, *C. & F.*, **13**, 440.
BLANC, M. V., GUEST, P. G., VON ELBE, G., & LEWIS, B., 1947, *J.C.P.*, **15**, 798.
BLEEKRODE, R., & NIEUWPOORT, W. C., 1965, *J.C.P.*, **43**, 3680.
BLINOV, V. J., & KHUDIAKOV, G. N., 1957, *Dokl. Akad. Nauk. S.S.S.R.*, **113**, 1094.
BLUMENTHAL, J. L., & SANTY, M. J., 1967, *11th Symposium*, p. 417.
BOERS, A. L., ALKEMADE, C. T. J., & SMIT, J. A., 1956, *Physica*, **22**, 358.
BOLLINGER, L. M., & WILLIAMS, D. T., 1949, *N.A.C.A. tech. note*, No. 1707, also *3rd Symposium*, p. 176.
BONE, W. A., & TOWNEND, D. T. A., 1927, *Flame and Combustion in Gases*, Longmans, Green & Co., London.
BONHOEFFER, K. F., & EGGERT, J., 1939, *Z. Angew. Photogr.*, **1**, 43.
BONNE, U., HOMANN, K. H., & WAGNER, H. G., 1965, *10th Symposium*, p. 503.
—— & WAGNER, H. G., 1965, *Ber. Bunsenges. phys. Chem.*, **69**, 20.
BOTHA, J. P., & SPALDING, D. B., 1954, *P.R.S.*, **225**, 71.
BOWDEN, F. P., & YOFFE, A. D., 1958, *Fast Reactions in Solids*, Butterworths, London.
BOWSER, R. J., & WEINBERG, F. J., 1972, *C. & F.*, **18**, 296.
—— —— 1974, *Nature*, **249**, 339.
—— —— 1976, *C. & F.*, **27**, 21.
BOYER, M. H., & FRIEBERTSHAUSER, P. E., 1957, *C. & F.*, **1**, 264.
BOYS, S. F., & CORNER, J., 1949, *P.R.S.*, **197**, 90.
BRADLEY, D., & HUNDY, G. F., 1971, *13th Symposium*, p. 575.
—— & IBRAHIM, S. M. A., 1975, *C. & F.*, **24**, 169.
—— & JESCH, L. F., 1972, *C. & F.*, **19**, 237.
—— & MATTHEWS, K. J., 1967, *11th Symposium*, p. 359.
—— & SHEPPARD, C. G. W., 1970, *C. & F.*, **15**, 323.
BRAGG, S., 1963, *J. Inst. Fuel*, **36**, 12.
BRAME, J. S. S., & KING, J. G., 1955, *Fuel, Solid, Liquid and Gaseous*, Arnold & Co., London, 5th ed.
BREISACHER, P., DEMBROW, D., & BERL, W. G., 1959, *7th Symposium*, p. 894.
BREWER, L., & SEARCY, A. W., 1951, *J. Amer. Chem. Soc.*, **73**, 5308.
BROIDA, H. P., 1951, *J.C.P.*, **19**, 1383.
—— & CARRINGTON, T., 1955, *J.C.P.*, **23**, 2202.
—— & KANE, W., 1952, *J.C.P.*, **20**, 1042.
—— & SHULER, K. E., 1957, *J.C.P.*, **27**, 933.
BRONFIN, B. R., 1975, *15th Symposium*, p. 935.
BROWN, B., & MCARTY, K. P., 1962, *8th Symposium*, p. 814.
BROWN, G. B., 1932, *Phil. Mag.*, **13**, 161.
—— 1935, *Proc. Phys. Soc.*, **47**, 703.
BROWN, R. L., EVEREST, D. A., LEWIS, J. D., & WILLIAMS, A., 1968, *J. Inst. Fuel*, **41**, 433.
BROWNE, W. G., PORTER, R. P., VERLIN, J. D., & CLARK, A. H., 1969, *12th Symposium*, p. 1035.
BRULE, G., MICHAUD, P., & BARASSIN, A., 1973, *C. & F.*, **21**, 33.
BRZUSTOWSKI, T. A., & GLASSMAN, I., 1964, *Heterogeneous Combustion, Progress in Astronautics and Aeronautics*, **15**, p. 41. Academic Press, New York.

BULEWICZ, E. M., EVANS, D. G., & PADLEY, P. J., 1975, *15th Symposium*, p. 1461.
—— JAMES, C. G., & SUGDEN, T. M., 1956, *P.R.S.*, **235**, 89.
BUNDY, F. P., & STRONG, H. M., 1949, *3rd Symposium*, p. 647.
BURGESS, M. J., & WHEELER, R. V., 1911, *J. Chem. Soc.*, **99**, 2013.
BURGOYNE, J. H., & COHEN, L., 1954, *P.R.S.*, **225**, 375.
—— & KATAN, L. L., 1947, *J. Inst. Petrol.*, **33**, 158.
—— & NEWITT, D. M., 1955, *Trans. Inst. Marine Eng.*, **67**, 255.
—— & RICHARDSON, J. F., 1949, *Fuel*, **28**, 1, 145 & 150.
—— —— 1950, *Fuel*, **29**, 93.
—— & ROBERTS, A. F., 1968, *P.R.S.*, **308**, 55 & 69.
—— —— & QUINTON, P. G., 1968, *P.R.S.*, **308**, 39 & 69.
—— & WEINBERG, F. J., 1954, *P.R.S.*, **224**, 286.
BURKE, S. P., & SCHUMANN, T. E. W., 1928, *Industr. Engng. Chem.*, **20**, 998.
BURT, R., & THOMAS, A., 1968, *P.R.S.*, **307**, 183.

CALCOTE, H. F., 1949, *3rd Symposium*, p. 245.
—— & JENSEN, D. E., 1966, *Ion Molecule Reactions in Flames*, Advances of Chemistry Series, No. 58, American Chemical Society, p. 291.
—— & KING, I. R., 1955, *5th Symposium*, p. 423.
—— KURZIUS, S. C., & MILLER, W. J., 1965, *10th Symposium*, p. 605.
CALDIN, E. F., 1946, *Proc. Phys. Soc.*, **58**, 341 & 350.
CALLEAR, A. B., 1965, *Applied Optics*, Supplement 2 of Chemical Lasers, p. 145.
CARNEVALE, E. H., WOLNIK, S., LARSON, G., CAREY, C., & WARES, G. W., 1967, *Phys. Fluids*, **10**, 1459.
CARRINGTON, T., 1959, *J.C.P.*, **31**, 1418.
CASSEL, H. M., DAS GUPTA, A. K., & GURUSWAMY, S., 1949, *3rd Symposium*, p. 185.
—— & LEIBMAN, I., 1958, personal communication.
CHARTON, M., & GAYDON, A. G., 1958, *P.R.S.*, **245**, 84.
CHEDAILLE, J., & BRAUD, Y., 1972, *Industrial Flames*, **1**, *Measurements in Flames*, Arnold, London.
CHIGIER, N. A., & DVORAK, K., 1975, *15th Symposium*, p. 573.
CHIPPET, S., & GRAY, W. A., 1978, *C. & F.*, **31**, 149.
CLARKE, A. E., HUNTER, T. G., & GARNER, F. H., 1946, *J. Inst. Petrol.*, **32**, 627.
—— ODGERS, J., & RYAN, P., 1962, *8th Symposium*, p. 982.
—— —— STRINGER, F. W., & HARRISON, A. J., 1965, *10th Symposium*, p. 1151.
CLOUSTON, J. G., GAYDON, A. G., & GLASS, I. I., 1958, *P.R.S.*, **248**, 429.
CLUSIUS, K., KÖLSCH, W., & WALDMAN, L., 1941, *Z. Elektrochem.*, **47**, 820 and *Z.p.C.*, **189**, 131.
COMBOURIEU, J., & MOREAU, R., 1969, *12th Symposium*, p. 1015.
—— —— HALL, A. R., & PEARSON, G. S., 1969, *C. & F.*, **13**, 596.
CONAN, H. R., & LINNETT, J. W., 1951, *T.F.S.*, **47**, 981.
CONWAY, J. B., SMITH, W. F. R., LIDDELL, W. J., & GROSSE, A. V., 1955, *J. Amer. Chem. Soc.*, **77**, 2026.
COOKSON, R. A., & KILHAM, J. K., 1963, *9th Symposium*, p. 257.
CORBEELS, R. J., 1970, *C. & F.*, **14**, 49.
CORLISS, C. H., & BOZMAN, W. R., 1962, *Nat. Bur. Standards Monograph 53*, Washington.
COTTON, D. H., FRISWELL, N. J., & JENKINS, D. R., 1971, *C. & F.*, **17**, 87.

COWARD, H. F., & WOODHEAD, D. W., 1949, *3rd Symposium*, p. 518.
CROS, J. C., BOUVIER, A., & CHEVALEYRE, J., 1971, *C. & F.*, **16**, 205.
CROSSWHITE, H. M., 1958, *Johns Hopkins Spectroscopic Report* No. 13 (Baltimore).
CULLIS, C. F., 1975, *Amer. Chem. Soc., Symp.*, **21**, *Petroleum Derived Carbons*, p. 348.
—— FISH, A., & GIBSON, J. F., 1965, *P.R.S.*, **284**, 108.
—— & FOSTER, C. D., 1977, *P.R.S.*, **355**, 153.
—— & FRANKLIN, N. H., 1964, *P.R.S.*, **280**, 139.
—— READ, I. A., & TRIMM, D. L., 1967, *11th Symposium*, p. 391.
CULSHAW, G. W., & GARSIDE, J. E., 1949, *3rd Symposium*, p. 204.
CUMMINGS, G. A. McD., 1958, *Nature*, **181**, 1327.
—— & HALL, A. R., 1965, *10th Symposium*, p. 1365.
—— & HUTTON, E., 1967, *11th Symposium*, p. 335.

DAILY, J. W., & KRUGER, C. H., 1977, *J. Quant. Spectrosc. Rad. Transfer*, **17**, 327.
D'ALESSIO, A., *et al.*, 1975, *15th Symposium*, p. 1427.
DALZELL, W. H., 1969, personal communication.
—— WILLIAMS, G. C., & HOTTEL, H. C., 1970, *C. & F.*, **14**, 161.
DAMKÖHLER, G., 1940, *Z. Elektrochem.*, **46**, 601.
—— & EDSE, R., 1943, *Z. Elektrochem.*, **49**, 178.
DASHCHUK, M., 1971, *13th Symposium*, p. 659.
DAVID, W. T., 1934, *Engineering*, **138**, 475.
DAVIES, D. A., 1960, *J. Sci. Instrum.*, **37**, 15.
DAY, M. J., DIXON-LEWIS, G., & THOMPSON, K., 1972, *P.R.S.*, **330**, 199.
—— STAMP, D. V., THOMPSON, K., & DIXON-LEWIS, G., 1971, *13th Symposium*, p. 705.
DECKKER, B. E. L., & SAMPATH, P., 1971, *13th Symposium*, p. 505.
DELANEY, R. M., & WEBER, A. H., 1962. St. Louis University report.
DE LEEUW, J. H., 1963, *Physico-Chemical Diagnostics of Plasmas*, p. 65. (Ed. T. P. Anderson, *et al.*) *Proc. Fifth Biennial Gas Dynamics Symposium*, Northwestern Univ. Press, Evanston.
DEPOY, P. E., & MASON, D. M., 1971, *C. & F.*, **17**, 107.
DE SOETE, G. G., 1971, *13th Symposium*, p. 735.
DE VOS, J. C., 1954, *Physica*, **20**, 690.
DIEDERICHSEN, J., & WOLFHARD, H. G., 1956, *P.R.S.*, **236**, 89.
DIEKE, G. H., & CROSSWHITE, H. M., 1962, *J. Quant. Spectr. Rad. Transfer*, **2**, 97.
DIXON-LEWIS, G., 1970, *P.R.S.*, **317**, 235.
—— ISLES, G. L., & WALMSLEY, R., 1973, *P.R.S.*, **331**, 571.
—— SUTTON, M. M., & WILLIAMS, A., 1970, *P.R.S.*, **317**, 227.
—— & WILLIAMS, A., 1963, *9th Symposium*, p. 576.
—— & WILSON, M. J. G., 1951, *T.F.S.*, **47**, 1106.
DREW, C. M., GORDON, A. S., & KNIPE, R. H., 1964, *Heterogeneous Combustion, Progress in Astronautics and Aeronautics*, **15**, p. 17. Academic Press, New York.
DROWART, J., BURNS, R. P., DE MARIA, G., & INGHRAM, M. G., 1959, *J.C.P.*, **31**, 1131.
DUBOIS, M., 1949, *C.R. Acad. Sci.*, Paris, **229**, 747.
—— 1950, *C.R. Acad. Sci.*, Paris, **231**, 217.

DUFF, R. E., KNIGHT, H. T., & WRIGHT, H. R., 1954, *J.C.P.*, **22**, 1618.
DUGGER, G. L., & HEIMEL, S., 1952, *N.A.C.A. tech. note*, No. 2624.
—— WEAST, R. C., & HEIMEL, S., 1955, *5th Symposium*, p. 589.
DUNNING, W. J., 1973, *Faraday Symposium, Chem. Soc.*, **7**, 7.
DURÃO, D. F. G., & WHITELAW, J. H., 1974, *P.R.S.*, **338**, 479.
DURIE, R. A., 1952a, *Proc. Phys. Soc.*, **A65**, 125.
—— 1952b, *P.R.S.*, **211**, 110.
DURST, F., MELLING, A., & WHITELAW, J. H., 1972, *C. & F.*, **18**, 197.
—— & WHITELAW, J. H., 1971, *J. Phys.*, **E4**, 804.

EBERIUS, K. H., HOYERMANN, K., & WAGNER, H. G., 1973, *14th Symposium*, p. 147.
ECHIGO, R., NISHIWAKI, N., & HIRATA, M., 1967, *11th Symposium*, p. 381.
ECKSTEIN, B. H., & VAN ARTSDALEN, E. R., 1958, *J. Amer. Chem. Soc.*, **80**, 1352.
EDELMAN, R. B., et al., 1973, *14th Symposium*, p. 399.
EDMONDSON, H., & HEAP, M. P., 1969a, *12th Symposium*, p. 1007.
—— —— 1969b, *C. & F.*, **13**, 322.
—— —— 1970, *C. & F.*, **15**, 179.
—— —— & PRITCHARD, R., 1970, *C. & F.*, **14**, 195.
EGERTON, A. C., EVERETT, A. J., & MOORE, N. P. W. 1953, *4th Symposium*, p. 689.
—— & POWLING, J., 1948, *P.R.S.*, **193**, 172 & 190.
—— & THABET, S. K., 1952, *P.R.S.*, **211**, 445.
EISEMAN, J. H., 1949, *J. Res. Nat. Bur. Stand.*, Washington, **42**, 541.
ELLIS, C., 1937, *The Chemistry of Petroleum Derivatives*, Reinhold, New York.
ELTENTON, G. C., 1947, *J.C.P.*, **15**, 455.
EMMONS, H. W., & SHUH-JING YING, 1967, *11th Symposium*, p. 475.
ENGELMAN, V. S., BARTOK, W., LONGWELL, J. P., & EDELMAN, R. B., 1973, *14th Symposium*, p. 755.
ENGLISH, P., & DINGLE, M. G. W., 1966, *J. Sci. Instrum.*, **43**, 121.
ERICKSON, W. D., WILLIAMS, G. C., & HOTTEL, H. C., 1964, *C. & F.*, **8**, 127.
ESSENHIGH, R. H., & CSABA, J., 1963, *9th Symposium*, p. 111.
EULER, J., & HÜPPNER, W., 1950, *Optik*, **6**, 332.
—— 1954, *Ann. Phys.* Leipzig, **14**, 145.
EYRING, H., GERSCHINOWITZ, H., & SUN, C. E., 1935, *J.C.P.*, **3**, 786.

FABELINSKII, I. L., 1968, *Molecular Scattering of Light*, Plenum Press, New York.
FAIRBAIRN, A. R., 1962, *P.R.S.*, **267**, 88.
—— & GAYDON, A. G., 1955, *5th Symposium*, p. 324.
FAIZULLOV, F. S., SOBOLEV, N. N., & KUDRYARTSEV, E. M., 1960, *Optics & Spectrosc.*, **4**, 311 & 400.
FAY, J. A., 1954, *J. Aero. Sci.*, **21**, 681.
FENIMORE, C. P., 1975, *C. & F.*, **25**, 85.
—— & JONES, G. W., 1957, *J. phys. Chem.*, **61**, 654.
—— —— 1963, *9th Symposium*, p. 597.
—— —— & MOORE, G. E., 1957, *6th Symposium*, p. 242.
FERGUSON, R. E., 1955, *J.C.P.*, **23**, 2085.
—— 1957, *C. & F.*, **1**, 431.
FERRISO, C. C., & LUDWIG, C. B., 1964, *J.O.S.A.*, **54**, 657.
—— —— & ACTON, L., 1965, *J.O.S.A.*, **56**, 171.

FEUGIER, A., & VAN TIGGELEN, A., 1965, *10th Symposium*, p. 621.
FIELD, M. A., 1969, *C. & F.*, **13**, 237.
FISSAN, H. J., 1971, *C. & F.*, **17**, 355.
FLETCHER, E. A., 1967, *11th Symposium*, p. 729.
FONS, W. L., 1961, *C. & F.*, **5**, 283.
FONTIJN, A., 1969, Meeting, Eastern Section of Combustion Institute, Morgantown, West Virginia.
—— & VREE, P. H., 1967, *11th Symposium*, p. 343.
FOX, J. S., & MIRSCHANDANI, I., 1974, *C. & F.*, **22**, 267.
FOX, M. D., & WEINBERG, F. J., 1962, *P.R.S.*, **268**, 222.
FRANK, H. H., & REICHARDT, H., 1936, *Naturwissenschaften*, **24**, 171.
FRIEDMAN, R., 1949, *3rd Symposium*, p. 110.
—— 1953, *4th Symposium*, p. 259.
—— & BURKE, E., 1954, *J.C.P.*, **22**, 824.
—— & CYPHERS, J. A., 1955, *J.C.P.*, **23**, 1875.
—— —— 1956, *J.C.P.*, **25**, 448.
—— & JOHNSTON, W. C., 1950, *J. appl. Phys.*, **21**, 791.
—— —— 1952, *J.C.P.*, **20**, 919.
—— & LEVY, J. B., 1963, *C. & F.*, **7**, 195.
—— & MACĚK, A., 1964, *9th Symposium*, p. 703.
—— —— 1965, *10th Symposium*, p. 731.
FRISTROM, R. M., 1956, *J.C.P.*, **24**, 888.
—— 1957, *6th Symposium*, p. 96.
—— 1963, *9th Symposium*, p. 560.
—— PRESCOTT, R., NEUMANN, R., & AVERY, W. H., 1953, *4th Symposium*, p. 267.
—— & WESTENBERG, A. A., 1957, *C. & F.*, **1**, 217.
—— —— 1965, *Flame Structure*, McGraw-Hill, New York.
FUHS, A., 1963, *Proc. Fifth Biennial Gas Dynamics Symp.*, Northwestern Univ. Press, Evanston.
FUIDGE, G. H., MURCH, W. O., & PLEASANCE, B., 1939, *Gas Wld.*, **111**, 350 and *Gas Journal*, **228**, 571.
FULLER, L. E., PARKS, D. J., & FLETCHER, E. A., 1969, *C. & F.*, **13**, 455.

GARNER, F. H., LONG, R., & ASHFORTH, G. K., 1949, *Fuel*, **28**, 272.
—— —— —— 1951, *Fuel*, **30**, 17.
GARNER, W. E., & JOHNSON, C. H., 1928, *J. Chem. Soc.*, p. 280.
GARSIDE, J. E., 1949, *Fuel*, **28**, 221.
GARTON, W. R. S., & RAJARATNAM, A., 1957, *Proc. Phys. Soc.*, **A70**, 815.
GAY, I. D., KERN, R. D., KISTIAKOWSKY, G. B., & NIKI, H., 1966, *J.C.P.*, **45**, 2371.
—— KISTIAKOWSKY, G. B., MICHAEL, J. V., & NIKI, H., 1965, *J.C.P.*, **43**, 1720.
GAY, N. R., AGNEW, J. T., WITZELL, O. W., & KARABELL, C. E., 1961, *C. & F.*, **5**, 257.
GAYDON, A. G., 1948, *Spectroscopy and Combustion Theory*, Chapman & Hall, London, 2nd ed.
—— 1966 *Brit. Coal Utilisation Research Assocn. Gazette*, No. 59, p. 1.
—— 1968, *Dissociation Energies and Spectra of Diatomic Molecules*, Chapman & Hall, London, 3rd ed.

GAYDON, A. G. 1974, *Spectroscopy of Flames*, Chapman & Hall, London, 2nd ed.
—— & HURLE, I. R., 1963, *The Shock Tube in High Temperature Chemical Physics*, Chapman & Hall, London.
—— —— & KIMBELL, G. H., 1963, *P.R.S.*, **273**, 291.
—— & KOPP, I., 1971, *J. Phys.*, **B4**, 752.
—— & WHITTINGHAM, G., 1947, *P.R.S.*, **189**, 313.
—— & WOLFHARD, H. G., 1947, *Disc. Faraday Soc.*, **2**, 161.
—— —— 1948, *P.R.S.*, **194**, 169.
—— —— 1949a, *Rev. Inst. Française du Pétrole*, **4**, 405.
—— —— 1949b, *P.R.S.*, **196**, 105.
—— —— 1949c, *P.R.S.*, **199**, 89.
—— —— 1949d, *3rd Symposium*, p. 504.
—— —— 1950a, *P.R.S.*, **201**, 561.
—— —— 1950b, *P.R.S.*, **201**, 570.
—— —— 1950c, *Fuel*, **29**, 15.
—— —— 1951a, *P.R.S.*, **205**, 118.
—— —— 1951b, *P.R.S.*, **208**, 63.
—— —— 1954, *Fuel*, **33**, 286.
GAYDON, B. G., 1976, *I.M.E.K.O. Congress VII*, London, paper AEN/163.
GERSTEIN, M., LEVINE, O., & WONG, E. L., 1951, *J. Amer. Chem. Soc.*, **73**, 418.
GILBERT, M., 1957, *6th Symposium*, p. 74.
GILMORE, F. R., BAUER, E., & MCGOWAN, J. W., 1969, *J. Quant. Spectr. Rad. Transfer*, **9**, 157.
GLICK, H. S., 1959, *7th Symposium*, p. 98.
—— & WURSTER, W. H., 1957, *J.C.P.*, **27**, 1224.
GLIWITZKY, W., 1936, *Z. Ver. dtsch. Ing.*, **83**, 687.
GOLDMANN, F., 1929, *Z.p.C.*, **B5**, 307.
GOLLAHALLI, S. R., & BRZUSTOWSKI, T. A., 1975, *15th Symposium*, p. 409.
GOLOTHAN, D. W., 1967, *Soc. Auto. Eng. Trans.*, **76**, 616.
GOODGER, E. M., 1977, *Combustion Calculations*, Macmillan, London.
GOODY, R. M., 1964, *Atmospheric Radiation*, Part I, Clarendon Press, Oxford.
GORDON, A. S., 1948, *J. Amer. Chem. Soc.*, **70**, 395.
GORDON, J. S., 1957, *WADC tech. report*, 57–33; *Astia Doc.*, 110735.
GOULDIN, F. C., & GLASSMAN, I., 1971, *13th Symposium*, p. 847.
GOUY, G., 1879, *Ann. chim. phys.*, p. 18 & 27.
GRAHAM, S. C., HOMER, J. B., & ROSENFELD, J. L. J., 1975, *P.R.S.*, **344**, 259.
GRAIFF, L. B., 1964, *Nature*, **203**, 856.
GRANT, A. J., & JONES, J. M., 1975, *C. & F.*, **25**, 153.
GRAY, B. F., & YANG, C. H., 1969, *C. & F.*, **13**, 20.
GRAY, K. L., LINNETT, J. W., & MELLISH, C. E., 1952, *T.F.S.*, **48**, 1155.
GRAY, P., HALL, A. R., & WOLFHARD, H. G., 1955a, *Nature*, **176**, 695.
—— —— —— 1955b, *P.R.S.*, **232**, 389.
—— & HOLLAND, S., 1970, *C. & F.*, **14**, 203.
—— & LEE, J. C., 1959, *7th Symposium*, p. 61.
—— —— LEACH, A. H., & TAYLOR, D. C., 1957, *6th Symposium*, p. 255.
—— MACKINVEN, R., & SMITH, D. C., 1967, *C. & F.*, **11**, 217.
—— & PRATT, M. W. T., 1957, *6th Symposium*, p. 183.
—— & SMITH, D. B., 1967, *Chem. Communications*, p. 146.
—— & YOFFE, A. D., 1950, *J. Chem. Soc.*, p. 3180.
—— —— 1955, *Chem. Rev.*, **55**, 1069.

GREEN, H. L., & LANE, W. R., 1965, *Particulate Clouds, Dusts, Smokes and Mists*, Spon, London.
GREENE, E. F., COWAN, G. R., & HORNIG, D. F., 1951, *J.C.P.*, **19**, 427.
—— TAYLOR, R. L., & PATTERSON, W. L., 1958, *J. phys. Chem.*, **62**, 238.
GRIEM, H. R., 1964, *Plasma Spectroscopy*, McGraw-Hill, New York.
GRIFFING, V., & LAIDLER, K. J., 1949, *3rd Symposium*, p. 432.
GRIFFITHS, E., & AWBERY, J. H., 1929, *P.R.S.*, **123**, 401.
GRISDALE, R. O., 1953, *J. appl. Phys.*, **24**, 1082.
GROSSE, A. V., & CONWAY, J. B., 1958, *Industr. Engng. Chem.*, **50**, 663.
—— & KIRSHENBAUM, A. D., 1955, *J. Amer. Chem. Soc.*, **77**, 5012.
GROVE, J. R., HOARE, H. F., & LINNETT, J. W., 1950, *T.F.S.*, **46**, 745.
GROVER, J. H., FALES, E. N., & SCURLOCK, A. C., 1963, *9th Symposium*, p. 21.
GUDERLEY, G., 1938, *Z. angew. Math. Mech.*, **18**, 285.
GUMZ, W., 1939, *Theorie und Berechnung der Kohlenstaubfeuerungen*, Springer, Berlin.
GÜNTHER, R., & JANISCH, G., 1972, *C. & F.*, **19**, 49.
GUPTA, A. K., SYRETH, N., & BEER, J. M., 1975, *15th Symposium*, p. 1367.
GURVICH, L. V., et al., 1962, *Thermodynamic Properties of Individual Substances*, Vol. 2, Akad. Nauk, U.S.S.R.
—— —— 1974, *Chemical Bond Energies, Ionisation Potentials and Electron Affinities*, Akad. Nauk, U.S.S.R.
GUYOMARD, F., 1952, *C.R. Acad. Sci.*, Paris, **234**, 67.

HAHNEMANN, H., & EHRET, L., 1943, *Z. tech. Phys.*, **24**, 228.
HALL, A. R., & DIEDERICHSEN, J., 1953, *4th Symposium*, p. 837.
—— MCCOUBREY, J. C., & WOLFHARD, H. G., 1957, *C. & F.*, **1**, 53.
—— & PEARSON, G. S., 1969, *12th Symposium*, p. 1025.
—— & WOLFHARD, H. G., 1956, *T.F.S.*, **52**, 1520.
—— —— 1957, *6th Symposium*, p. 190.
HALSTEAD, C. J., & JENKINS, D. R., 1969, *12th Symposium*, p. 979.
HALSTEAD, M. P., KIRSCH, L. J., PROTHERO, A., & QUINN, C. P., 1975, *P.R.S.*, **346**, 515.
—— PROTHERO, A., & QUINN, C. P., 1971, *P.R.S.*, **322**, 377.
HANSEN, G., 1940, *Z. Instrum. Kde.*, **60**, 325.
HARNED, B. W., & GINSBURG, N., 1958, *J.O.S.A.*, **48**, 178.
HARRIS, M. E., GRUMER, J., VON ELBE, G., & LEWIS, B., 1949, *3rd Symposium*, p. 80.
HARRISON, A. J., & WEINBERG, F. J., 1971, *P.R.S.*, **321**, 95.
HARRISON, G. R., LORD, R. C., & LOOFBOUROW, J. R., 1949, *Practical Spectroscopy*, Blackie, London.
HARRISON, P. L., 1959, *7th Symposium*, p. 913.
HARSHMAN, R. C., 1957, *Jet Propulsion*, **29**, 398.
HARTLEY, H., 1932, *Inst. Gas Engr. Trans.*, p. 466.
HARTMANN, I., 1948, *Industr. Engng. Chem.*, **40**, 752.
—— 1954, *The Scientific Monthly*, **79**, 97.
—— NAGY, J., & BROWN, H. R., 1943, *U.S. Bureau of Mines Rep. Investigations*, No. 3722.
—— —— & JACOBSON, M., 1951, *U.S. Bureau of Mines Rep. Investigations*, No. 4835.
HASTIE, J. W., 1973, *C. & F.*, **21**, 187.

REFERENCES

HAWKSLEY, P. G. W., 1952, *Brit. Coal Utilizn, Res. Assn. Mon. Bull.*, **16**, 117 & 181.
HAYHURST, A. N., & KITTELSON, D. B., 1974, *P.R.S.*, **338**, 155 & 175.
—— & VINCE, I. M., 1977, *Nature*, **266**, 524.
HEALD, M. A., & WHARTON, C. B., 1965, *Plasma Diagnostics with Microwaves*, Wiley & Sons, New York.
HEATH, G. A., & PEARSON, G. S., 1967, *11th Symposium*, p. 967.
HEIMEL, S., & WEAST, R. C., 1957, *6th Symposium*, p. 296.
HELLER, C. A., & GORDON, A. S., 1955, *J. phys. Chem.*, **59**, 773.
HENKEL, M. J., HUMMEL, H., & SPAULDING, W. P., 1949, *3rd Symposium*, p. 135.
HENNING, F., & TINGWALDT, C., 1928, *Z. Phys.*, **48**, 805.
HERLAN, A., 1978, *C. & F.*, **31**, 297.
HERZBERG, G., 1945, *Infra-red and Raman Spectra*, Van Nostrand, New York.
—— 1950, *Molecular Spectra and Molecular Structure. I. Diatomic Molecules*, Van Nostrand, New York.
—— 1966, *Electronic Spectra of Polyatomic Molecules*, Van Nostrand, New York.
—— 1971, *The Spectra and Structure of Simple Free Radicals*, Cornell University Press, Ithica.
HIBBARD, R. R., & PINKEL, B., 1951, *J. Amer. Chem. Soc.*, **73**, 1622.
HINCK, E. C., SEAMANS, T. F., VANPEE, M., & WOLFHARD, H. G., 1965, *10th Symposium*, p. 21.
HIRSCHFELDER, J. O., 1963, *9th Symposium*, p. 553.
—— CURTISS, C. F., & CAMPBELL, D. E., 1953, *J. phys. Chem.*, **57**, 403.
HO, C. M., JAKUS, K., & PARKER, K. H., 1976, *C. & F.*, **27**, 113.
HOFMANN, U., & WILM, D., 1936, *Z. Elektrochem.*, **42**, 504.
HOLLANDER, T. J., 1968, *AIAA Journal*, **6**, 395.
—— KALFF, P. J., & ALKEMADE, C. T. J., 1963, *J.C.P.*, **39**, 2558.
HOMANN, K. H., 1967a, *C. & F.*, **11**, 265.
—— 1967b, *11th Symposium*, p. 254.
—— & WAGNER, H. G., 1967, *11th Symposium*, p. 371.
—— —— 1968, *P.R.S.*, **307**, 141.
HONG, N. S., JONES, A. R., & WEINBERG, F. J., 1977, *P.R.S.*, **353**, 77.
HÖRMANN, H., 1935, *Z. Phys.*, **97**, 539.
HOTTEL, H. C., 1959, *Fire Research Abs.*, **1**, 41.
—— & BROUGHTON, F. P., 1932, *Industr. Engng. Chem., Anal. Edn.*, **4**, 166.
—— & HAWTHORNE, W. R., 1949, *3rd Symposium*, p. 254.
—— & MANGELSDORF, H. G., 1935, *Trans. Amer. Inst. Chem. Eng.*, **31**, 517.
—— & SAROFIM, A. F., 1967, *Radiative Transfer*, McGraw-Hill, New York.
—— WILLIAMS, G. C., & BAKER, M. L., 1957, *6th Symposium*, p. 398.
HOWARD, J. B., 1969, *12th Symposium*, p. 877.
—— & ESSENHIGH, R. H., 1967, *11th Symposium*, p. 399.
—— WERSBORG, B. L., & WILLIAMS, G. C., 1973, *Faraday Symposium, Chem. Soc.*, **7**, 109.
HSIEH, M. S., & TOWNEND, D. T. A., 1939, *J. Chem. Soc.*, p. 332.
HÜBNER, H. J., & KLÄUKENS, H., 1941, *Ann. Phys.*, Leipzig, **39**, 33.
HUDDLESTONE, R. H., & LEONARD, S. L. (editors), 1965, *Plasma Diagnostic Techniques*, Academic Press, New York.
HURLE, I. R., 1964, *J.C.P.*, **41**, 3911.
—— PRICE, R. B., & PYE, D. B., 1968, *Nature*, **219**, 849.

HURLE, I. R., PRICE, R. B., SUGDEN, T. M., & THOMAS, A., 1968, *P.R.S.*, **303**, 409.
—— SUGDEN, T. M., & NUTT, G. B., 1969, *12th Symposium*, p. 387.

IBIRICU, M. M., & GAYDON, A. G., 1964, *C. & F.*, **8**, 51.

JAGGERS, H. C., & VON ENGEL, A., 1971, *C. & F.*, **16**, 275.
JAIN, D. C., 1964, *J. Quant., Spectr. Rad. Transfer*, **4**, 427.
JAMES, H., & GREEN, P. D., 1971, *J. Phys.*, **D4**, 738.
JANAF (Joint Army Navy Air Force), 1965, *Thermochemical Tables*, Dow Chemical Co., PB 168370.
—— 1966, First addendum, PB 168370-1.
—— 1967, Second addendum, PB 168370-2.
—— 1971, *Thermochemical Tables*, Nat. Bur. Stand., Washington D.C.
JANISCH, G., & GÜNTHER, R., 1973, *Combustion Institute, European Symposium*, p. 689.
JENKINS, D. R., 1966, *P.R.S.*, **293**, 493.
—— 1968, *P.R.S.*, **303**, 453 & 467; **306**, 413.
—— 1969, *P.R.S.*, **313**, 551.
JENSEN, D. E., 1968, *C. & F.*, **12**, 261.
—— 1974, *P.R.S.*, **338**, 375.
—— & JONES, G. A., 1974, *J.C.P.*, **60**, 4321.
—— —— 1978, *C. & F.*, **32**, 1.
—— & PADLEY, P. J., 1966, *T.F.S.*, **62**, 2132 & 2140.
—— —— 1967, *11th Symposium*, p. 351.
JESSEN, P. F., 1971, *Gas Council tech. note*, LRSTN 207.
—— & GAYDON, A. G., 1969, *12th Symposium*, p. 481.
JOHN, R. R., & SUMMERFIELD, M., 1957, *Jet Propulsion*, **27**, 169.
JOHNSON, T. R., & BEER, J. M., 1973, *14th Symposium*, p. 639.
JONES, A., & PADLEY, P. J., 1975, *C. & F.*, **25**, 1.
JONES, A. R., 1973, *Combustion Institute, European Symposium*, p. 376.
—— LLOYD, S. A., & WEINBERG, F. J., 1978, *P.R.S.*, **360**, 97.
—— SCHWAR, M. J. R., & WEINBERG, F. J., 1971, *P.R.S.*, **322**, 119.
—— & WONG, W., 1975, *C. & F.*, **24**, 139.
JONES, F. J., BECKER, P. M., & HEINSOHN, R. J., 1972, *C. & F.*, **19**, 351.
JONES, G. W., LEWIS, B., FRIAUF, J. B., & PERROTT, G. St. J., 1931, *J. Amer. Chem. Soc.*, **53**, 869.
—— et al., 1944, *U.S. Bureau of Mines Rep. Investigations*, No. 3755.
JONES, H., 1977, *P.R.S.*, **353**, 459.
JONES, J. M., & ROSENFELD, J. L. J., 1972, *C. & F.*, **19**, 427.
JOST, W., 1944, *Z.p.C.*, **A193**, 332.
—— 1946, *Explosion and Combustion Processes in Gases*, McGraw-Hill, New York.

KALLEND, A. S., & NETTLETON, M. A., 1966, *Petrochemie*, **19**, 354.
KANDYBA, V. V., 1948, *Izv. Akad. Nauk. S.S.S.R.*, **12**, 387.
KANURY, A. M., 1975, *Introduction to Combustion Phenomena*, Gordon & Breach.
KAPILA, A. K., & LUNDFORD, G. S. S., 1977, *C. & F.*, **29**, 167.
KARL, G., KRAUS, P., & POLANYI, J. C., 1967, *J.C.P.*, **46**, 224.
KARLOVITZ, B., 1959, *7th Symposium*, p. 604.

KARLOVITZ, B., DENNISTON, D. W., & WELLS, F. E., 1951, *J.C.P.*, **19**, 541.
――― ――― & KNAPSHAEFER, D. H., 1953, *4th Symposium*, p. 613.
KASKAN, W. E., 1953, *4th Symposium*, p. 575.
――― 1965, *10th Symposium*, p. 41.
KAUFMAN, F., & KELSO, J. R., 1955, *J.C.P.*, **23**, 1702.
KAVELER, H. H., & LEWIS, B., 1937, *Chem. Rev.*, **21**, 421.
KAYE, G. W. C., & LABY, T. H., 1966, *Physical and Chemical Constants*.
KENT, J. H., 1970, *C. & F.*, **14**, 279.
KERKER, M., 1969, *The Scattering of Light*, Academic Press, New York.
KHITRIN, L. N., 1936, *J. Tech. Phys. U.S.S.R.*, **3**, 1048.
――― MOIN, P. B., SMIRNOV, D. B., & SHEVCHUK, V. U., 1965, *10th Symposium*, p. 1285.
KILHAM, J. K., 1949, *3rd Symposium*, p. 733.
――― & GARSIDE, J. E., 1949, *Gas Research Bd. Comm.*, **47**.
KIMURA, I., 1965, *10th Symposium*, p. 1295.
――― & UKAWA, H., 1962, *8th Symposium*, p. 521.
KINBARA, T., & NAKAMURA, J., 1955, *5th Symposium*, p. 285.
――― & NODA, K., 1971, *13th Symposium*, p. 333.
――― ――― 1975, *15th Symposium*, p. 993.
KING, I. R., 1957, *J.C.P.*, **27**, 817.
――― & CALCOTE, H. P., 1955, *J.C.P.*, **23**, 2203.
KIRK, R. E., & OTHMER, D. F., 1964, *Encyclopaedia of Chemical Technology*, **4**, 244, Wiley and Sons, New York.
KISTIAKOWSKY, G. B., & TABBUTT, F. D., 1959, *J.C.P.*, **30**, 577.
KLÄUKENS, H., & WOLFHARD, H. G., 1948, *P.R.S.*, **193**, 512.
KLEIN, G., 1957, *Phil. Trans. Roy. Soc.*, **A249**, 389.
KNEWSTUBB, P. F., 1965, *10th Symposium*, p. 623.
――― & SUGDEN, T. M., 1958, *Nature*, **181**, 474 & 1261.
KOHN, H., 1914, *Ann. Phys.* Leipzig, **44**, 749.
KOLODTSEV, K., & KHITRIN, L., 1936, *J. Tech. Phys. U.S.S.R.*, **3**, 1034.
KOSTKOWSKI, H. J., & BROIDA, H. P., 1956, *J.O.S.A.*, **46**, 246.
KOVASZNAY, L. S. G., 1956, *Jet Propulsion*, **26**, 485.
KRIER, H., & WRONKIEWICZ, J. A., 1972, *C. & F.*, **18**, 159.
KRYGSMAN, C., 1938, *Physica*, **5**, 918.
KUEHL, D. K., 1962, *8th Symposium*, p. 510.
KUHN, G., & TANKIN, R. S., 1968, *J. Quant. Spectr. Rad. Transfer*, **8**, 1281.
KUMAGAI, S., SAKAI, T., & OKAJIMA, S., 1971, *13th Symposium*, p. 779.
KUNUGI, M., & JINNO, H., 1959, *7th Symposium*, p. 942.
KURLBAUM, F., 1902, *Phys. Z.*, **3**, 187.
KYDD, P. H., & FOSS, W. I., 1965, *10th Symposium*, p. 101.

LADENBURG, R., & BERSHADER, D., 1955, *Physical Measurements in Gas Dynamics and Combustion*, Oxford University Press, p. 47.
LAFFITTE, P., et al., 1967, *11th Symposium*, p. 941.
LALOS, G. T., 1951, *J. Res. Nat. Bur. Stand.*, Wash., **47**, 179.
LAND, T., 1963, *Industrial and Process Heating*, **3**, 4, see also *C. & F.*, 1964, **8**, 66.
LANDER, C., 1942, *Proc. Inst., Mech. Eng.*, **148**, 81.
LANDOLT, H., 1856, *Pogg. Ann.*, **99**, 389.
LAPP, M., GOLDMAN, L. M., & PENNEY, C. M., 1972, *Science*, **175**, 1112.
――― & PENNEY, C. M., 1974, *Laser–Raman Gas Diagnostics*, Plenum Press.

LAUD, B. B., & GAYDON, A. G., 1971, *C. & F.*, **16**, 55.
LAW, C. K., 1978, *C. & F.*, **31**, 285.
LAWTON, J., & WEINBERG, F. J., 1969, *Electrical Aspects of Combustion*, Clarendon Press, Oxford.
LEARNER, R. C. M., 1962, *P.R.S.*, **269**, 311.
—— & GAYDON, A. G., 1959, *Nature*, **183**, 242.
LEFEBVRE, A. H., & REID, R., 1966, *C. & F.*, **10**, 355.
LEVY, A., 1965, *P.R.S.*, **283**, 134.
—— DROEGE, J. W., TIGHE, J. J., & FOSTER, J. F., 1962, *8th Symposium*, p. 524.
—— & WEINBERG, F. J., 1959, *C. & F.*, **3**, 229.
LEVY, J. B., & FRIEDMAN, R., 1962, *8th Symposium*, p. 663.
LEWIS, B., 1954, *Selected Combustion Problems*, Butterworths, London, p. 177.
—— 1973, *14th Symposium*, p. 1116.
—— & VON ELBE, G., 1939, *J.C.P.*, **7**, 197.
—— —— 1943, *J.C.P.*, **11**, 75.
—— —— 1961, *Combustion, Flames and Explosions of Gases*, Academic Press, New York, 2nd ed.
—— & GRUMER, J., 1948, *Industr. Engng. Chem.*, **40**, 1123.
LEYER, J. C., & MANSON, N., 1971, *13th Symposium*, p. 551.
LIEB, D. F., & ROBLEE, L. H. S., 1970, *C. & F.*, **14**, 285.
LIGHTHILL, M. J., 1952, *P.R.S.*, **211**, 564.
—— 1954, *P.R.S.*, **222**, 1.
—— 1962, *P.R.S.*, **267**, 147.
LINNETT, J. W., PICKERING, H. S., & WHEATLEY, P. J., 1951, *T.F.S.*, **47**, 974.
—— & SIMPSON, C. J. S. M., 1957, *6th Symposium*, p. 20.
LITCHFIELD, E. L., & BLANC, M. V., 1958, personal communication.
LLOYD, S. A., & WEINBERG, F. J., 1975, *Nature*, **257**, 367.
LOCKWOOD, F. C., & ODIDI, A. O., 1973, *Combustion Institute, European Symposium*, p. 507.
LOEB, L. B., 1960, *Basic Processes of Gaseous Electronics*, University of California Press.
LONGWELL, J. P., & WEISS, M. A., 1955, *Industr. Engng. Chem.*, **47**, 1634.
LOSHAEK, S., FEIN, R. S., & OLSEN, H. L., *J. Acoust. Soc. Amer.*, **31**, 605.
LOVACHEV, L. A., 1971, *C. & F.*, **17**, 275.
LÜCK, K. C., & MÜLLER, F. J., 1977, *J. Quant. Spectr. Rad. Transfer*, **17**, 403.
LURIE, H. H., & SHERMAN, G. W., 1933, *Industr. Engng. Chem.*, **25**, 404.

MACÉK, A., FRIEDMAN, R., & SEMPLE, J. M., 1964, *Heterogeneous Combustion, Progress in Astronautics and Aeronautics*, **15**, p. 3. Academic Press, New York.
—— & SEMPLE, J. M., 1969, *Combustion Sci. Techn.*, **1**, 181.
—— —— 1971, *13th Symposium*, p. 859.
MACFARLANE, J. J., HOLDERNESS, F. H., & WHITCHER, F. S. E., 1964, *C. & F.*, **8**, 215.
MACH, L., 1892, *Z. Instrum. Kde*, **12**, 275.
MACHE, H., 1943, *Forsch. Ing. Wes.*, **14**, 77.
—— & HEBRA, A., 1941, *S. B. Ost. Akad. Wiss. Abt.*, IIa, **150**, 157.
MACLEAN, D. I., & WAGNER, H. G., 1967, *11th Symposium*, p. 871.
MADGEBURG, H., & SCHLEY, U., 1966, *Z. angew. Phys.*, **20**, 465.
MANSON, N., 1945, *C.R. Acad. Sci.*, Paris, **220**, 734.
—— 1948, *C.R. Acad. Sci.*, Paris, **226**, 230.

MANTON, J., VON ELBE, G., & LEWIS, B., 1953, *4th Symposium*, p. 358.
MARKSTEIN, G. H., 1949, *J.C.P.*, **17**, 428.
—— 1951, *J. Aero. Sci.*, **18**, 199.
—— 1959, *7th Symposium*, p. 289.
—— 1963, *9th Symposium*, p. 137.
—— 1975, *15th Symposium*, p. 1285.
—— & SOMERS, L. M., 1953, *4th Symposium*, p. 527.
MARLOW, D. G., NISEWANGER, C. R., & CADY, W. M., 1949, *J. appl. Phys.*, **20**, 771.
MARR, G. V., 1957, *Canad. J. Phys.*, **35**, 1265.
MARRONE, P. V., 1966, *UTIAS* (Toronto) Report 113.
MARSEL, J., & KRAMER, L., 1959, *7th Symposium*, p. 906.
MARSH, P. A., VOET, A., MULLENS, T. J., & PRICE, L. D., 1971, *Carbon*, **9**, 797.
MASON, W. E., & WILSON, M. J. G., 1967, *C. & F.*, **11**, 195.
MASSEY, H. S. W., 1949, *Rep. Phys. Soc. Prog. Phys.*, **12**, 248.
MATSUI, K., KOYAMA, A., & UEHARA, K., 1975, *C. & F.*, **25**, 57.
MATTHEWS, C. S., & WARNECK, P., 1969, *J.C.P.*, **51**, 854.
MATTON, G., & FOURÉ, C., 1957, *6th Symposium*, p. 757.
MAYO, P. J., & WEINBERG, F. J., 1970, *P.R.S.*, **319**, 351.
MEDALIA, A. I., & HECKMAN, F. A., 1969, *Carbon*, 7, 567.
MELAERTS, W., DE SOETE, G., BERTRAND, J. N., & VAN TIGGELEN, A., 1960, *Bull. Soc. Chim. Belg.*, **69**, 95.
MENTALL, J. E., & NICHOLLS, R. W., 1965, *Proc. Phys. Soc.*, **86**, 873.
MERER, A. J., 1967, *Canad. J. Phys.*, **45**, 4103.
MERTENS, J., & POTTER, R. L., 1958, *C. & F.*, **2**, 181.
MIE, G., 1908, *Ann. Phys.* Leipzig, **25**, 377.
MILLER, D. R., EVERS, R. L., & SKINNER, G. B., 1963, *C. & F.*, **7**, 137.
MILLER, E., & SETZER, H. J., 1957, *6th Symposium*, p. 164.
MILLER, W. J., 1968, *Ionization in Combustion Processes, Oxidation and Combustion Reviews*, Elsevier, Amsterdam.
—— 1973, *14th Symposium*, p. 307.
MILLIKAN, R. C., 1962, *J. phys. Chem.*, **66**, 794.
MILLS, R. M., 1968, *C. & F.*, **12**, 513.
MINCHIN, S. T., 1935, *Proc. Wrld. Petrol Congr.*, **2**, 738.
—— 1949, *J. Inst. Fuel.*, **22**, 299.
MINKOFF, G. J., & TIPPER, C. F. H., 1962, *Chemistry of Combustion Reactions*, Butterworths, London.
MITCHELL, A. C. G., & ZEMANSKY, M. W., 1934, *Resonance Radiation and Excited Atoms*, Cambridge (reprinted 1961).
MIZUTANI, Y., 1972, *C. & F.*, **19**, 203.
MOORHOUSE, J., WILLIAMS, A., & MADDISON, T. E., 1974, *C. & F.*, **23**, 203.
MORGAN, F. H., & DANFORTH, W. E., 1950, *J. appl. Phys.*, **21**, 112.
MORLEY, C., 1976, *C. & F.*, **27**, 189.
MORRISON, M. E., & SCHELLER, K., 1972, *C. & F.*, **18**, 3.
MULLANEY, G. J., 1958, *Rev. sci. Instrum.*, **29**, 87.
MULLINS, B. P., 1955, *Spontaneous Ignition*, Butterworths, London.
MUNDAY, G., UBBELOHDE, A. R., & WOOD, I. F., 1968, *P.R.S.*, **306**, 179.
MUNTZ, E. P., 1962, *Phys. of Fluids*, **5**, 80.
MURRAY, R. C., & HALL, A. R., 1951, *T.F.S.*, **47**, 743.
MYERSON, A. L., TAYLOR, F. R., & FAUNCE, B. G., 1957, *6th Symposium*, p. 154.

NAKAKUKI, A., 1973, *C. & F.*, **20**, 135.
NEEDHAM, D. P., & POWLING, J., 1955, *P.R.S.*, **232**, 337.
NETTLETON, M. A., & STIRLING, R., 1967, *P.R.S.*, **300**, 62.
NEUBERT, H., 1943, *Z. tech. Phys.*, **24**, 180.
NEWITT, D. M., & THORNES, L. S., 1937, *J. Chem. Soc.*, p. 1656.
NICHOLLS, R. W., 1956, *Proc. Phys. Soc.*, **A69**, 741.
NOLTINGK, B. E., ROBINSON, N. E., & GAYDON, B. G., 1975, *J. Inst. Fuel*, **48**, 127.
NURUZZAMAN, A. S. M., & BEER, J. M., 1971, *Combustion Sci. Techn.*, **3**, 17.
—— HEDLEY, A. B., & BEER, J. M., 1971, *13th Symposium*, p. 787.

OLSEN, H. L., 1949, *3rd Symposium*, p. 663.
ONSAGER, L., & WATSON, W. W., 1939, *Phys. Rev.*, **56**, 474.
OPPENHEIM, A. K., URTIEW, P. A., & WEINBERG, F. J., 1966, *P.R.S.*, **291**, 279.
ORNSTEIN, L. S., & BRINKMAN, H., 1931, *K. Akad. Amsterdam*, **34**, 33 & 489.

PADLEY, P. J., & SUGDEN, T. M., 1958, *P.R.S.*, **248**, 248.
—— —— 1959, *7th Symposium*, p. 235.
—— —— 1962, *8th Symposium*, p. 164.
PALMER, H. B., & BEER, J. M., 1974, *Combustion Technology, Some Modern Developments*, Academic Press, New York.
PALMER, K. N., 1959, *7th Symposium*, p. 497.
—— 1973, *Dust Explosions and Fire*, Chapman & Hall, London.
PALM-LEIS, A., & STREHLOW, R. A., 1969, *C. & F.*, **13**, 111.
PANDYA, T. P., & SRIVASTAVA, N. K., 1975, *Combustion Sci. Techn.*, **11**, 165.
—— & WEINBERG, F. J., 1963, *9th Symposium*, p. 587.
—— —— 1964, *P.R.S.*, **279**, 544.
PARKER, K. H., & GUILLON, O., 1971, *13th Symposium*, p. 667.
PARKER, W. G., & WOLFHARD, H. G., 1950, *J. Chem. Soc.*, p. 2038.
—— —— 1953, *4th Symposium*, p. 420.
—— —— 1956, *Fuel*, **35**, 323.
PATANKAR, S. V., & SPALDING, D. B., 1973, *14th Symposium*, p. 605.
PAYNE, K. G., & WEINBERG, F. J., 1959, *P.R.S.*, **250**, 316.
—— —— 1962, *8th Symposium*, p. 207.
PEETERS, J., 1973, *Combustion Institute, European Symposium*, p. 245.
—— & MAHNEN, G., 1973, *Combustion Institute, European Symposium*, p. 53.
PEGG, R. E., & RAMSDEN, A. W., 1966, *Proc. International Clean Air Conference*, London. Part 1, Paper VI/1.
PENNER, S. S., 1951, *J.C.P.*, **19**, 272.
—— 1959, *Quantitative Molecular Spectroscopy and Gas Emissivities*, Addison-Wesley, Reading, Mass.
PERGAMENT, H. S., & CALCOTE, H. F., 1967, *11th Symposium*, p. 597.
PERRIN, F., 1930, *Chem. Rev.*, **7**, 231.
PICKERING, H. S., & LINNETT, J. W., 1951a, *T.F.S.*, **47**, 985.
—— —— 1951b, *T.F.S.*, **47**, 1101.
PILLOW, M. E., 1952, personal communication.
PITZ, R. W., CATTOLICA, R., ROBBEN, F., & TALBOT, L., 1976, *C. & F.*, **27**, 313.
PLACE, E. R., & WEINBERG, F. J., 1966, *P.R.S.*, **289**, 192.
—— —— 1967, *11th Symposium*, p. 245.
POLANYI, J. C., 1963, *J. Quant. Spectr. Rad. Transfer*, **3**, 471.

POLLARD, F. H., & WYATT, R. M. H., 1950, *T.F.S.*, **46**, 281.
POLYMEROPOULOS, C. E., & DAS, S., 1975, *C. & F.*, **25**, 247.
PORTER, G., 1953, *4th Symposium*, p. 248.
PORTER, J. W., 1967, *C. & F.*, **11**, 501.
PORTER, R. P., 1970, *C. & F.*, **14**, 275.
—— CLARK, A. H., KASKAN, W. E., BROWNE, W. G., 1967, *11th Symposium*, p. 907.
POTTER, A. E., 1960, *Prog. Combustion Sci. Techn.*, **1**, 145.
—— HEIMEL, S., & BUTLER, J. N., 1962, *8th Symposium*, p. 1027.
POTTER, R. L., & WAYMAN, D. H., 1958, *C. & F.*, **2**, 129.
POWELL, A., 1963, *J. Acoust. Soc. Amer.*, **35**, 405.
POWLING, J., 1949, *Fuel*, **28**, 25.
—— & SMITH, W. A. W., 1958, *C. & F.*, **2**, 157.
POWNALL, C., & SIMMONS, R. F., 1971, *13th Symposium*, p. 585.
PRANDTL, L., 1942, *Ströumungslehre*, Vieweg, Brunswick, p. 370.
PRICE, R. B., HURLE, I. R., & SUGDEN, T. M., 1969, *12th Symposium*, p. 1093.
PRITCHARD, R., EDMONDSON, H., & HEAP, M. P., 1972, *C. & F.*, **18**, 13.
PROTHERO, A., 1969, *C. & F.*, **13**, 399.
PUTNAM, A. A., & BROWN, D. J., 1974, *Combustion Technology, Some Modern Developments* (ed. Palmer & Beer), Academic Press, New York, p. 128.
—— & DENNIS, W. R., 1956, *J. Acoust. Soc. Amer.*, **28**, 246 & 260.

QUINN, H. F., 1950, *Canad. J. Res.*, **A28**, 411.

RAEZER, S. D., & OLSEN, H. L., 1962, *C. & F.*, **6**, 227.
RAO, K. V. L., & LEFEBVRE, A. H., 1976, *C. & F.*, **27**, 1.
RASBASH, D. J., ROGOWSKI, Z. W., & STARK, G. W. V., 1956, *Fuel*, **35**, 94.
RAY, S. K., & LONG, R., 1964, *C. & F.*, **8**, 139.
RAYLEIGH, LORD, 1882, *Phil. Mag.*, **13**, 340 and *Scientific Papers*, II, p. 101.
—— 1896, *Theory of Sound*, Macmillan, London.
REED, S. B., 1967, *C. & F.*, **11**, 177.
—— 1971, *C. & F.*, **17**, 105.
REKERS, R. G., & VILLARS, D. S., 1956, *J.O.S.A.*, **46**, 534.
RIBNER, H. S., 1959, *J. Acoust. Soc. Amer.*, **31**, 245.
—— 1960, *UTIAS* (Toronto) Report 67.
—— 1967, *UTIAS* (Toronto) Report 128.
RICHARDSON, E. G., 1923, *Proc. Phys. Soc.*, **35**, 47.
RICHMOND, J. K., & SINGER, J. M., et al., 1957, *6th Symposium*, p. 303.
ROBSON, K., & WILSON, M. J. G., 1969, *C. & F.*, **13**, 626.
ROPER, F. G., 1977, *C. & F.*, **29**, 219.
—— 1978, *C. & F.*, **31**, 251.
ROSHKO, A., 1976, *AIAA 14th Aerospace Meeting*, p. 1.
ROSSER, W. A., INAMI, S. H., & WISE, H., 1963, *C. & F.*, **7**, 107.
ROSSINI, F. D., et al., 1947, *Nat. Bur. Stand. Circular*, 461.
RÖSSLER, F., & BEHRENS, H., 1950, *Optik*, **6**, 145.
ROTH, W., 1958, *J.C.P.*, **28**, 668.
—— & BAUER, W. H., 1955, *5th Symposium*, p. 710.
—— & RAUTENBERG, T. H., 1956, *J. phys. Chem.*, **60**, 379.
ROZLOVSKII, A. I., & ZAKAZNOV, V. F., 1971, *C. & F.*, **17**, 215.
RUMMEL, K., & VEH, P. O., 1941, *Arch. Eisenhüttenwesen*, **14**, 489.

SACHSE, H., & BARTHOLOMÉ, E., 1949, *Z. Elektrochem*, **53**, 183.
SAHA, M. N., & SAHA, H. K., 1934, *A Treatise on Modern Physics*, Vol. I, Allahabad and Calcutta.
SAHNI, O., 1969, *J. Phys.*, **D2**, 471.
SALOOJA, K. C., 1972, *Nature*, **240**, 350.
—— 1973, *Combustion Institute, European Symposium*, p. 400.
SANDERSON, J. A., CURCIO, J. A., & ESTES, D. V., 1948, *Phys. Rev.*, **74**, 1221.
SATO, A., et al., 1975, *C. & F.*, **24**, 35.
SAYERS, J. F., TEWARI, G. P., WILSON, J. R., & JESSEN, P. F., 1971, *J. Inst. Gas Engineers*, **11**, 322.
SCHACK, A., 1925, *Z. tech. Phys.*, **6**, 530.
SCHARDIN, H., 1942, *Ergebn. exakt. Naturw.*, **20**, 303.
SCHMIDT, H., 1909, *Ann. Phys.* Leipzig, **29**, 998.
SCHMITZ, R. A., 1967, *C. & F.*, **11**, 49.
SCHOFIELD, K., & SUGDEN, T. M., 1965, *10th Symposium*, p. 589.
SCHOLEFIELD, D. A., & GARSIDE, J. E., 1949, *3rd Symposium*, p. 102.
SCHORPIN, S. N., 1950, *Izv. Akad. Nauk. U.S.S.R.*, **7**, 995.
SCHWAR, M. J. R., PANDYA, T. P., & WEINBERG, F. J., 1967, *Nature*, **215**, 239.
—— & WEINBERG, F. J., 1969, *Nature*, **221**, 357; *P.R.S.*, **311**, 469.
SEMENOFF, N., 1940, *Progr. Phys. Sci. U.S.S.R.*, **24**, 433 and *N.A.C.A. tech. note*, 1026.
SENFTLEBEN, H., & BENEDICT, E., 1919, *Ann. Phys.*, Leipzig, **60**, 297.
SHCHELKIN, K. J., 1943, *J. Tech. Phys. U.S.S.R.*, **13**, 520.
SHEINSON, R. S., & WILLIAMS, F. W., 1973, *Combustion Institute, European Symposium*, p. 707.
SHETINKOV, E. S., 1959, *7th Symposium*, p. 583.
SHIRODKAR, A., 1933, *Phil. Mag.*, **15**, 426.
SIDDALL, R. G., & MCGRATH, I. A., 1963, *9th Symposium*, p. 102.
SILLA, H., & DOUGHERTY, T. J., 1972, *C. & F.*, **18**, 65.
SILVERMAN, S., 1949, *3rd Symposium*, p. 498.
SIMMONDS, W. A., & WILSON, M. J. G., 1951, *Gas Research Bd. Comm.*, **61**.
SIMMONS, F., 1967, *AIAA Journal*, **5**, 778.
SIMMONS, R. F., & WOLFHARD, H. G., 1955, *T.F.S.*, **51**, 1211.
—— —— 1957a, *Jet Propulsion*, **27**, 44.
—— —— 1957b, *Z. Elektrochem.*, **61**, 601.
SIMON, D. M., 1951, *J. Amer. Chem. Soc.*, **73**, 422.
—— 1959, *7th Symposium*, p. 413.
SIMPSON, C. J. S. M., & LINNETT, J. W., 1957, *6th Symposium*, p. 149.
SINGER, J. M., 1953, *4th Symposium*, p. 352.
—— & VON ELBE, G., 1957, *6th Symposium*, p. 127.
SJÖGREN, A., 1973, *14th Symposium*, p. 919.
SKIRROW, G., & WOLFHARD, H. G., 1955, *P.R.S.*, **232**, 78 & 577.
SMITH, D., & AGNEW, J. T., 1957, *6th Symposium*, p. 83.
SMITH, E. C. W., 1940a, *P.R.S.*, **174**, 110.
—— 1940b, *Inst. Gas Engr.*, No. 237.
SMITH, F. A., & PICKERING, S. F., 1929, *J. Res. Nat. Bur. Stand.*, Wash., **3**, 65.
—— —— 1936, *J. Res. Nat. Bur. Stand.*, Wash., **17**, 7.
SMITH, S. J., & BRANSCOMB, L. M., 1955, *J. Res. Nat. Bur. Stand.*, Wash., **55**, 165.
SMITH, S. R., & GORDON, A. S., 1956, *J. phys. Chem.*, **60**, 759.

SMITH, T. J. B., & KILHAM, J. K., 1963, *J. Acoust. Soc. Amer.*, **35**, 715.
SMITHELLS, A., & INGLE, H., 1892, *Trans. Chem. Soc.*, **61**, 204.
SMOOT, L. D., HECKER, W. C., & WILLIAMS, G. A., 1976, *C. & F.*, **26**, 323.
SNELLEMAN, W., 1967, *C. & F.*, **11**, 453.
SNYDER, W. T., 1962, *8th Symposium*, p. 573.
SOBOLEV, N. N., 1949, *J. Exp. Theor. Phys. U.S.S.R.*, **19**, 25.
—— & SHCHETININ, T. I., 1950, *J. Exp. Theor. Phys. U.S.S.R.*, **20**, 356.
SORENSON, S. C., SAVAGE, L. D., & STREHLOW, R. A., 1975, *C. & F.*, **24**, 347.
SPALDING, D. B., 1954, *Fuel*, **33**, 255.
—— 1955a, *Fuel*, **34**, Supplement, p. S.100.
—— 1955b, *Some Fundamentals of Combustion*, Butterworths, London.
—— 1956a, *Fuel*, **35**, 347.
—— 1956b, *Phil. Trans. Roy. Soc. London*, **A249**, 1.
—— & JAIN, V. K., 1962, *C. & F.*, **6**, 265.
—— STEPHENSON, P. L., & TAYLOR, R. G., 1971, *C. & F.*, **17**, 55.
SPENCE, D., & MCHALE, E. T., 1975, *C. & F.*, **24**, 211.
SPENCE, K., & TOWNEND, D. T. A., 1949, *3rd Symposium*, p. 404.
SPIERS, H. M., 1955, *Technical Data on Fuels*, 5th edition, Brit. Nat. Comm., World Power Conference, London.
SPOKES, G. N., & GAYDON, A. G., 1958, *Nature*, **180**, 1114.
STAIR, R., SCHNEIDER, W. E., & JACKSON, J. K., 1963, *Appl. Optics*, **2**, 1151.
STARR, W. L., 1965, *J.C.P.*, **43**, 73.
STEERE, N. V., 1967, *Handbook of Laboratory Safety*, The Chemical Rubber Co., Cleveland.
STEINBERGER, R., & SCHAAF, V. P., 1958, *J. phys. Chem.*, **62**, 280.
STEINLE, H., 1939, *Z. angew. Mineralogie*, **2**, 28.
STENHOUSE, I. A., & WILLIAMS, D. R. (AERE, Harwell), 1978. Informal presentation at Institute of Physics meeting held at British Gas Corporation
STRAUSS, W. A., & EDSE, R., 1959, *7th Symposium*, p. 377.
STREET, J. C., & THOMAS, A., 1955, *Fuel*, **34**, 4.
STREHLOW, R. A., 1968, *Fundamentals of Combustion*, International Textbook Co., Scranton.
—— & SAVAGE, L. D., 1978, *C. & F.*, **31**, 209.
—— & STUART, J. G., 1953, *4th Symposium*, p. 329.
STRENG, A. G., & GROSSE, A. V., 1957, *6th Symposium*, p. 264.
—— STOKES, C. S., & STRENG, L. A., 1958, *J.C.P.*, **29**, 458.
STRICKER, W., 1976, *C. & F.*, **27**, 133.
STRONG, H. M., BUNDY, F. P., & LARSON, D. A., 1949, *3rd Symposium*, p. 641.
—— —— 1954, *J. appl. Phys.*, **25**, 1521, 1527 & 1531.
STULL, V. R., & PLASS, G. N., 1960, *J.O.S.A.*, **50**, 121.
SUMMERFIELD, M., REITER, S. H., KEBELY, V., & MASCALO, R. W., 1955, *Jet Propulsion*, **25**, 377.

TANFORD, C., 1947, *J.C.P.*, **15**, 433.
—— & PEASE, R. N., 1947a, *J.C.P.*, **15**, 431.
—— —— 1947b, *J.C.P.*, **15**, 861.
TAYLOR, R. L., & BITTERMAN, S., 1969, *Rev. Mod. Phys.*, **41**, 26.
TELLER, E., 1937, *J. phys. Chem.*, **41**, 109.
TESNER, P. A., 1959, *7th Symposium*, p. 546.
—— 1962, *8th Symposium*, p. 627.

TESNER, P. A., 1973, *Faraday Symposium, Chem. Soc.*, **7**, 104.
—— SNEGIRIOVA, T. D., & KNORRE, V. G., 1971, *C. & F.*, **17**, 253.
—— et al., 1971, *C. & F.*, **17**, 279.
TEWARI, G. P., & WEINBERG, F. J., 1967, *P.R.S.*, **296**, 546.
THABET, S. K., 1951, *Ph.D. Thesis*, London.
THOMAS, A., 1962, *C. & F.*, **6**, 46.
—— & WILLIAMS, G. T., 1966, *P.R.S.*, **294**, 449.
THOMAS, D. B., 1962, *J. Res. Nat. Bur. Stand.*, Wash., **66C**, 255.
THOMAS, D. L., 1968a, *C. & F.*, **12**, 541.
—— 1968b, *C. & F.*, **12**, 569.
THOMAS, N., GAYDON, A. G., & BREWER, L., 1952, *J.C.P.*, **20**, 369.
THORP, C. E., 1955, *Bibliography of Ozone Technology*, Vol. 2, Armour Research Foundation, Chicago.
—— LONG, R., & GARNER, F. H., 1951, *Fuel*, **30**, 266.
THRING, M. W., 1962, *The Science of Flames and Furnaces*. Chapman & Hall, 2nd ed.
TOEPLER, A., 1866, *Ann. Phys.* Leipzig, **128**, 126.
TOLLMIEN, W., 1926, *Z. angew. Math. Mech.*, **6**, 468.
—— 1931, *Handbuch der Exp. Phys.*, IV, Pt. 1, p. 241.
TOONG, T. Y., SALANT, R. F., STOPFORD, J. M., & ANDERSON, G. Y., 1965, *10th Symposium*, p. 1301.
TOPPS, J. E. C., & TOWNEND, D. T. A., 1946, *T.F.S.*, **42**, 345.
TOURIN, R. H., 1961, *J.O.S.A.*, **51**, 175.
—— 1966, *Spectroscopic Gas Temperature Measurement*, Elsevier, Amsterdam.
TOWNEND, D. T. A., 1927, *P.R.S.*, **116**, 637.
—— GARSIDE, J. A., & CULSHAW, G. W., 1941–8, *Inst. Gas. Engr. Comm.*, **220**, 244, 245, 255, 274, 306, 325 & 346.
TRAVERS, B. E. L., & WILLIAMS, H., 1965, *10th Symposium*, p. 657.
TROPSCH, H., & EGLOFF, G., 1935, *Industr. Engng. Chem.*, **27**, 1063.
TSE, R. S., MICHAUD, P., & DELFAU, J. L., 1978, *Nature*, **272**, 153.
TSUJI, H., & TAKENO, T., 1965, *10th Symposium*, p. 1327.
—— & YAMAOKA, I., 1967, *11th Symposium*, p. 979.
—— —— 1971, *13th Symposium*, p. 723.

UBEROI, M. S., 1954, *J.C.P.*, **22**, 1784.
UNSÖLD, A., 1955, *Physik der Sternatmosphären*, Springer, Berlin.

VAN DER HULST, H. C., 1957, *Light Scattering by Small Particles*, Wiley, New York.
VAN DER POLL, A. N. J., & WESTERDIJK, T., 1941, *Z. tech. Phys.*, **22**, 29.
VANPEE, M., HINCK, E. C., & SEAMANS, T. F., 1965, *C. & F.*, **9**, 393.
—— & SEAMANS, T. F., 1967, *11th Symposium*, p. 931.
VASILIEVA, I. A., DEPUTATOVA, L. V., & NEFEDOV, A. P., 1974, *C. & F.*, **23**, 305.
VEAR, C. J., HENDRA, P. J., & MACFARLANE, J. J., 1972, *J.C.S. Chem. Comm.*, p. 381.
VENUGOPALAN, M., 1971 (Editor), *Reactions under Plasma Conditions*, Wiley-Interscience.
VERKAMP, F. J., HARDIN, M. C., & WILLIAMS, J. R., 1967, *11th Symposium* p. 985.

REFERENCES

VON ELBE, G., & LEWIS, B., 1959, *7th Symposium*, p. 342.
—— & MENTSER, M., 1945, *J.C.P.*, **13**, 89.
VON ENGEL, A., & COZENS, J. R., 1963, *Proc. Phys. Soc.*, **82**, 85.
VON KARMAN, T., & PENNER, S. S., 1954, *Selected Combustion Problems*, Butterworths, London, p. 5.

WATERMEIER, L. A., 1957, *J.C.P.*, **27**, 1118.
WATSON, J. H. L., 1948, *Analyt. Chem.*, **20**, 576.
WATSON, R., & FERGUSON, W. R., 1965, *J. Quant. Spectr. Rad. Transfer*, **5**, 595.
WAYMAN, D. H., & POTTER, R. L., 1957, *C. & F.*, **1**, 321.
WEBSTER, J. M., WEIGHT, R. P., & ARCHENBOLD, E., 1976, *C. & F.*, **27**, 395.
WEINBERG, F. J., 1955, *Fuel*, **34**, S.84.
—— 1956a, *Fuel*, **35**, 359.
—— 1956b, *P.R.S.*, **235**, 510.
—— 1957, *P.R.S.*, **241**, 132.
—— 1963, *Optics of Flames*, Butterworths, London.
—— 1968, *P.R.S.*, **307**, 195.
—— 1975, *15th Symposium*, p. 1; *Prog. Energy Combust. Sci.*, **1**, 17.
—— & WILSON, J. R., 1970, *P.R.S.*, **314**, 175.
—— —— 1971, *P.R.S.*, **321**, 41.
—— WONG, W. W. Y., 1975, *P.R.S.*, **345**, 379.
—— & WOOD, N. B., 1959, *J. Sci. Instrum.*, **36**, 227.
WELTMANN, R. N., & KUHNS, P. W., 1951, *N.A.C.A. tech. note*, 2580.
WENAAS, E. P., & MCCHESNEY, J., 1970, *C. & F.*, **15**, 85.
WESTENBERG, A. A., & FRISTROM, R. M., 1965, *10th Symposium*, p. 473.
WESTERMANN, I., 1930, *Erzbergbau*, **28**, 613.
WHARTON, C. B., 1963, *Physico-chemical Diagnostics of Plasmas* (Ed. T. P. Anderson *et al.*), *Proc. Fifth Biennial Gas Dynamics Symposium*, Northwestern University Press, Evanston, p. 14.
WHARTON, W. W., VIOLETT, T. D., & MILLER, E., 1957, *6th Symposium*, p. 173.
WHATLEY, A. T., & PEASE, R. N., 1954, *J. Amer. Chem. Soc.*, **76**, 1997.
WHITTAKER, A. G., & WILLIAMS, H., 1957, *J. phys. Chem.*, **61**, 388.
—— —— & RUST, P. M., 1956, *J. phys. Chem.*, **60**, 904.
WIDHOPF, G. F., & LEDERMAN, S., *A.I.A.A. Jour.*, **9**, 309.
WIESE, W. L., 1965, *Plasma Diagnostic Techniques*, Academic Press, New York.
WILKINS, J., *et al.*, 1977, *9th Annual Offshore Technology Conference*, p. 123.
WILLIAMS, A., 1973, *C. & F.*, **21**, 1.
WILLIAMS, H., 1959, *7th Symposium*, p. 269.
WILSON, R. H., 1953, *Publ. Astronom. Soc. Pacific*, **65**, 295.
WILSON, R. P., & WILLIAMS, F. A., 1971, *13th Symposium*, p. 833.
WILSON, W. E., 1965, *10th Symposium*, p. 47.
WOHL, K., GAZLEY, C., & KAPP, N., 1949, *3rd Symposium*, p. 288.
—— KAPP, N., & GAZLEY, C., 1949, *3rd Symposium*, p. 3.
WOLFHARD, H. G., 1939, *Z. Phys.*, **112**, 107.
—— 1943, *Z. tech. Phys.*, **24**, 206.
—— 1955, *Fuel*, **24**, 60.
—— 1956, *Selected Combustion Problems, II.* Butterworths, London, p. 328.
—— & BRUSZAK, A. E., 1960, *C. & F.*, **4**, 149.
—— & HINCK, E., 1967, *11th Symposium*, p. 589.
—— & PARKER, W. G., 1949a, *Proc. Phys. Soc.*, **B62**, 523.

WOLFHARD, H. G., & PARKER, W. G., 1949b, *Proc. Phys. Soc.*, **A62**, 722.
—— —— 1950, *Fuel*, **29**, 325.
—— —— 1952, *Proc. Phys. Soc.*, **A65**, 2.
—— —— 1955, *5th Symposium*, p. 718.
—— & SEAMANS, T. F., 1962, *Fire Research Abstracts and Reviews*, **4**, 92.
—— & STRASSER, A., 1958, *J.C.P.*, **28**, 172.
WOOD, B. J., WISE, H., & ROSSER, W. A., 1957, *J.C.P.*, **27**, 807.
WRIGHT, F. J., 1970, *C. & F.*, **15**, 217.

YOFFE, A. D., 1953, *Research*, **6**, Supplement, p. 11.
YUMLU, V. S., 1967, *C. & F.*, **11**, 190.
—— 1968, *C. & F.*, **12**, 14.

ZEHNDER, L., 1891, *Z. Instrum. Kde*, **11**, 275.
ZELDOVICH, Y. B., 1947, *Z. tech. Phys., U.S.S.R.*, **17**, 3.
—— 1949, *Z. tech. Phys. U.S.S.R.*, **19**, 1199.
—— & FRANK-KAMENETSKY, D. A., 1938a, *C.R. Acad. Sci. U.S.S.R.*, **19**, 693. & 699.
—— —— 1938b, *J. Phys. Chem. U.S.S.R.*, **12**, 100.
—— —— 1938c, *Acta Physicochim. U.S.S.R.*, **9**, 341.
—— & SEMENOFF, N., 1940, *J. Exp. Theor. Phys. U.S.S.R.*, **10**, 1116 & 1427.
—— & SHAULOV, Y. K., 1946, *J. Phys. Chem. U.S.S.R.*, **20**, 1359.
ZICKENDRAHT, H., 1941, *Helv. Phys. Acta*, **14**, 132.

Author Index

Author Index

Acton, L., 263
Adams, G. K., 374, 380, 383
Addecott, K. S. B., 209, 210
Adomeit, G., 28
Agnew, J. T., 25, 83, 217, 218
Agnew, W. G., 25, 195
Agoston, G. A., 176
Akita, K., 175
Alkemade, C. T. J., 282, 355
Allen, C. W., 349, 350
Anacker, F., 277
Anagnostou, E., 164
Andersen, J. W., 46, 47, 68, 70, 79, 80
Anderson, G. Y., 188
Anderson, T. P., 347
Andrews, G. E., 65, 79, 81, 136, 142, 145
Archenhold, E., 55
Arden, E. A., 379
Arthur, J. R., 208, 209, 215
Ashforth, G. K., 63, 64
Atallah, S., 176
Aten, C. F., 225
Avery, W. H., 97
Awbery, J. H., 273, 283, 317
Ay, J., 130, 359

Babcock, W. R., 194
Badami, G. N., 72
Baddiel, C. B., 385
Baden, H. C., 388
Baker, K. L., 194
Baker, M. L., 129
Baker, R. J., 45
Ballal, D. R., viii, 27, 142, 144, 145
Ballinger, P. R., 25
Balwanz, W. W., 342-4, 367, 368
Barassin, A., 321
Barnes, M. H., 30
Barr, J., vi, 150, 153, Plate 8
Bartholomé, E., 66, 79-81, 87, 88, 99, 108-13, 117, 209, 339, 385
Bartok, W., 131
Bauer, E., 244
Bauer, W. H., 388

Beams, J. W., 69
Beattie, I. R., 311
Bechert, K., 120
Becker, P. M., 124, 164
Beer, J. M., vii, 170-2, 181, 267
Behrens, H., vi, 34, 37, 189, 202, 217, 258, 259, 296, 381, Plates 4, 5
Bell, J. C., 321
Benedict, E., 258
Benedict, W. S., 263
Bennett, R. G., 249
Berl, E., 30
Berl, W. G., 389, 390
Berlad, A. L., 17
Bershader, D., 53
Bertholet, 224
Bertrand, J. N., 284
Billig, F. S., 400
Bitterman, S., 244
Blair, D. W., 366
Blanc, M. V., 16, 373
Bleekrode, R., 220
Blinov, V. J., 175
Blumenthal, J. L., 395
Boers, A. L., 282
Bollinger, L. M., 14, 90, 143, Plate 7
Bone, W. A., v, 4, 29, 196, 206, 217
Bonhoeffer, K. F., 242
Bonne, U., 203, 214, 221, 222, 228
Botha, J. P., 73, 80
Bourke, P. J., 45
Bouvier, A., 305
Bowden, F. P., 381
Bowser, R. J., 210, 233, 359, 360
Boyer, M. H., 378
Boys, S. F., 117
Bozman, W. R., 299
Bradley, D., 65, 78-80, 136, 142, 145, 321, 322
Bragg, S., 177, 181, 182, 187
Brame, J. S. S., 175
Braud, Y., vii, 2, 42, 316-8
Breisacher, P., 389, 390
Brewer, L., 220, 308, 390, 395

Brinkman, H., 307
Broatch, D., Plate 15
Broida, H. P., vi, 38, 103, 298, 303, 305, 389
Bronfin, B. R., 403
Broughton, F. P., 295, 296, 316
Brown, B., 397
Brown, D. J., 181
Brown, G. B., 190–2
Brown, H. R., 393
Brown, R. L., 338
Browne, W. G., 105, 124
Brule, G., 321
Bruszak, A. G., 19
Brzustowski, T. A., 171, 396
Bulewicz, E. M., 101, 209, 210
Bundy, F. P., 273, 274
Bunsen, R., 7, 269
Burgess, M. J., 30
Burgoyne, J. H., vi, 96, 97, 172–6, 312
Burke, E., 96
Burke, S. P., 147, 148, 150, 168
Burns, R. P., 217
Burt, R., 195
Butler, J. N., 164

Cady, W. M., 315
Calcote, H. F., 347, 352, 358–70
Caldin, E. F., 396
Callear, A. B., 248, 251
Campbell, D. E., 121
Carey, C. R., 279, 315
Carnevale, E. H., 279, 315
Carrington, T., 303
Cassel, H. M., 394
Cattaneo, A. G., 194
Cattolica, R., 300, 301
Charton, M., 101, 257
Chedaille, J., vii, 2, 42, 316–8
Chevaleyre, J., 305
Chigier, N. A., 45
Clark, A. H., 105, 124
Clarke, A. E., 129, 197–9
Clouston, J. G., 257, 258
Clusius, K., 36
Cohen, L., 172, 173
Combourieu, J., 381, 382
Commins, B. T., 215
Conan, H. R., 72
Conway, J. B., 390, 393, 396
Cookson, R. A., 18
Corbeels, R. J., 73
Corliss, C. H., 299
Corner, J., 117
Cotton, D. H., 209, 210
Cowan, G. R., 257

Coward, H. F., 198, 199
Cozens, J. P., 322
Cros, J. C., 305
Crosswhite, H. M., 298, 302
Csaba, J., 173
Cullis, C. F., 23, 224, 225, 227, 235, 385
Culshaw, G. W., 65, 80, 81, Plate 13
Cummings, G. A., 363, 373, 381
Curcio, J. A., 296
Curtiss, C. F., 121
Cyphers, J. A., 88, 113, 284

Dalby, F. W., 249
Daily, J. W., 274
d'Alessio, A., 213, 222, 228
Dalzell, W. H., 213, 260, 261
Damköhler, G., 55, 90, 106, 133, 136, 326
Danforth, W. E., 317
Das Gupta, A. K., 394
Das, S., 172
Dashchuk, M., 136
David, W. T., 18, 283, 284, 318
Davies, D. A., 317
Davy, H., 196, 224
Day, M. J., 115, 122
Deckker, B. E. L., 188
Delaney, R. M., 315
De Leeuw, J. H., 346
Delfau, J. L., 359
De Maria, G., 217
Dembrow, D., 389, 390
Dennis, W. R., 186
Denniston, D. W., 126, 136
Depoy, F. E., 25
Deputatova, L. V., 274
De Soete, G. G., 27, 142, 284
Desty, D. H., 170
De Vos, J. C., 271
Diederichsen, J., 84, 186
Dieke, G. H., 298, 302
Dingle, M. G. W., 279
Dixon-Lewis, G., 92, 97–102, 115, 121–3, 312
Dougherty, T. J., 321
Drew, C. M., 396
Droege, J. W., 115
Drowart, J., 217
Dubois, M., 192
Duff, R. E., 391
Dugger, G. L., 82
Dunning, W. J., 229
Durão, D. F. G., 45, 188
Durie, R. A., 208, 382
Durst, F., 45, 136
Dvorak, K., 45, 51

AUTHOR INDEX

Eberius, K. H., 124
Echigo, R., 196
Eckstein, B. H., 388
Edelman, R. B., 131, 154
Edmondson, H., 16, 55, 74, 80, 126, 127
Edse, R., 83, 326, 332, 374
Egerton, A. C., vi, 19, 25, 29, 30, 35–7, 72, 100
Eggert, J., 242
Egloff, G., 224, 225
Ehert, L., 42, 80, 192–4
Eiseman, J. H., 7, 8, 10
Ellis, C., 236
Eltenton, G. C., 186
Emmons, H. W., 176
Engelman, V. S., 131
English, P., 279
Erickson, W. D., 260
Essenhigh, R. H., 170, 173
Estes, D. V., 296
Euler, J., 51, 276
Evans, D. G., 209, 210
Everest, D. A., 338
Everett, A. J., 19
Evers, R. L., 89
Eyring, H., 247

Fabelinskii, I. L., 260
Fairbairn, A. R., 231, 277
Faizullov, F. S., 277
Fales, E. N., 135
Faunce, B. G., 377
Fay, J. A., 150
Fein, R. S., 46, 47, 68, 70, 79, 80, 192
Fenimore, C. P., 124, 130, 131, 202–4, 207, 213
Ferguson, R. E., 223
Ferguson, W. R., 308
Ferriso, C. C., 263
Féry, C., 269
Feugier, A., 361
Field, M. A., 173
Fish, A., 23
Fissan, H. J., 294
Fletcher, E. A., 30, 79, 80, 400
Fons, W. L., 176
Fontijn, A., 364
Foss, W. I., 128, 129
Foster, C. D., 23
Foster, J. F., 115
Fouré, C., 318
Fox, J. S., 366
Fox, M. D., 135, 137–9, Plate 7
Frank, H. H., 374
Frank-Kamenetsky, D. A., 115, 117
Franklin, N. H., 224

Friauf, J. B., 274, 284, 290
Friebertshauser, P. E., 378
Friedman, R., 17, 19, 88, 113, 165, 284, 339, 364, 395, 397
Fristrom, R. M., vii, 41, 56, 65, 93, 97, 103, 104, 284
Friswell, N. J., 209, 210
Fuhs, A., 348
Fuidge, G. H., 8
Fuller, L. E., 79, 80

Garner, F. H., 63, 64, 80, 197–9, 215
Garner, W. E., 87
Garside, J. E., vi, 65, 80, 81, 168, 318, Plate 13
Garton, W. R. S., 299
Gay, I. D., 225
Gay, N. R., 217, 218
Gaydon, B. G., 189
Gayhart, E. L., 389
Gazley, C., 150, 166, 167
Gerschinowitz, H., 247
Gerstein, M., 78, 80, 84, 86
Gibson, J. F., 23
Gilbert, J. A. S., 215
Gilbert, M., 70, 83
Gilmore, F. R., 240
Ginsburg, N., 300
Glass, I. I., 257, 278
Glassman, I., 396, 398
Glick, H. S., 225, 257
Gliwitzky, W., 395
Gohrbandt, W., vi
Goldman, L. M., 310
Goldmann, F., 36
Gollahalli, S. R., 171
Golothan, D. W., 210
Goodger, E. M., 332
Goody, R. M., 294
Gordon, A. S., 162, 224, 381, 396
Gordon, J. S., 335
Gouldin, F. C., 398
Gouy, G., 60
Graham, S. C., 226, 235
Graiff, L. B., 186
Grant, A. J., 188
Gray, B. F., 25
Gray, K. L., 76, 80
Gray, P., 77, 119, 377–9, 387
Gray, W. A., 261
Green, H. L., 260
Green, P. D., 162
Greene, E. F., 225, 257
Griem, H. R., 298, 348
Griffing, V., 160
Griffiths, E., 273, 283, 317

AUTHOR INDEX

Grisdale, R. O., 215, 229, 237
Grosse, A. V., 383, 386, 390, 393, 396
Grove, J. R., 70
Grover, J. H., 135
Grumer, J., 8, 80
Guderley, G., 186
Guest, P. G., 16
Guillon, O., 136
Gumz, W., 233, 234
Günther, R., 16, 74
Gupta, A. K., 181
Guruswamy, S., 394
Gurvich, L. V., 216, 333–5, 351, 358
Guyomard, F., 149

Hahnemann, H., vi, 42, 80, 192–4
Hall, A. R., 117, 186, 374–382, 387, Plate 13
Halstead, C. J., 101
Halstead, M. P., 25, 189
Hansen, G., 52
Hardin, M. C., 25
Harned, B. W., 300
Harris, M. E., 80
Harrison, A. J., 129, 366
Harrison, G. R., 239
Harrison, P. L., 398
Harshman, R. C., 388
Hartley, H., 242
Hartmann, I., 393–5
Hastie, J. W., 104, 105
Hawksley, P. G. W., 260
Hawthorne, W. R., 150, 152
Hayhurst, A. N., 130, 132, 356
Heald, M. A., 342
Heap, M. P., 16, 55, 74, 80, 126, 127
Heath, G. A., 381
Hebra, A., 65, 80
Hecker, W. C., 104, 123
Heckman, F. A., 261
Hedley, A. B., 172
Heinsohn, R. J., 164
Heimel, S., 82, 164
Heller, C. A., 381
Hendra, P. J., 310
Henkel, M. J., 372, 373
Henning, F., 273, 275, 276
Herlan, A., 228
Herzberg, G., 241
Hibbard, R. R., 86
Hinck, E. C., 84, 244, 286
Hirata, M., 196
Hirschfelder, J. O., 121
Ho, C. M., 320
Hoare, H. F., 70
Hofmann, U., 229

Hofsaess, M., 81
Holderness, F. H., 203, 205
Holland, S., 119
Hollander, T. J., 355, 357
Homann, K. H., 202, 205, 212–5, 221, 222, 228
Homer, J. B., 226, 235
Hong, N. S., 44
Hörmann, H., 46
Hornig, D. F., 257
Hottel, H. C., 129, 150, 152, 176, 213, 260, 263–6, 295, 296, 316
Howard, J. B., 173, 210, 213, 232
Hoyermann, K., 124
Hsieh, M. S., 22
Hübner, H. L., 45, Plate 6
Huddlestone, R. H., 341, 347
Hummel, H., 372, 373
Hundy, G. F., 78
Hunter, T. G., 197–9
Hüppner, W., 51
Hurle, I. R., 4, 28, 180, 181, 225, 248, 275–9, 282, 283, 363
Hutton, E., 363

Ibiricu, M. M., 164, 165, 208
Ibrahim, S. M. A., 321
Inami, S. H., 89
Inghram, M. G., 217
Ingle, H., 12, 13, 36
Isles, G. L., 122

Jackson, J. K., 276
Jacobson, M., 393
Jaggers, H. C., 366
Jahn, G., 110
Jain, D. C., 308
Jain, V. K., 154
Jakus, K., 320
James, C. G., 101
James, H., 162
Janisch, G., 16, 74
Jenkins, D. R., 101, 209, 210, 251
Jensen, D. E., 123, 124, 230, 234, 235, 352–7
Jesch, L. F., 322
Jessen, P. F., 26, 161, 218, 220, 235, 305, 308, 309
Jinno, H., 320
John, R. R., 88, 141
Johnson, C. H., 87
Johnson, T. R., 267
Johnston, W. C., 18, 19
Jones, A., 308
Jones, A. R., 32, 44, 45, 213
Jones, F. J., 164

AUTHOR INDEX

Jones, G. A., 124, 125
Jones, G. W., 124, 202–4, 207, 274, 284, 373, 390
Jones, H., 188
Jones, J. M., 150, 188
Jost, W., v, 36, 148

Kalff, P., 355
Kallend, A. S., 173
Kandyba, V. V., 293
Kane, W., 38
Kanury, A. M., vii, 399
Kapila, A. K., 25
Kapp, N., 150, 166, 167
Karabell, C. E., 217, 218
Karl, G., 248
Karlovitz, B., 126, 136, 142
Kaskan, W. E., 105, 189, 274
Katan, L. L., 175
Kaufman, F., 373
Kaveler, H. H., 271
Kaye, G. W. C., 250
Kebely, V., 140
Kelso, J. R., 373
Kent, J. H., 318
Kerker, 260
Kern, R. D., 225
Khitrin, L. N., 16, 75, 83
Khudiakov, G. N., 175
Kilham, J. K., 18, 181, 318
Kimbell, G. H., 28
Kimura, I., 61, 188
Kinbara, T., 254, 255, 358
King, I. R., 347, 358
King, J. G., 175
Kirchhoff, 269
Kirk, R. E., 236
Kirsch, L. J., 25, 189
Kirschenbaum, A. D., 383
Kistiakowsky, G. B., 225, 305
Kittelson, D. B., 356
Klaükens, H., 45, 71, 96, 98, 319
Klein, G., 121
Knapshaefer, D. H., 126
Knewstubb, P. F., 223, 361
Knight, H. T., 391
Knipe, R. H., 396
Knorre, V. G., 231
Kohn, H., 274
Kolodtsev, K., 83
Kolsch, W., 36
Kopp, I., 252
Kostkowski, H. J., 305
Kovasznay, L. S. G., 141
Koyama, A., 233
Kramer, L., 386

Kraus, P., 248
Krier, H., 171
Kruger, C. H., 274
Krygsman, C., 276
Kudryartsev, E. M., 277
Kuehl, D. K., 66, 82
Kuhn, H., 281
Kuhns, P. W., 315
Kumagai, S., 171
Kunugi, M., 320
Kurlbaum, F., 269, 283
Kurzius, S. C., 359
Kydd, P. H., 128, 129

Laby, T. H., 250
Ladenburg, R., 53
Laffitte, P., 381
Laidler, K. J., 160
Lalos, G. T., 316
Land, T., 317
Lander, C., 263–5
Landolt, H., 166
Lane, W. R., 260
Lapp, M., 310
Larson, D. A., 273
Larson, G., 279, 315
Laud, B. B., 220, 228
Law, C. K., 171
Lawton, J., vii, 16, 58, 210, 340, 347, 358, 365
Leach, A. H., 387
Learner, R. C. M., 301, 308
Lederman, S., 310
Lee, J. C., 77, 387
Lefebvre, A. H., 27, 90, 136, 142–4, Plate 21
Leibman, I., 394
Lennard-Jones, J. E., 216
Leonard, S. L., 341, 347
Levine, O., 78, 80, 86
Levy, A., 31, 42, 46, 72, 115
Levy, J. B., 165, 339, 382
Lewis, B., v, vi, 4, 5, 8, 16, 25, 28, 33, 36, 46, 63, 66, 77–81, 85, 125, 142, 190, 271, 274, 284–6, 390, Plate 6
Lewis, J. D., 338
Leyer, J. C., 188
Liddell, W. J., 390
Lieb, D. F., 223
Lighthill, M. J., 177, 181, 182
Lindsey, A. J., 215
Linnett, J. W., 30, 70–6, 80, 81, 88, 98, 112
Litchfield, E. L., 373
Lloyd, S. A., 32
Lockwood, F. C., 320
Loeb, L. B., 347
Long, R., 63, 64, 215

AUTHOR INDEX

Longwell, J. P., 128, 129, 131
Loofbourow, S., 239
Lord, R. C., 239
Loshaek, S., 192
Lovachev, L. A., 31, 32
Lück, K. C., 300
Ludwig, C. B., 263
Lundford, G. L. S., 25
Lurie, H. H., 283
Lwakabamba, S. B., 136, 142, 145

McArty, K. P., 397
McChesney, J., 194
McCoubrey, J. C., 374
Macék, A., 364, 395, 397, 398
Macfarlane, J. J., 203, 205, 310
McGowan, J. W., 244
McGrath, I. A., 259, 296
Mach, L., 51
McHale, E. T., 115
Mache, H., 14, 65, 80
Mackinven, R., 77
Maclean, D. I., 106
Maddison, T. E., 26
Madgeburg, H., 276
Mahnen, G., 124
Maier, E., 389
Mangelsdorf, H. G., 263-6
Mannkopff, R., 277
Manson, N., 75, 188
Manton, J., 77, 80, 81
Marchand, 224
Markstein, G. H., 36, 37, 189, 267
Marlow, D. G., 315
Marr, G. V., 308
Marrone, P. V., 307
Marsel, J., 386
Marsh, P. A., viii, 215, 237, Plate 22
Mascalo, R. W., 140
Mason, D. M., 25
Mason, W. E., 174
Massey, H. S. W., 246
Matsui, K., 233
Matthews, C. S., 359
Matthews, K. J., 322
Matton, G., 318
Mayo, P. J., 210-2, 232
Méker, G., 9
Medalia, A. I., 261
Melaerts, W., 284
Melling, A., 136
Mellish, C. E., 76, 80
Mentall, J. E., 308
Mentser, M., 75
Merer, A. J., 220
Mertens, J., 378

Michael, J. V., 225
Michaud, P., 321, 359
Mie, G., 258, 396
Miller, D. R., 89
Miller, E., 377
Miller, W. J., 358, 359, 361
Millikan, R. C., 205
Mills, R. M., 115
Minchin, S. T., 169, 197
Minkoff, G. J., 5
Mirschandani, I., 366
Mitchell, A. C. G., 249
Mizutani, Y., 143, 144
Moin, P. B., 16
Moore, G. E., 203, 204, 207
Moore, N. P. W., 19
Moorhouse, J., 26
Moreau, R., 381, 382
Morgan, F. H., 317
Morley, C., 130
Morrison, M. E., 89, 100
Mullaney, G. J., 315
Mullens, T. J., 215, 237, Plate 22
Müllins, B. P., vi, 23, 150, Plate 8
Müller, F. J., 300
Muller-Dethlefs, K., 196
Munday, G., 207
Muntz, E. P., 307
Murch, W. O., 8
Murray, R. C., 117, 387
Myerson, A. L., 377

Nagy, J., 393
Nakakuki, A., 176
Nakamura, J., 358
Napier, D. H., 215
Needham, D. P., 379
Nefedov, A. P., 274
Nettleton, M. A., 173
Neubert, H., 46
Neumann, R., 97
Newitt, D. M., 24, 172
Newman, B. H., 383
Nicholls, R. W., 308
Nieuwpoort, W. C., 220
Niki, H., 225
Nisewanger, C. R., 315
Nishiwaki, N., 196
Noda, K., 254, 255
Noltingk, B. E., 189
Nuruzzaman, A. S. M., 171, 172
Nutt, C. W., 209, 210
Nutt, G. B., 363

Odgers, J., 129
Odidi, A. O., 320

AUTHOR INDEX

Okajima, S., 171
Olsen, H. L., 53, 78, 80, 81, 192, 312
Ong, R. S. B., 359
Onsager, L., 168
Oppenheim, A. K., 54
Ornstein, L. S., 307
Othmer, D. F., 236

Padley, P. J., 209, 210, 253, 256, 308, 352–4, 357
Palmer, H. B., vii, 170
Palmer, K. N., 19, 395
Palm-Leis, A., 144
Pandya, T. P., 46, 55, 162, 163
Parker, K. H., 136, 320
Parker, W. G., vi, 109, 130, 154–8, 162, 203, 207, 220, 225–9, 258, 275, 296, 372–4, 388, 392, 395, 397, Plate 15
Parks, D. J., 79, 80
Patankar, S. V., 170
Patterson, W. L., 225
Payne, K. G., 75, 210
Pearson, G. S., 381, 382
Pease, R. N., 88, 110, 113, 118, 388
Peeters, J., 124, 359, 361
Pegg, R. E., 210
Penner, S. S., 121, 262, 293, 294, 302, 310
Penney, C. M., 310
Pergament, H. S., 368–70
Perrin, F., 244
Perrott, G. St. J., 274, 284, 390
Pickering, H. S., 74, 111, 112
Pickering, S. F., 38, 63
Pillow, M. E., 308
Pinkel, B., 86
Pitz, R. W., 300, 301
Place, E. R., 164, 210–3, Plate 15
Plass, G. N., 258
Pleasance, B., 8
Polanyi, J. C., 248
Pollard, F. H., 376
Polymeropoulos, C. E., 172
Porter, G., 227
Porter, J. W., 186
Porter, R. P., 105, 124, 321
Potter, A. E., 19, 164
Potter, R. L., 378
Powell, A., 181
Powling, J., 30, 37, 42, 72, 100, 379, 380
Pownall, C., 115, 116
Prandtl, L., 37, Plate 5
Pratt, M. W. T., 379
Prescott, R., 97
Price, L. D., 215, 237
Price, R. B., 180, 181
Pritchard, R., 74

Prothero, A., 25, 189, 333
Putnam, A. A., 181, 186
Pye, D. B., 180

Quinn, C. P., 25, 189, 293
Quinn, H. F., 293
Quinton, P. G., 175

Raezer, S. D., 78, 80, 81
Rajaratnam, A., 299
Ramsden, A. W., 210
Rao, K. V. L., 27
Rasbash, D. J., 175
Rautenberg, T. H., 376
Ray, S. K., 215
Rayleigh, Lord, 183, 190
Read, I. A., 225
Reed, S. B., 16, 126, 127
Reichardt, H., 374
Reid, R., 90, 136, 143, 144, Plate 21
Reiter, S. H., 140
Rekers, R. G., 381
Renich, W. T., 389
Ribner, H. S., 182
Richardson, E. G., 183
Richardson, J. F., 174
Richmond, J. K., 135
Robben, F., 300, 301
Roberts, A. F., 175, 176
Robins, A. B., 383
Robinson, N. E., 189
Roblee, L. H. S., 223
Robson, K., 154
Rogowski, Z. W., 175
Roper, F. G., 149
Rosenfeld, J. L. J., 150, 226, 235
Roshko, A., 134, 135, 168
Rosser, W. A., 89, 171
Rossini, F. D., 335
Rössler, F., 258, 259, 296
Roth, W., 376, 388, 390
Rozlovskii, A. I., 19
Rummel, K., 228
Rust, P. M., 381
Ryan, P., 129
Ryanson, P. R., 25

Sachse, H., 81, 109, 113, 209, 339
Saha, H. K., 349
Saha, M. N., 349
Sahni, O., 322
Sakai, T., 171
Salant, R. F., 188
Salooja, K. C., 209, 210
Sampath, P., 188
Sander, 184

AUTHOR INDEX

Sanderson, J. A., 296
Santy, M. J., 395
Sarofim, A. F., 260, 264
Sato, A., 316
Savage, L. D., 34, 127
Sayers, J. F., 26
Schaaf, V. P., 380
Schack, A., 259
Schardin, H., 48, 49
Scheller, K., 89, 100
Schley, U., 276
Schmidt, H., 293
Schmitz, R. A., 154
Schneider, W. E., 276
Schofield K., 355
Scholefield, D. A., 168
Schorpin, S. N., 124
Schumann, T. E. W., 147, 148, 150, 168
Schwar, M. J. R., 44, 45, 54, 55, 315
Scrivener, J., 380
Scurlock, A. C., 135
Seamans, T. F., 41, 84, 386
Searcy, A. W., 395
Semenoff, N., 116
Semple, J. M., 395, 398
Senftleben, H., 258
Setzer, H. J., 377
Shaulov, Y. K., 380
Shchelkin, K. J., 136, 141
Shchetinin, T. I., 296
Sheinson, R. S., 25
Shen, F. C. T., 366
Shepard, C. E., 320
Sheppard, C. G. W., 322
Sherman, G. W., 283
Shetinkov, E. S., 141
Shevchuk, V. U., 16
Shirodkar, A., 314
Shuler, K. E., 298, 389
Sichel, M., 130, 359
Siddall, R. G., 259, 296
Siedentopf, Plate 5
Silla, H., 321
Silverman, S., 293, 294
Simmonds, W. A., vi, Plate 3
Simmons, F., 306
Simmons, R. F., 89, 115, 116, 385, 386, 391
Simon, D. M., 76, 86
Simpson, C. J. S. M., 30, 88
Singer, J. M., 65, 79, 135
Sjögren, A., 200
Skinner, G. B., 89
Skirrow, G., 383
Smirnov, D. B., 16
Smit, J. A., 282

Smith, D., 83
Smith, D. B., 77, 119
Smith, E. C. W., 22, 203, 231, 256
Smith, F. A., 38, 63
Smith, S. R., 162
Smith, T. J. B., 181
Smith, W. A. W., 380
Smith, W. F. R., 390
Smithells, A., 12, 13, 36
Smoot, L. D., 104, 123
Snegiriova, T. D., 231
Snelleman, W., 280
Snyder, W. T., 90, Plate 7
Sobolev, N. N., 276, 296, 298
Somers, L. M., 189
Sorensen, S. C., 34
Spalding, D. B., 73, 80, 87, 121–6, 154, 170, 176
Spaulding, W. P., 372–3
Spence, D., 115
Spence, K., 25, 130
Spiers, H. M., 34
Spokes, G. N., 202, 219, 220, 228
Springer, R. W., 347
Srivastava, N. K., 163
Stair, R., 276
Stamp, D. V., 115
Stark, W. V., 175
Starr, W. L., 248
Steere, N. V., 34, 175
Steinberger, R., 380
Steinle, H., 37
Stenhouse, I. A., 311
Stephenson, P. L., 121, 123
Stirling, R., 173
Stokes, C. S., 386
Stokes, G., 196
Stopford, J. M., 188
Strasser, A., 375
Strause, W. A., 83, 373
Street, J. C., 201, 205, 208
Strehlow, R. A., 4, 34, 76, 80, 81, 127, 144
Streng, A. G., 386
Streng, L. A., 386
Stricker, W., 310
Stringer, F. W., 129
Strong, H. M., 273, 274
Stuart, J. G., 76, 80, 81
Stull, V. R., 258
Sugden, T. M., 84, 101, 180, 181, 223, 253, 256, 352, 355, 363
Summerfield, M., 140, 141
Sun, C. E., 247
Sutton, M. M., 97, 98
Syred, N., 181

AUTHOR INDEX

Tabbutt, F. D., 305
Takeno, T., 186, 187
Talbot, L., 300, 301
Tanford, C., 88, 110, 113, 117, 118
Tankin, R. S., 281
Taylor, D. C., 387
Taylor, F. R., 377
Taylor, R. G., 121, 123
Taylor, R. L., 225, 244
Teclu, 12
Teller, E., 248
Tesner, P. A., 230, 231
Tewari, G. P., 19, 25
Thabet, S. K., 25, 29, 30, 35-7, 72, Plates 3, 5
Thomas, A., 177-81, 195, 201, 205, 208, 214, 228
Thomas, D. B., 214
Thomas, D. L., 279, 281
Thomas, N., 307, 390
Thompson, K., 115, 122
Thornes, L. S., 24
Thorp, C. E., 215, 386
Thring, M. W., v, 2, 170, 264, 267, 320
Tighe, J. J., 115
Tingwaldt, C., 273, 275, 276
Tipper, C. F. H., 5
Toepler, A., 48, 182
Tollmein, W., 42, 192
Toong, T. Y., 188
Topps, J. E. C., 25
Tourin, R. H., 263, 294, 298
Townend, D. T. A., v, 4, 22-5, 29, 65, 130, 196, 207, 217, Plate 13
Travers, B. E. L., 347
Trimm, D. L., 225
Tropsch, H., 224, 225
Tse, R. S., 359
Tsuji, H., 165, 186, 187

Ubbelohde, A. R., 207
Ubbelohde, L., 81
Uberoi, M. S., 57
Uehara, K., 233
Ukawa, H., 61
Unsöld, A., 239, 249
Urtiew, P. A., 54

Van Artsdalen, E. R., 388
Van der Hulst, H. C., 260
Van der Poll, A. N. J., 71
Vanpee, M., 84, 386
Van Tiggelen, A., 284, 361
Vasilieva, I. A., 274
Vear, C. J., 310
Veh, P. O., 228

Venugopalan, M., 364
Verkamp, F. J., 25
Verlin, J. D., 124
Villars, D. S., 381
Vince, I. M., 130, 132
Violett, T. D., 377
Voet, A., 215, 237, Plate 22
Von Elbe, G., v, vi, 4, 5, 8, 16, 25, 28, 33, 36, 46, 63, 66, 75-81, 125, 142, 190, 284-6, Plate 6
Von Engel, A., 322, 366
Von Kărmăn, T., 121
Vree, T. H., 364

Wagner, H. G., 106, 124, 203, 205, 214, 215, 221, 222, 228
Waldman, L., 36
Walmsley, R., 122
Warder, D. C., 347
Wares, G. W., 179, 315
Warneck, P., 359
Warshawsky, I., 320
Watermeier, L. A., 119
Watson, J. H. L., 230
Watson, R., 308
Watson, W. W., 168
Wayman, D. H., 378
Weast, R. C., 82
Weber, A. H., 315
Webster, J. M., 55
Weight, R. P., 55
Weinberg, F. J., vii, 16, 18, 28, 32, 42-55, 58, 69-72, 75, 96-9, 137-9, 162-4, 170, 210-3, 232, 233, 312-5, 317, 333, 340, 347, 358-60, 365, 366, Plates 7, 15
Weiss, M. A., 128, 129
Wells, F. E., 126, 136
Weltmann, R. N., 315
Wenaas, E. P., 194
Werner, G., 30
Wersborg, B. L., 213
Westenberg, A. A., vii, 41, 65, 93, 103, 284
Westerdijk, T., 71
Westermann, T., 392
Wharton, C. B., 342, 345
Wharton, W. W., 377
Whatley, A. T., 388
Wheatley, P. J., 75
Wheeler, R. V., 30
Whitcher, F. S. E., 203, 205
Whitelaw, J. H., 45, 136, 188
Whittaker, A. G., 381
Whittingham, G., 207, 233
Wiberley, S. E., 388
Widhopf, G. F., 310
Wiese, W. L., 348

Wilkins, J., 170
Williams, A., 26, 97, 98, 101, 102, 170, 338
Williams, D. R., 311
Williams, D. T., 14, 90, 143, Plate 7
Williams, F. A., 397
Williams, F. W., 25
Williams, G. A., 104, 123
Williams, G. C., 129, 213, 260
Williams, G. T., 177, 179, 181
Williams, H., 347, 348, 381
Williams, J. R., 25
Wilm, D., 214, 229
Wilson, J. R., 26, 28, 99
Wilson, M. J. G., 97, 99, 154, 174, 312, Plate 3
Wilson, R. H., 276
Wilson, R. P., 397
Wilson, W. E., 114
Wise, H., 89, 171
Witzell, O. W., 217, 218
Wohl, K., vi, 150, 166, 167, Plates 9, 10
Wolnik, S., 279, 315
Wong, E. L., 78, 80, 86

Wong, W., 213
Wong, W. W. Y., 44
Wood, B. J., 171
Wood, I. F., 207
Wood, N. B., 53
Woodhead, D. W., 198
Wright, F. J., 200
Wright, H. R., 391
Wronkiewicz, J. A., 171
Wurster, W. H., 257
Wyatt, R. M. H., 376

Yamaoka, I., 165
Yang, C. H., 25
Ying, S. J., 176
Yoffe, A. D., 377, 378, 381
Yumlu, V. S., 73, 87
Yumoto, T., 175

Zakaznov, V. F., 19
Zehnder, L., 52,
Zeldovich, Y. B., 115–7, 154, 380
Zemansky, M., 249
Zickendraht, H., 192

Subject Index

ns to
Subject Index

With Definitions of Symbols and Values of Physical Constants and Conversion Factors

In this fourth edition we have moved towards S.I. units (e.g. joules and newtons rather than ergs and dynes) but have still found it necessary to use customary units such as kcal for energy and torr and atm instead of newton m^{-2} for pressure. Conversion factors between the various units are therefore given here.

Under each letter symbols and physical constants are given first and then text subjects. Greek letters are given at the end, after Z. Physical constants are listed under the usual symbols (e.g. velocity of light under c). For symbols and constants the first page reference to its use is often included. Bold type for page numbers indicates the more important references.

Å, angstrom unit, 10^{-10} m
Abel inversion, 274, 295, **298**
Absorption coefficient, carbon particles, 292, **295**
   measurement, **292**
Absorption spectrum, diffusion flames, 162, **220**
   premixed flames, 219
Acetylene, burning velocity, 86, **88**, 108
   decomposition flame, 224, 390
   feather, 39
   pyrolysis, 222, 224
   soot precursor, **227**
Acetylene flame, 11, 12, 17, 34, 390, Plate 4
   carbon formation in, 34, 218
   equilibrium composition, **218**, 337
   temperature, 218, 338
Activation energy, 23, 92, 120, 129, 230, 244
Additives, effect on carbon formation, **207**
   effect on ionisation, **352**
   effect on S_u, **86**, **113**
Adiabatic flame temperature, **323**, **329**
Adiabatic heating in fast gas stream, 319
Aerosols, flames of, 172
Afterburning, 102, 129, 282, 284
   in rocket motors, 369

Air entrainment, 7, 39, 61, 169, 175, Plate 3
Alcohols, smoke point, 197, **199**
Alkali metals in flames, 89, 165, 209, 353
   ionisation by, 352
Alkaline earth metals, ionisation, 209, 355
Alpha-particle method for T, 314
Aluminium, flames of, 392, 394–7, Plate 17
Ammonia, diffusion flame, 156, **158**
   flame with nitric oxide, 372
   flame with O_2, spectrum, 158, 305, Plate 18
   gas composition of flame, 336
Ammonium perchlorate, 381, 382
Anemometry, hot-wire, 77
   laser-Doppler, **43**, 90
Angle, cone of flame, 57, **63**
Aperture stop, 270, 280
Area method of measuring S_u, 59, **63**, 72, 90
Aromatics in soot formation, 215, 222, 226, **228**
Atoms, free, *see* Oxygen, Hydrogen
   persistence of free, 100, 256, 282
atm (atmosphere pressure) = 101·325 kN m^{-2} = 760 torr
Attachment energies, 352
Attenuation, microwave, **340**, 367

SUBJECT INDEX

B.T.U., B.t.u. (British thermal unit) = 0·252 kcal = 1055·06 joule
Background sources for reversal T, 270, 275
Back pressure, 55, 74, Plate 6
Barium, smoke reduction by, 209, 210
Base, diffusion flames, premixing in, 39, 161
  premixed flames, 61
Batswing burner, 169
Benzene, absorption, 162, 221, Plate 11
  as intermediate to soot, 221, 228
  flame, 34, 38, 185, Plate 4
Beryllium flame, 395
Bibliography, **406**
Black-body radiation, **238**
Blow off, 8, 15, 17, 39, 127, 164
Boltzmann constant, 243; *see also* Maxwell–Boltzmann law
Bomb, spherical for S_u, 77
Boron, combustion, 392, Plate 17
Boron hydrides, flames, **388**
Boudouard equilibrium, **216**, 332
Boundary layer, 42, 66
Brightness temperature, 271, 292
  tungsten, λ dependence, 272
Brightness-emissivity method, **292**, 315
Brush, turbulent flame, 14, 90, **133**, 138
Bubble method, for S_u, **75**
Bunsen flame, 2, **7**, 12, 38
  stability, **15**
Buoyancy, effect on diffusion flames, 154
  effect at limits, 31, 33
Burner diameter, 8, 10, **17**, 19, 65, 168
Burner length, 8, **14**, 41
Burners, flat flame, **72**, 155, 162
Burning velocity, S_u, 19, 58, 106
  at flammability limits, 30, 73
  boron hydrides, 389
  comparison of methods, **79**
  effect of additives, **86**, 113
  effect of electric fields, 366
  effect of mixture strength, **86**, 111
  effect of radical concentration, **107**
  effect of sound, 194
  effect of turbulence, 90, **136**, 143, Plate 21
  for thermal propagation, 117
  of droplets and dusts, 172, 394
  of flames with halogens, 383
  of hydrazine, 387
  of hydrocarbons, **86**
  pressure dependence, 77, **82**
  temperature dependence, **81**, 110
  variation across flame, 68
  with nitrogen oxides, 373–7

c, velocity of light, $2 \cdot 997\ 925 \times 10^8$ m s^{-1}
c_1, first radiation constant, hc^2, $5 \cdot 88 \times 10^{-17}$ W m^2 or $5 \cdot 88 \times 10^{-6}$ erg cm^2 s^{-1}, 239
c_2, second radiation constant, $1 \cdot 4388 \times 10^{-2}$ m K, 238
c_p, specific heat at constant pressure, 95
c_v, specific heat at constant volume
C_2, concentration, 218, 231
  formation, 255
  rotational temperature, 305–6
  spectrum, 104, 161, 165, 219, 241, 384, Plate 20
  soot nucleation by, **231**
  vibrational temperature, 308
C_3, 219, 220, 230
cal, thermochemical calorie, 4·184 joule
Calculation of final flame temperature, 323
Candle flame, **154**, 258, Plates 6, 12
Candoluminescence, 256
Caps, flame, 29
Carbon, *see also* Soot
  arc, as background, 275, **276**
  burning of, 391
  effect of electric field, 164, 210, Plate 15
  formation mechanism, **226**
  in flames, 34, 160, **195**, Plate 4
Carbon black, 227, **236**, Plate 22
Carbon dioxide, emissive power, **263**
Carbon monoxide flame, 11, 322
  S_u, 87, 113
  spectrum, 241, 255
  with oxides of nitrogen, 374, 376
Carbon particles, combustion, 233–4, 392, Plate 17
  emissivity, **258**
  growth, **213**
  light scattering by, 196, **257**
  size, 213, 260
  structure, **213**, Plate 22
Carbon subnitride flame, 390
Carbon tetrachloride, effect on S_u, 88
Catalytic effects, on S_u, 68, 87
  on wires, 318
Cellular flames, **36**, 76, 189, Plate 5
CH, formation, 254
  rotational temperature, 305
  spectrum, 22, 104, 161, 165, 219, 241, 384, Plate 20
  translational temperature, 300
CH$_2$ radicals, 227
CH$_3$ radicals, 104, 220
C$_2$H, 227, 230, 235, 255
C$_3$H$_3^+$, 359

SUBJECT INDEX

Chains of carbon particles, 260, Plate 15
Channel process, 236
Chemi-ionisation, 358–62
Chemiluminescence, **252**, 282, 363
Chlorine, flames with, 385, Plate 17
Chlorine dioxide, flames with, 381, 382
Chlorine-hydrogen flames, composition and T, 339
Chlorine trifluoride, flames with, 384, Plate 20
CHO^+, 359, 363
Chopper, light, 220, 281, 293
Chromium lines for reversal T, 275
Chugging, 187
cm^{-1}, wave number (1 eV = 8065·7 cm^{-1})
CN, vibrational temperature, 307
Coagulation of carbon particles, 229
Coal dust, **173**, 195, 281, 393
Coanda burner, **169**
Coated wires, 318
Coherent anti-Stokes Raman spectroscopy, 311
Coherent turbulence, 135, 168
Collision, complexes, 246, 248
   cross-sections, 248, **250**, 341
   frequency, 250
   frequency, of electrons, 341
   processes, 243, **245**
Colour filters, 296
Colour temperature, 258, **295**, 395
Combustion chamber oscillations, 186
Complex flame structure, effect on reversal temperature, 273, 294
Composition, calculation of, **324**
   diffusion flames, 152, **159**
   for various mixture strengths, **336**
   interconal gases, 13
   typical flame gases, **339**
Computer handling of data, 121, 332
Concentration, free radicals, **100**, 110
Condensation and soot formation, 227, 228
Conductivity, thermal, 95, 117, 125
Cone angle, 57, **63**, 66, 72, 90
Continuous spectra, **162**, 220, 241
   background sources, **275**
   in pyrolysis, 162, 220
Contour of sodium lines, 273, 293
Cool flames, 24, 72, 130, 189, 253, 377, Plate 3
Counter-flow diffusion flame, **162**, 220
Convection effects, on diffusion flames, 154
   on limits, 31, 33
Cross correlation, of flicker, 189
Cross-section, collision, 248, **250**, 341

Curvature, effect on S_u, 58, 76, 77
Curve of growth, 297
Cusp formation, 137
Cyanogen flame, carbon in, 197, 207, 390
   temperature, 307, 338, 339

D, diffusion coefficient, 94, **118**, 148
Dead space, 56, 127
Decomposition flames, 117, 121, 224, 372, 279, 390
Detonation, 4, 19, 28, 181, 365, 391
   temperature measurement, 274, 278
Deuterium flames, 36, 119
Diacetylene in soot formation, 221–7
Diameter, of burner, 8, 10, 16, 19, 65, 168
   of tube, 78
Diesel engines, sooting, 209
Diffusion coefficient, effect of T, 118, 148
Diffusion flames, 2, 38, **146**, Plates 8, 9
   carbon in, 150, **197**
Diffusion, of H atoms, 110, 118
   selective thermal, 34, **36**, 166, 202
   theory of propagation, **115**
Diluents, effect on S_u, **86**
Dissociation in burnt gases, 324
Distortion, optical, 46, 162, 314
Divergence of flow lines, 57, 63, 64
Doppler effect, 43, 274, **299**
Double-beam line-reversal systems, 278
Droplets, combustion, **170**, 200, 228, 366
   evaporation, 170, 228
   minimum ignition energy, 27
Dusts, flames of, 173, 392–5

e, electronic charge, 4·80298 × 10^{-10} e.s.u.; 1·6021 × 10^{-19} coulomb
eV, electron volt, 1·6021 × 10^{-19} joule; 1 ev = 11 605 K.
E_λ, emissivity at wavelength λ, 239
Egerton–Powling burner, 37, **72**, 90
Electric field, effect on soot, 164, **210**, 233, Plate 15
   effect on flames, 194, **365**
Electron, affinities, 352
   microscopy of particles, 213, 215, Plates 15, 22
   temperature, **320**, 345
Electronic excitation, abnormal, 165, **288**, 363
   processes, **245**
Electrons, concentration, **340**, 349
Emissive power, **262**, 265
Emissivity, measurement, **292**, 296
   of flames, 240, 263–7
   of tungsten, **272**, 275
   particles, 258

Energy, noise, 181
Engines, combustion in, 4, 25, 130, 189, 195
Enthalpy, excess, **125**
  values for gases, 323, 333, **335**
Entrainment of air, 7, 39, 61, 169, 175
Equilibrium, carbon, 35, 201, **216**
  constants, 325, **334**
  departures from, 101, 110, 238, **256**, 288
  ionisation, **349**
Equipartition of energy, **243**
Equivalence ratio, 35, 86
Erg, 10^{-7} joule
Ether flame, 25, Plate 3
Ethylene, burning velocity, **81**
Ethylene flames, 10, 12, Plate 1
  diffusion, 156, **160**
  gas composition, 331
Ethylene oxide flame, 30
Ethyl nitrate flame, 378
Excess enthalpy, **125**
Excitation, by collision, 243, **245**
  temperature, 269, 288
Expansion ratio, 74, 75, 77, 178
Explosion flames, **4**, 205, 323, 394
Explosives, *see* Propellants
Extracts from soot particles, 215, 228

Final flame temperature, at limits, 29
  calculation, **323**, **329**
  relation to S_u, **108**, 113
Fire detection, 254
Fire extinguishers, 89, 174
Fire whorl, 176
Flame caps, 29
Flame monitoring, 189
Flame stretch, 16, **126**, 140
Flame thrust, **55**, **74**
Flame traps, 19, 66
Flammability limits, **29**, 36, 113, 142, 388, 394
Flash back, 8, 10, 15, 17
Flash point, 34, **175**, 176
Flat-flame burner, counter flow, **162**, 220
  diffusion, **155**, **162**, 220
  Egerton–Powling, **72**, 90, Plate 3
  porous plate, 73
Flickering, 169, **188**
Flow, alternation laminar to turbulent, 14
Flow-line divergence, 57, 63, 64
Flow meters, **43**
Flow patterns, **40**, 58, 66, 193
Fluorescence, of flames, 196, 220
  quenching, 246, **257**, 303

Fluorine, flames with, 2, 208, **383**
Formaldehyde, flames, 11, 197, 302
  spectrum, 24, 255, 378, 380, 384
Formation, heats of, **335**
Four-grating interferometer, 53
Free atoms, persistence, 100, 256, 282
Free radicals, concentration, **100**, 110
Freezing of equilibria, 201
Fulvenes and soot formation, 219
Furnace black, 236
Furnace flames, 170, 267, 320

Gas flow, measurement, **43**
Gouy's method of measuring S_u, 59
Graphite, dust, flames, Plate 17
  formation in flames, 214, 228, 229
Grating interferometer, 53
Gravity effects, 154, 171
Green flames, 22, 203
Grid-generated turbulence, 15, 90, 132

h, Planck's constant, $6·6256 \times 10^{-34}$ Js, 238
Halides, effect of S_u, 89, 114
Halogens, effect on C_2, CH and soot, 208
  effect on ionisation, 351
  flames with, 208, **383**
Heat content, various gases, 323, **335**
Heat of formation, 335
Heat transfer, 99, 119, 155, 173, 316
  in fast flows, 316, 319
  radiative, **261**
  to droplets and particles, 171, 259
  to wires, **316**
Height, diffusion flames, **147**, 166
  to carbon luminosity, 155, **197**
  to turbulence, **166**, 190, Plate 13
High-energy fuels, **372**, **386**
High-temperature flames, 364, 390
H_3O^+, 359, 362
Holography, **54**, 314
Homogeneous reactor, **128**
Hot wire, anemometery, 78
  ignition by, 27
  temperature measurement, 291, **315**
Hydrazine flames, 117, 119, 121, **386**
  with nitric oxide, 372
Hydrocarbons, burning velocities, **86**
  pyrolysis, **224**
  tendency to smoke, **198**
Hydrogen atoms, concentration, **101**, 110, **112**, 325, 336–9
  diffusion of, 110, 118
  effect on S_u, 110
  persistence of excess, **100**, 256, 282
Hydrogen bromide, 115, 116

SUBJECT INDEX

Hydrogen flame, 8, 36, 87
  burning velocity, 121
  gas composition, 339
  reaction zone, **100**
  with nitric oxide, **373**
  with nitrogen dioxide, 376
Hypergolic fuels, 118, 399

Ignition, **23**, **25**, 173
  energy, minimum, **25**, 125, 142, 373, 393
  point, **99**, 114
  temperature, 23, 393
Inclined slit, 70, 96, **312**
Incomplete combustion, 256, 284
Indium line, reversal temperature, 274, 277
Induction period, **24**, 28, 99
Infra-red radiation from flames, 242, **261**, 294
Inhibitors, **89**, 100, 165, 171
Interconal gases, composition, **13**
  luminosity, 13
  temperature, **284**
Interferometry, **52**, 312
Intermediate products to soot, **219**
Internal combustion engine, 4, 25, 130, 189, 195
Inverted flames, 16, 34, 127
Ion recombination, 352, 361
Ions, as soot nuclei, 229, **232**
Ionisation, **340**
  and soot formation, 210, 212, 229
  potentials, 350, **358**
  temperature measurement by, **320**
Iron carbonyl in flames, 129, 289, Plates 16, 18
Iron lines, for line-ratio method, 298
  for reversal temperature, 274, 288
Iso-intensity method, 302, 305
Isotope tracer studies, **223**, 225

J, joule
J, mechanical equivalent of heat, 4·184 J cal^{-1}
Jet flow pattern, 41
Jet noise, 181
Jets, sensitive, 191

k, thermal conductivity, 95
k, Boltzmann constant, 1·38054 × 10^{-23} J K^{-1} or 1·38054 × 10^{-16} erg K^{-1}
Karlovitz number, 126
kcal, kilocalorie, 4184 J
Kirchhoff's law, 258, 269, 280, 292
Knock, 5, 25, 113, 189
Kurlbaum method, **283**

L_e, Lewis number, 125
Lampblack, 236
Laminar flames, **10**, 14
Langmuir probes, 321, **345**
Laser-Doppler anemometry, **43**, 90
Laser-Raman spectroscopy, 54, **308**
Lasers, **53**, 300
  ignition by, 28
Latent energy, 256, 257, 283
Lead tetraethyl, antiknock, 190
  effect on S_u, 89
  for reversal T, 274, Plate 16
Lean-limit flames, 30
Le Chatelier's rule, 30, 389
Length of burner, 8, **14**, 41
Lewis number, 124, **125**
Lifetime, excited states, 250
Lifted flames, 153, 168
Light chopper, 220, 281, **293**
Light scattering, 46, **257**, 280
Limits of flame propagation, 29, 36, 73
Limits of flammability, 17, 26, **29**, 36, 73, 388, 394
Line-ratio method for T, **297**
Liquid pools, burning, **175**
Lithium in flames, 101, 354
  for reversal T, 274
Location of flame front, **69**
Longwell reactor, **128**
Low-pressure flames, 17, **19**, 82, 287, Plate 2
Luminosity, alcohol flames, 199
  hydrogen flames, 100
  onset, 199, **201**
Luminosity of reaction zone, 107
  optical distortion of, 46, 162
Luminous flames, 11, 34, **196**
  temperature measurement, 254, 283, 292, 295
Luminous zone, 107, 114
  carbon, 34, 151, Plates 4, 5
Lycopodium dust flames, 173

m_e, mass of electron, 9·1091 × 10^{-28} g
Mach–Zehnder interferometer, **52**
Magnesium, flames of, 394, 397
Mantles of flames, 11, 39, 200
Mass spectrometry, 104, 124, 215, **222**, 358–361
Matrix for uniform flow, 42, 72
Maxwell–Boltzmann distribution law, 120, **243**, 249, 288, 302
Méker burner, 9
Meniscus-shaped flames, 153
Metal alkyls, flames of, 386, 399
Metal atoms in flames, 209, 353

Metal oxide particles, 257
Metal oxides, properties, 393
Metal powders, flames or, 393
Metal salts, effect on S_u, 89
Metals for thermocouples, 317
Methane flames, 8, 26, 29, 30, 338, 339
   temperature, 285–7, 338–9
   values of S_u, 81, 86
Methane, pyrolysis, 222, 235
   reactions, 104, **123**
Methyl alcohol flame, 11, 197, 302
Methyl bromide in flames, 89, 114, 127
Methyl nitrate flames, 379, Plate 19
Microreversability, 243, 246
Micro-wave attenuation, **340**, 367
Mie theory, 258
Minimum final flame temperature, 29
Minimum ignition energy, **25**, 125, 142, 373, 393
Mists, see Sprays
Mixed fuels, 30, 87, **174**
Mixing length, turbulent, 132
Mixture strength, effect on gas composition, 331, **336**
   effect on luminosity, 11, 35, Plate 1
   effect on S_u, **86**
   for soot formation, 35, **201**, 216
Mobilities of ions, 365
Mole number, **329**
Momentum of gas jet, 7

n, refractive index, 51
n, Loschmidt's number, $2{\cdot}6870 \times 10^{19}$ cm^{-3} at S.T.P.
N, newton (unit of force), kg m s^{-2}; 10^5 dynes
N, Avogadro's number, $6{\cdot}0225 \times 10^{23}$ mole^{-1}
Natural gas, 26, see also Methane
Negative ions, 351, 361
NH spectrum, 158, 305, 374–6
NH$_2$ spectrum, 158, 374, 375
Nitrates and nitrites, flames of, **378**
Nitric acid, flames with, **378**
Nitric oxide, concentration, 130, 244, 325
   decomposition flame, 117, **372**
   flames with, 12, 109, 130, **373**
   formation, 130
   ionisation, 350, 358, 363
Nitrogen dioxide, flames with, 109, 376, Plates 17, 19
Nitrous oxide, flames with, 12, 109, 130, 305, Plate 18
NO$_x$, **130**
Noise of flames, **177**
Nozzle burner, 15, 42, 66

Nozzle, flow pattern in, 42, 61, 66
   method for S_u, **65**
Nucleation of soot, **229**

O$_2$, absorption spectrum, 159, 161
   Schumann–Runge bands, 158, 241, 275, 375, 385, Plate 18
Octane number, 113, **190**
OH, absorption spectrum, 103, 158
   concentration, 101, **103**, 112, 330
   predissociation, 252, 257, 308
   radicals, effect on S_u, 106
   rotational temperature, 254, 301–5, Plate 16
   spectrum, 241, Plate 16
   vibrational intensity distribution, 254, 257, 308, Plate 16
Oil sprays, **170**, 172, 200, 229
Open-top flames, 38
Opposed-jet flame, 162
Optical distortion in flames, 46, 162, 314
Orientation of graphite layers, 229, Plate 22
Oscillating flames, **186**, 188, 189
Overhang of flame, 56
Oxides, heats of formation and melting points, 393
Oxygen atoms, concentration, 101, 252, 325–31
   effect on S_u, 112
   in NO$_x$ formation, 131
Ozone flames, 386

Parabolic velocity profile, 15, **41**, 63
Particle radiation, 196, **257**
Particle-track photography, **46**, **66**, 175
Peclet number, $\rho c u l / \lambda$, 19, 142
Perchloric acid flames, 381
Peroxides, 24
Phase-contrast electron microscopy, 215, 237, Plate 22
Phase-shift, microwave, 340, 344
Photography of flames, **45**
Pinking in engines, 25
Pitot head, 42, 55
Planck radiation law, **238**
Plasma frequency, 341
Platinum thermometry, 317
Pollution, 130, 195
Polyacetylenes, 221, 225, 255
Polyhedral flames, **36**, 202
Polymerisation of hydrocarbons, 227, 228
Pools, burning of liquid, 175
Porous-plate burner, 73
Porous-sphere burner, 65
Positive ions, 357

SUBJECT INDEX

Potential energy curves, 246
Potential flow pattern, 193
Prandtl number, $\nu\rho c\lambda$
Predissociation, 246
  of OH, 252, 257, 308
Preheating, effect on sooting, 205
  effect on S_u, **81**, 117
  zone, **94**, 114, 165, 312
Premixed flames, 2, **7**
  soot formation in, **200**
Premixing at base of diffusion flame, 39, 161
Pressure difference across flame front, **55**, 75
Pressure effect, on burning velocity, 77, **82**
  on diffusion flames, 151, Plate 8
  on final flame temperature, 337
  on ignition, 26
  on limits, 20
  on OH rotational temperature, 303
  on stability, **17**
Pressure pulses, 4, 177
Probes, soot collection on, 215, Plate 15
  Langmuir, 321, **345**
  sampling with, 100, 219
Products, sampling of, 13, 219
Prompt NO_x, 131
Propagation, 79, **92**, 115, **120**
  effect of direction, 33, 36
  effect of radiation, 124, 173, 385
  explosions, **4**, 79
  mechanism, 92, 115, **120**
  rate determining process, 119
  turbulent flames, **132**
Propane, burning velocity, **80**
  cool flame, 24
Propellant flames, 372, **280**
Pyrolysis, **224**
  continuous spectrum, 162, 200, Plate 11

Q, heat of reaction, 94
Quartz coated wires, 318
Quenching, **16**, 78, 365
  at burner rim, 15
  distance, 8, **17**, 25, 27, 125, 373
  of fluorescence, 246, **251**, 303

R, gas constant, 8·3143 J K^{-1} mole^{-1} or 8·3143 × 10^7 erg K^{-1} mole^{-1}, 243
R_e, Reynolds number, **14**
$r/R = 0.4$ method for S_u, 63
Radiation, amount from flames, 242, **261**
  constants, 238, 239
  depletion, 244, 281
  heat loss by, 124, 385, 391

Radiation, loss from hot wire, 316
  processes, **238**
  propagation by, 124, 173
Radiative equilibrium, **249**
Radiative lifetime, **250**
Radicals, concentration related to S_u, **100**, **110**
  propagation by diffusion of, **118**
Raman scattering, 308
Rate determining process in propagation, 119
Rate of reaction, 92, 106, 122, 123, **128**
Rayleigh scattering, 258, **260**, 309
Reaction rate, 92, 106, **128**
  in H_2 flames, **122**
  methane, **123**
  overall, 128
  relation to S_u, 92
Reaction zones, diffusion flames, **157**
  multiple, 376–89, Plate 19
  premixed flames, **92**, 95–7
  reversal temperature, **287**
  thickness, 21, 61, **106**, 120
Recombination of ions, 352
Recovery factor, wire in fast flow, 319
Rectified flat flames, **72**
References, **406**
Reflection loss correction, for reversal T, 272, 281
Refraction, deflection of light by, 46, 281
  for temperature measurement, **312**
Refractive index, 51, 311
Relaxation times, 244, 257
Resistance thermometry, 315
Reversal of spectrum lines, 268
Reversal temperature, **268**
Revised tube method for S_u, 78
Reynolds number, R_e, **14**, 132, 144
Rim of burner, effects at, 16, 58, 127
Rockets motors, ionisation in, **366**
  noise, **187**
  temperature, 254, **301**

S_u, burning velocity, 19, **58**
Safety, 8, 33, 66, 175, 402
Saha equation, **349**
Sampling of products, 13, 100, 219, 221
Scattering of light, by particles, 257, 281
  by soot, 196, **257**, Plate 15
Schlieren photography, **47**, 71, 153, Plates 9, 10, 12
  for S_u, **71**
Schmidt method for T, 293
Screaming (rocket noise), 187
Self-absorption, 297, 302
Sensitive flames, 190

448 SUBJECT INDEX

Separator, Smithells, **12**, Plate 2
Shadow photography, 47, 51, **69**, 166
Shape, diffusion flames, **147**
  effect of sound, 193
  inner cone, 14, 55
Shock-diamonds, ionisation, 367–9
  temperature, 274
Shock-tube studies, ignition, 28, 173
  pyrolysis, 225
  temperature, 274, 277
Silica coatings, 318
Singing flames, **182**
Slow combustion, 24
Smithells separator, **12**, 35, 38, Plate 2
Smoke, **195**; *see* Soot and Carbon
Smoke point, 155
Soap-bubble method for S_u, **75**
Sodium line, contour, 273, 293
  reversal method for T, 252, **269**
  temperature, results, **284**
  use for brightness-emissivity, 294
Solid fuels, **391**
Solid particles, radiation from, **257**
Solid propellants, 381
Sonic thermocouples, 316, 320
Soot, **195**, **213**, **226**; *see also* Carbon
  height to, 155, 197
  nature of particles, **213**, Plate 22
Sound, effect on diffusion flames, **190**
  effect on premixed flames, **192**
  temperature from velocity of, **315**
Sources, background, 270, **275**
Spalding burner, 75, 117
Spark ignition, 16, **25**, 142
Specific heats, 323
Spectrum, absorption, 162, **220**
  cool flames, 24, 377
  diffusion flames, **158**, 211, 220, Plate 11
  effect of electric field, 211
  flames with boranes, 389
  flames with halogens, 383–5
  flames with oxides of nitrogen, 373–82, Plates 18, 19
  hydrazine flame, 387
  line breadth, 299, 300
  low-pressure flames, 22, 105
Spectrum-line reversal T, 161, 252, **268**
Spherical bomb method for S_u, **77**
Spherical detonations, 28
Spiral burner, 32
Spontaneous thermal ignition, **23**, 28, 99, 113, 373
  of dusts, 392
Sprays, flames of, **170**, 172, 200, 229, Plate 12
Stability, bunsen flames, 8, **15**

Stability, regions, **17**
Standard lamps, 275
Stark-effect line broadening, 321, **348**
Starting vortex, 194, Plate 13
Static temperature, 319
Stationary flames, 2
Statistical weights, 243, 350
Stefan–Boltzmann law, 240
Stirred reactor, **128**
Stoichiometry, 7, 35
Stop, aperture for line reversal, 270, 280
Striking back, *see* Flash back
Suction pyrometer, 316, 318, 320
Sulphur trioxide, effect on sooting, 207, 233
Supersonic combustion, 399
Surface catalysis, 18, 68, 260, 318
Surface combustion, 173, 175, 233

t, time, sec
T, temperature, K
Temperature, brightness-emissivity, **292**, 315
  calculation adiabatic flame, **323**, 329
  candle flame, 155
  course through flame front, **95**, 114, 314
  distribution in flames, **284**, 298
  effect on sooting, 199, **205**
  effect on S_u, **81**, 108
  electron, **320**, 345
  excitation, 288
  final flame, at limit, 29
  flame, values, **338**, **339**
  hot wires, 291, **315**
  ignition, 23
  in reaction zone, **287**
  laser-Raman scattering, **308**
  line-ratio method, **297**
  measurement, **268**
  refractive-index method, **311**
  relation to S_u, 81, **108**, 110
  rise in fast flow, 319
  rotational, **301**
  spectrum-line reversal, 161, **268**
  time-resolved study, **277**, 320
  translational, **299**
  vibrational, 307, 309
Thallium line for reversal T, 274, 288
Thermal conductivity, 95, 117, 125
Thermal ignition, **23**, 113
Thermal theory of flame propagation, **115**
Thermocouple measurements, 96, **315**, 320
Thermocouples, fast response, 320
  sonic, 316, 320
Thickness of reaction zone, 21, 61, 72, **106**, 120

SUBJECT INDEX

Thin diffusion flames, 169
Time-resolved study of T, **277**, 320
Tip of flame, 22, 34, **56**, 61, 138
Titanium combustion, 398
Toepler's schlieren system, 48
Tollmien–Schlichting instability, 188
torr, unit of pressure, 1 mm Hg, 133·322 $N\,m^{-2}$
Transition, premixed to diffusion flames, 3, **38**
 turbulent to laminar flow, 14
Transition probabilities, 299, 301, 307
Translational temperatures, **299**, 322
Tubes, propagation in, 4, 33, **78**
Tungsten, emissivity, **272**
 lamp, 270, **275**
Turbulence, effect on S_u, **90**, **136**, 143, Plate 21
 intensity, 132
 minimum ignition energy, 142
 onset of, 10, 42
 relation to noise, 180
Turbulent diffusion flames, 39, **166**, 200, Plates 7, 9, 13
Turbulent premixed flames, 14, **132**, Plates 10, 21
 light emission, 135, 140, 141
 propagation mechanism, **132**
Two-line method for T, see Line ratio
Two-path method for emissivity, 293
Two-stage ignition, **25**, 130, 189, 377

u', turbulent r.m.s. fluctuation velocity, 132
U, reaction velocity, 94, 106
Uniform flow pattern, 42

Vaporisation, of droplets, 170, 172
 of liquids, 175
Vibrating flames, 21, 78, **184**, Plate 14
Vibrational energy relaxation, 244, 257, 285
Vibrational temperature distribution of OH, 254, 257, 308
Vibrational temperature, 282, **307**, 309
Vinyl acetylene, soot precursor, 221–7

Vitiated flames, 150
Vortex, starting, 194, Plate 13
Vortices, 41, 134, 153, 175, 187, **188**, 191

Wall effects, 18, 23, 56, 78
Water vapour, effect on S_u, 87, 90
 emissive power, **264**
 spectrum, 242
Weinberg–Wood grating interferometer, 53
Wicks, flames on, 154, **174**
Wien's displacement law, 240
Wien's radiation law, **239**, 271, 278
Wires, catalytic heating, 318
 heating in fast flow, 319
 ignition by, 27
 temperature measurements with, 291, **315**
Wolfhard–Parker flat flame, **155**, 220
Wrinkled flame front, 90, **135**, Plate 7

Xenon lamp, 277
X-rays for temperature measurement, 314

Yield, light, 242

Zero gravity effects, 154, 171
Zirconia lamp, 276
Zirconium flames, 392, 398

α, exponent in expression for emissivity, 259, 295
α-particles, for T measurement, 314
γ, ratio of specific heats, c_p/c_v
δ_{pr}, thickness of preheating zone, 95
δ_r, thickness or reaction zone, 95
λ, mixture strength, actual O_2/stoichiometric O_2, 35
λ, wavelength (in angstroms)
ν, kinematic viscosity, 14, 140
ν, frequency
ρ, density
σ, Stefan–Boltzmann constant, 5·6697 $10^{-8}\,W\,m^{-2}K^{-4}$
σ, collision cross-section, 250
ϕ, equivalence ratio, 35, 86

QD	Gaydon, Alfred Gordon.
516	Flames, their structure, radiation,
G28	and temperature / A. G. Gaydon and H.
1978	G. Wolfhard. -- 4th ed. -- London :
	Chapman & Hall ; New York : Wiley,
	1979.
	xiii, 449 p., [23] pages of plates :
	ill. (some col.), tables ; 24 cm.

QND

"A Halsted Press book."
Includes indexes.
Bibliography: p. 406-426.
ISBN 0-470-26481-0

1. Flame.   I. Wolfhard, Hans G.   II. Title.

MUNION ME   820910   820907 CStoC
C001097   MT /EW   A*   82-B7999
                                 78-16087